KT-555-827

OCTOPUS BIG BOOK OF NATURE

octopus

OCTOPUS
BIG BOOK
OF
NATURE

Foreword by Dr Maurice Burton

Zoology
Edited by John Rostron Ph.D.
Illustrated by Jaromír Knotek,
Libuše Knotková and Petr Rob
Written by Luděk Dobroruka
Translated by Olga Kuthanová

Botany
Edited by Mark Lambert B.Sc.
Illustrated by Edita Plicková
Written by Zdenka Podhajská
Translated by Olga Kuthanová

Mineralogy
Edited by Keith Lye B.A., F.R.G.S.
Illustrated by Ladislav Pros
Written by Jaroslav Bauer
Translated by Olga Kuthanová

English version first published 1981 by Octopus Books Limited
59 Grosvenor Street, London WI
Second edition published 1982

© Artia 1981, Prague

ISBN 0 7064 1275 3

Printed in Czechoslovakia by TSNP Martin
1/19/01/51-02

Contents

Foreword

by Dr Maurice Burton

The Octopus *Big Book of Nature* gives a comprehensive, fully-illustrated account of the natural world. Today, the scientific study of nature is known as natural history, three branches of which are zoology, botany and mineralogy. Together these sciences explore the earth and its life forms.

The meaning of 'natural history' has changed over the years. Anyone who takes the trouble to look it up may find several different definitions for this term. The reader may perhaps come across an ancient dictionary of the nineteenth century which defines natural history as the 'unintelligent study of biology'. This is slightly amusing since the person who studies natural history is called a naturalist. Charles Darwin was a naturalist. He was also the greatest scientist in the field of biology. Nobody could accuse him of having approached his subject unintelligently.

The mystery is solved if we go back to the sixteenth century. The words 'zoology', 'botany' and 'mineralogy' had not been invented at that time. Anyone who studied the natural world would have been interested in everything vegetable, animal or mineral. The studies he made were embraced by the term 'natural history'. It was not until the second half of the seventeenth century, when the Royal Society in London was founded, that the sciences of zoology, botany and mineralogy were given their names.

Apart from clearing up the meanings of words, this tells us something about the history of these sciences. Thus, in the sixteenth century, one man could know all there was to be known about the animals, plants and minerals throughout the world. In those days anyone who collected these things did so as a matter of storing curiosities, or curios, as they are sometimes called. There were no museums, precious few books and very few people who could in any sense be called scientists.

With the foundation of the Royal Society came the need for organizing the sciences. Collections of specimens were being brought from all over the world and men began specializing in one or other of the sections of what had formerly been called, broadly, natural history. Some studied animals only, others studied plants, others minerals. They called themselves zoologists, botanists and mineralogists. Moreover, they wrote about their discoveries and about what they saw, and they published these in scientific papers. The natural sciences in the modern sense had been born but they still had a long way to go.

Progress they did, during the eighteenth and nineteenth centuries, but the progress was slow. The pace quickened a little with the dawn of the twentieth century but to nothing like the speed we have seen in the last 30 years. Before this time, anyone deciding to make zoology a career was regarded as a little unusual, if not peculiar. Professional botanists were more understandable but they were still looked upon as a trifle odd.

There were already far-sighted people who realized that in the near future the world would have far greater and more urgent need of biological information. This has indeed happened and in the 30 years following the end of World War II the study of zoology, more especially, has expanded almost beyond belief.

Suddenly, it seems, the word 'conservation' is mentioned everywhere: in the press, on the radio, indeed on almost everybody's lips. Television has brought animals we had never known, or even heard of, into our homes through the medium of the screen. Underlying this interest in conservation is the realization that the world's resources in forests and other forms of plant life, minerals and animals are getting dangerously low and that something must be done about it. The only things that are increasing in number are human beings and their domestic animals and plants, as well as the poisonous litter man is strewing around the universe.

Simultaneously, books on animals have become popular. Whereas in the early years of this century one could go into a bookshop and have difficulty in buying a book on animals, now the shelves are filled with them. Moreover, printing in colour has made tremendous strides so that the animals are no longer depicted in black and white.

The general reader no longer demands books filled with animal stories but encyclopedias which give comprehensive coverage of the natural sciences. People want to know about natural history in breadth and depth; and publishers have responded to this new interest. The Octopus *Big Book of Nature* is the latest, although not the last, and certainly not the least, of the encyclopedic books being offered to satisfy this new appetite for knowledge.

Zoology

Animals occupy almost every part of the earth's surface from the depths of the sea to the edges of the ice, from the lush forests of the tropics almost to the top of the highest mountains. However it is clear that animals are not distributed at random over the world; different groups are adapted to different habitats. Less obvious is the fact that in similar habitats in different parts of the world we find distinct groups of animals. Two different factors apply here. Firstly the different major habitats of the earth, which are called biomes, are determined largely by latitude, and secondly the areas of land and sea in the earth cut across these biomes so that, for instance, the tropical forests of the Amazon basin are widely separated from those of West Africa.

First let us look at the major biomes which animals occupy. The most obvious distinction between high and low latitudes is that it is cold near the poles and warm at the equator. The reason for this is simple. At the equator, the sun warms up the earth, its rays passing through the atmosphere as it does so. Near the poles, the sun's rays strike much more obliquely and their available heat has to pass through more atmosphere and spread itself over a greater area of the earth's surface. In the equatorial regions the air, warmed directly by both the sun and indirectly by the land, rises. As it does so, the air cools and drops any moisture it is carrying as rain. This regular and reliable source of rainfall allows the growth of dense forests and thus we find that rain forest is confined to a broad zone about the equator.

The air rising at the equator must be replaced, and indeed it is, for winds blow from higher latitudes towards the equator to do so. As these

Zoology

winds (called the trade winds) blow towards warmer areas they pick up moisture and in those latitudes where they blow all the year round, we find deserts. Thus, straddling the Tropics of Cancer and Capricorn we find the Sahara, the Kalahari and the Great Australian Deserts.

Further north, the wind systems are more complex. The air which rose above the equator descends in the middle latitudes and some air blows towards the equator as the trade winds, whereas other air blows polewards as the Westerlies. There it meets a mass of cold air centred on the ice-caps. Where they meet, they generate great rotating masses of air called cyclones or depressions, bringing with them rain. These latitudes then, between about 40 and 55 degrees, have rain all the year round and are characteristic of much of Europe, including England, Wales and Ireland and the New England states of America. The typical vegetation of such areas is deciduous forests. Further north, low temperatures become important factors and the deciduous forest is replaced by coniferous forests such as those of Canada, Scotland and Scandinavia. Further north still, the temperature prevents all tree growth and the landscape is one of tundra, with only low-growing shrubs and herbs. This is the characteristic vegetation of most land areas north of the Arctic Circle.

In addition to the effects of latitude just described, there are three other important factors. The first is that the earth tilts on its axis some 23 1/2 degrees. The effect of this is that these zones tend to move north and south with the seasons so that there is, in general, a broad band of transition between these zones. Many of these are distinct biomes in their own right. The zone between the desert and the tropical rain forest is called savannah. Although it is warm all the year round, the rainfall is seasonal and cannot sustain a great deal of tree growth. The typical vegetation consists of grasslands with scattered trees, although some areas (called forest savannah) can sustain open woodland. Between the desert and the deciduous-forest zone is a band which receives cyclonic rain in winter, but is dried by the trade winds in summer. The characteristic vegetation of this region, which is called maquis, consists of thorny evergreen shrubs. Maquis is found, for example, around the Mediterranean, in South Africa and southern Australia.

The second modifying factor is the distance from the sea, or continentality. As this increases, the rainfall decreases so that the climate becomes drier and cannot sustain forest. Thus the rain for-

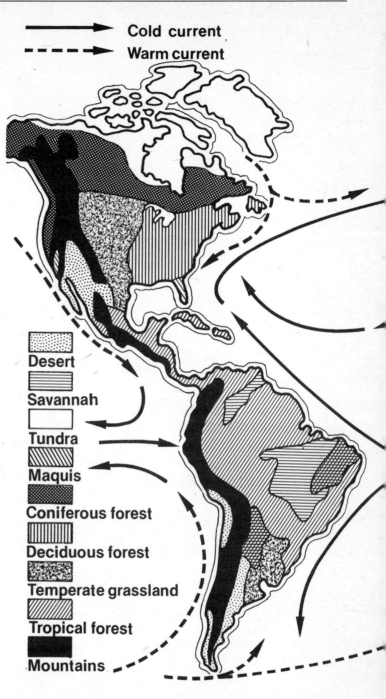

Cold current
Warm current

Desert
Savannah
Tundra
Maquis
Coniferous forest
Deciduous forest
Temperate grassland
Tropical forest
Mountains

est of West Africa gives way to savannah in East Africa. The deciduous woodland of temperate latitudes gives way to grassy plains, called steppes in Europe and Asia, prairies in North America and pampas in South America. Coniferous forest gives way to a sparsely-forested grassland known as taiga, found over much of Siberia.

The third modifier is altitude. As one ascends a mountain it gets colder — mountain vegetation in the tropics is often similar to that of lower altitudes in higher latitudes, thus even Mount Kenya on the equator is capped by tundra! These biomes each have a characteristic fauna, within any one continent. Thus in Africa monkeys are

typical inhabitants of the rain forest, gazelles occupy the savannah, jerboas inhabit the desert and species of gerbils inhabit the maquis in the north. However if we compare the faunas of the Amazon and West African rain forest we find marked differences. The two groups of monkeys are quite different, as are the porcupines. We must look for the origin of these distinctions elsewhere. What is not obvious as we look at the earth today is that the continents have not always been in their present positions. This has had a profound effect on the evolution and distribution of terrestrial animals in the past, but since the continents have been in their present position for about 30 mil-

The effects of latitude on the characteristic biomes is best seen along the eastern Atlantic coastline, from the tundra of Norway to the tropical forest of West Africa. In East and southern Africa there are large areas of forest savannah.

lion years it has not had any significant effect on their present-day distributions. To account for these we must look at the natural barriers to animal emigration.

The most important of these barriers is the sea. Thus Australia has been isolated from all the other continents for about 60 million years. The present-day fauna of lizards, snakes, birds, marsupials and mice arrived from over the sea, but

Zoology

the lungfishes and monotremes, and some of the reptiles, were present in Australia when it was part of a much larger land mass comprising Australia, South America, Africa and India. This supercontinent is called Gondwanaland and the other, northern, supercontinent consisting of North America and most of Asia is called Laurasia.

South America met North America about 60 to 70 million years ago. According to one theory, following this contact, several groups of mammals and other animals colonized South America. The most important of these were the ancestral opossums, the ancestors of the present-day edentates and two groups of primitive ungulates. The physical contact with North America was brief, so that for much of the past 60 million years South America has been isolated. During this time it developed a highly distinctive fauna, but in addition there were some colonizations from the north, across the sea. Many of these were birds, presumably including the ancestors of some of the distinctive present-day South American birds such as the cock-o-the-rock and the oven birds. However a number of mammals also managed the crossing. These included the ancestors of the American monkeys and the South American rodents. These animals had 20—30 million more years in isolation until, recently (no more than three million years ago) the present isthmus of Panama was formed. This resulted in a great interchange of faunas between the north and south; for example, armadillos and opossums colonized North America whilst the raccoon-like coatis and the deer colonized the South. The present-day South American fauna is thus the result of a long period of isolation, followed by a recent admixture.

Australia and South America are the two most distinctive regions of the world from a zoological point of view. Zoogeographers, who study the geographical distribution of animals, call these continents Notogea (meaning Southern Earth) and Neogea (meaning New Earth). The rest of the land is called Arctogea (Northern Earth) in contrast. Within Arctogea there is nevertheless some separation into distinct regions. The principal division is into the warm, more-or-less tropical southern regions, and the more temperate northern regions. The two southern regions are the Ethiopian or Afrotropical region comprising Africa south of the Sahara and the Oriental region comprising India, Bangladesh, South-east Asia and southern China. The northern region, comprising Alaska, Canada, the USA and northern

Mexico, Europe, North Africa and the rest of Asia, is called the Holarctic, though it is frequently divided into the Nearctic (New World) and Palaearctic (Old World) regions. These regions each have their characteristic faunas. The Afrotropical region for instance has hippopotamuses, giraffes and ostriches; the Oriental region has tarsiers, gibbons and peacocks. These southern regions are separated from the northern ones by two factors. One is climate. The warmth-loving animals of the south-China forests do not stray too far north. The other is the physical barriers. The boundary between the Palaearctic and Afrotropical region is the Sahara desert. Between the Palaearctic and the Oriental regions are the Thar desert in the west and the Himalayas in the mid-north, but in the east there is no physical barrier.

The Holarctic region is divided into New World and Old World sub-regions, the point of contact being the Bering Straits. These are now shallow seas but over the past few million years they have been dry land more than once, the most recent time being only 20—30 thousand years ago, during the last Ice Age. The only animals which could cross the land connection at this time were those well adapted to cold. Thus we find that high-Arctic animals such as the snowy owl and the arctic fox are identical all around the pole. Those found further south have different races (as for instance the red deer or wapiti) or closely related species (such as the elk or moose and the beavers).

Finally, let us look at the distribution of animals in time. The earliest records we have of the existence of animal life are found in rocks over 600 million years old. These animals were probably rather like modern jellyfish and soft-bodied worms. It is not until the Cambrian period 600 million years ago that there is an abundance of

The Australian region, being so distinct, is frequently referred to as Notogea. Similarly, the Neotropical region is called Neogea. Between the Australian and Oriental regions lie the islands which show transition between the two regions.

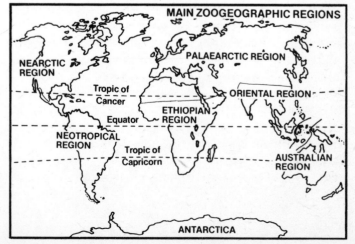

14

fossil evidence. This is because at this time many animal groups evolved hard, readily-preserved skeletons. The most important of these were the molluscs and the trilobites. All of these were invertebrates — animals without backbones, or any other type of bone for that matter. It was not for another 100 million years that the first vertebrates appeared, in the Silurian period. These were primitive fishes, many of which were armoured with bone. The fishes became the dominant animals in the sea during the Devonian period.

Meanwhile the land was being colonized, first by the plants and then by the first animals, primitive insects and scorpions, in the Silurian period 420 million years ago. The vertebrates did not crawl out until the Devonian period, perhaps 50 million years later. These were the first amphibians. For a further sixty-odd million years, these amphibians ruled the land, feeding on the large dragonflies, cockroaches and fish. The Carboniferous period saw the emergence of the first reptiles, about 300 million years ago. These rapidly diversified, much more rapidly than the amphibians had done. By the end of the Palaeozoic period they were the dominant form of life on land and included among their number forms which were recognizably the ancestors of the mammals. The Mesozoic period, between 220 and 70 million years ago, is called the Age of Reptiles. It saw the days of the dinosaurs and the great marine reptiles, but it also saw the emergence of the mammals 180 million years ago and the birds some 20 million years later. At the end of the Mesozoic period, for some reason not yet explained, the great reptiles disappeared leaving the land to the mammals and the air to the birds. In the sea, however, a new type of fish, the teleosts, had arisen and these today rule the seas.

The last great era, the Tertiary or Cenozoic, is often called the 'Age of Mammals' for during this period, which began 70 million years ago, the mammals increased in both number and diversity. An important factor in their radiation was that the continents were now spread in five separate blocks over the earth's surface. Each continent was, in its turn, colonized by mammals and within each they diversified so that different groups adapted to a different way of life. Some became herbivores, some carnivores and others remained less specialized. This is called adaptive radiation and has also resulted in the evolution of groups in diverse parts of the globe, but in similar environments. Thus we find kangaroo-rats in North America and jerboas in North Africa — both looking very similar and living in similar desert habitats but belonging to quite different groups of rodents. This phenomenon is called convergent evolution; it is also seen in the emergence of porcupine-like animals in both the Old and New Worlds. A related phenomenon is parallel evolution which is shown by the monkeys. The Old World and New World monkeys did not evolve from widely different ancestors, but from members of one family distributed throughout the northern hemisphere. Thus when they both colonized the warm, tropical rain forests, their evolution took parallel courses resulting in the two similar families of monkeys that we see today.

The last million years or so have seen four great ice ages. During these periods the north polar icecaps extended southwards to cover much of northern North America and Europe, so that the animal life in these regions was forced south. The southern continents were much less affected. This variable climatic regime was probably responsible for the great number of animal extinctions which occurred at this time. Today only the tropical regions retain their former diversity.

The major groups of backboned animals have arisen gradually over the past 400 million years. Each in turn has had a long period as an underling before becoming the dominant group, and in many cases a period of decline.

EMERGENCE OF MAJOR ANIMAL GROUPS		
PERIOD	FIRST APPEARANCE OF GROUP	DOMINANT GROUP
Tertiary	Rise of Mammals	Age of Mammals _70 million years ago_
Cretaceous	Extinction of Dinosaurs _120 million years ago_	Age of Reptiles
Jurassic	First Birds First Mammals	Age of Reptiles
Triassic		Age of Reptiles
Permian		Age of Reptiles
Carboniferous	First Reptiles _300 million years ago_	Age of Amphibians
Devonian	First Amphibians	Age of Fishes
Silurian	First Amphibians	Age of Fishes
Ordovician	First Bony Fishes	Age of Fishes
Cambrian	Fossils first common _600 million years ago_	
Pre-Cambrian		

The Simplest Animals

The earliest animals probably consisted of a single cell and may have been not unlike some of our present-day single-celled animals which are collectivelly called the Protozoa. These animals, despite their simple structure are surprisingly diverse, ranging from simple sluggish blobs of jelly like *Amoeba* to the highly active and complex ciliates. One of the most remarkable features of some protozoans, however, is their similarity to single-celled plants. Some organisms, such as *Euglena,* have both the green pigment chlorophyll, so characteristic of the plant kingdom, and a flagellum — a whiplike

life-styles. One simple style was adopted by the ancestors of the sponges; each cell remained more or less unspecialized but by mutual co-operation a new source of food — that of single-celled plants drifting in the sea — could be exploited.

In most many-celled animals, the cells have become specialized; groups of similar cells form tissues with specific functions. The simplest such animals are the coelenterates. Each individual consits of a hollow sac made up of two layers of cells separated by a jelly-like material.

The outer layer is responsible for

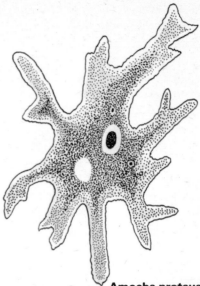
Amoeba proteus

pseudopodium and enclosed in a cavity or vacuole. Secreted digestive juices then decompose the organic material inside the vacuole and the undigested remnants are left behind when the amoeba moves on. Reproduction is simple. The cell nucleus divides into two more or less equal parts, and then the cytoplasm does likewise resulting in two independent individuals each containing a nucleus surrounded by cytoplasm. *Amoeba proteus* is a harmless animal found in mud or damp moss. Many amoebas, however, are parasitic on higher animals including man. One example is *Entamoeba histolytica* which causes amoebic dysentery.

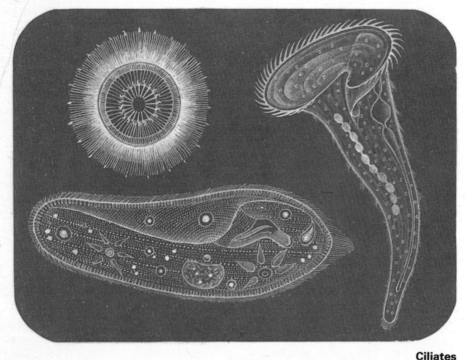
Ciliates

process which enables it to move about rapidly.

The body of a protozoan consists of a mass of jelly-like, granular material called cytoplasm inside an enclosing membrane. The cytoplasm surrounds the nucleus which governs the activities of the cell. If a protozoan is cut in two, the part containing the nucleus will usually survive and grow but the other part will die as soon as it has exhausted its food reserves.

Single-celled animals are nevertheless limited in possible life-styles. When the first animal cells began to remain together in groups, this opened up many new

the detection and catching of food and the inner layer for digesting it. The jelly-like layer between is frequently very highly developed, particularly in the jellyfish.

Amoeba proteus

Amoebas are among the simplest of all animals. A large amoeba measures up to 1 mm (0.04 in) and can be seen with the naked eye. The body continually changes shape as the cytoplasm shifts, making protrusions called pseudopodia or 'false feet' which enable it to move as well as feed. Its prey, for instance bacteria and slow-moving protozoans, is engulfed by the

Ciliates

Ciliates are protozoans whose body is enclosed by a firm membrane covered with rows of tiny hair-like processes called cilia. It is these that gave the whole phylum its name. In some the cilia cover the entire body while in others they cover it only in certain parts like combs. Some ciliates move about freely and actively catch prey, others are permanently fixed. Some species are parasitic on fish. One is parasitic in the intestines of pigs and causes loose, bloody stools. The disease may even be transmitted to man.

The ciliate most often used for demonstration purposes in the school laboratory is *Paramecium caudatum,* which may just about be seen with the naked eye. Paramecia are very important animals for they regulate the number of bacteria and serve as an indication of the organic pollution of bodies of water. Aquarists who have at one time or another examined a drop of water from their aquarium through a microscope are well acquainted with *Stentor coeruleus.* As a rule stentors are attached to aquatic plants with the narrow end of their body but often they cast themselves off from their moorings and float freely in the water.

A very unpleasant visitor in aquariums is *Trichodina domerguei* which is parasitic on the skin and gills of fish, where it feeds on the damaged epithelial cells. Such damaged skin is then readily attacked by moulds which cause serious diseases.

Detail of gemmule

Pond sponge

Pond sponge *Spongilla lacustris*
Sponges are primitive animals which grow under water, especially in the sea, attached to the bottom or other solid objects. Some species are found in freshwaters. A characteristic feature of sponges is the skeleton, often very decorative, composed of calcareous or siliceous spicules and organic matter called spongin which makes a resilient fibrous network. The pond sponge is very plentiful in still and slow-flowing waters. Though it is a very primitive animal — one that might not even be considered an animal by many — it has sepa-

rate sexes. It also reproduces asexually by means of overwintering buds called gemmules.

Brown hydra *Hydra oligactis*
Whereas the body of the sponge is composed of identical cells, the body of coelenterates is composed of cells differing in both shape and function. These form simple tissues. The hydra also possesses a nervous system which transmits stimuli to various parts of the body. Most coelenterates are marine but some are found also in freshwaters. Of these, hydras are the most well-known. Even though they are

plentiful, they readily escape notice because when disturbed they contract into a tiny ball. The brown hydra can be distinguished from other species by its long tentacles, which are up to five times as long as its body. Hydras are easily kept in a home aquarium, where they will feed on water fleas and also on fish fry.

Freshwater jellyfish
Craspedacusta sowerbyi
The word jellyfish is sure to evoke visions of the sea, for the sea is inhabited by a great many species. Some are beautifully coloured, others have extraordinary shapes and remarkably long arms. The freshwater jellyfish has none of these characteristics — it measures 2 cm (0.8 in) at the most and has a translucent jelly-like body. This interesting animal, which has been introduced into European waters, is native to North America and was probably introduced to Europe together with aquatic plants. On sunny days it floats near the surface but otherwise it generally rests on the bottom, rising slowly to the surface and then slowly drifting down again with arms outspread and mouth open.

Brown hydra

Freshwater jellyfish

Worms

The word 'worm' is used for a wide variety of invertebrates — and even for a few vertebrates! It is applied to animals with a long, slender, generally soft body usually, but not always, without any appendages. Nowadays, of course, zoologists recognize that this includes a large number of different animal phyla. Some of these (such as the flatworms) are very simple and primitive but some (such as the segmented worms) have a highly complex body organization.

Many worms are parasitic, living inside the bodies of other animals. These include flatworms such as the flukes and tapeworms and also the round worms and thorny-headed worms.

Free-living worms include the planarians (which are flatworms), proboscis-worms, peanut-worms, spoonworms, arrow-worms and segmented worms. Most of these are marine except for the planarians and segmented worms which include freshwater and land-living species.

Having soft bodies with no internal skeleton, the locomotion of worms depends on either the action of cilia or, in the more advanced worms, the presence of a body cavity containing water

Medicinal leech

under pressure. This is called a hydrostatic skeleton and its action can be clearly seen in an earthworm if it is allowed to crawl along inside a glass tube.

Medicinal leech
Hirudo medicinalis
The medicinal leech is not nearly as plentiful as other kinds of leeches but is doubtless one of the best-known. It feeds only on blood and in former times was commonly used in medicine. Bleeding patients by putting leeches on their bodies was the commonest treatment in centuries past. The medicinal leech

pierces the skin with its sharp jaws and injects into the wound a substance called hirudin which prevents the coagulation of blood. Leeches are used even today, not as blood-suckers, but as a source of hirudin which is used by the pharmaceutical industry as an important component in the preparation of certain drugs.

Common European earthworm
Lumbricus terrestris
If we take a look at the body of an earthworm we will see that it is composed of segments. And if we were to dissect it we would find

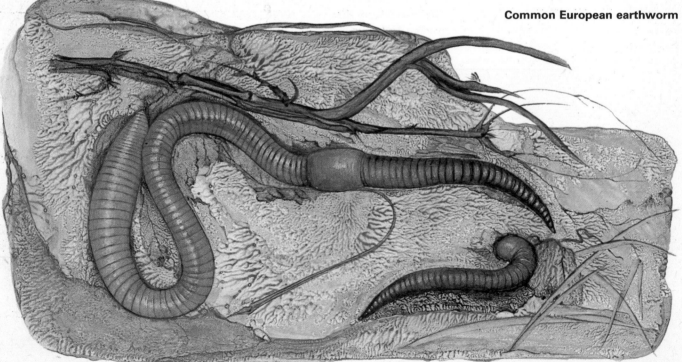

Common European earthworm

that the segmentation is not only external but also internal. This is why the phylum to which the earthworm belongs is called the segmented worms. Earthworms burrow in the soil and their presence is easily recognized by the small mounds they make. These mounds, called worm casts, are composed of soil depleted of humus and nutrients which has passed through the worm's digestive tract and has

approximately 10 tonnes of soil in the form of worm casts! This greatly contributes to, and is very important in aerating and fertilizing the soil. The larger lugworm is taken by fishermen as bait for fish.

Milky planarian
Dendrocoelum lacteum
and Land planarian
Bipalium kewensis
Planarians are typical representa-

extraordinary animals now living. For a long time they were believed to be a connecting link between the segmented worms and the arthropods, for they exhibit characteristics of both phyla. Velvet worms are found only in the tropical rain forests of South Africa, Australasia and South America. They are absolutely defenceless against drought, drying out twice as fast as an earthworm and four times as

Milky planarian

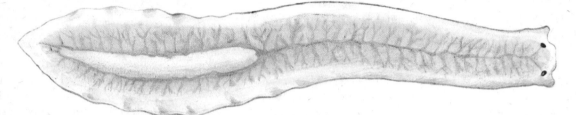

been left on the surface. Soil contains great numbers of earthworms. Most of them may be found in meadows, where they have ample food and little to disturb them. One acre may contain as many as three million earthworms comprising a total weight of about 800 kg (15 cwt). In a single year they move

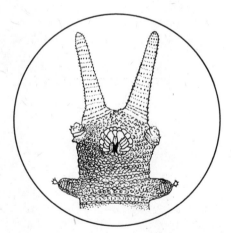

tives of the flatworm phylum. Their body is flat, soft and unsegmented and their anatomy indicates that they retain many simple and primitive characteristics. Most planarians are aquatic. Numerous species are found in various places, in still as well as flowing waters, chiefly under stones. The milky planarian is common in all types of freshwater.

The moist soil of the tropics is inhabited by several species of land planarians. These are generally much larger than their aquatic relatives and sometimes brightly coloured. The land planarian which is up to 15 cm (6 in) long was introduced into greenhouses throughout the world but bears the name of the place where it was discovered by science – London's famed Kew Gardens.

South African velvet worm
Peripatus capensis
Velvet worms are among the most

fast as a caterpillar of the same size. This is due to the structure of the respiratory system, for the velvet worm's breathing tubes terminate in many large openings on the body surface.

Apart from a few exceptions, velvet worms bear living young. The embryos develop inside the mother's womb for a period of 13 months. As offspring are produced every year, during one month of the year the mother carries two sets of embryos inside her body – one that has just begun to develop and a second which is about to be born. Even today scientists are not in agreement as to the classification of these worms. Most modern scientists, however, reject the theory of their relationship to the segmented worms and place them close to the land arthropods, that is, the centipedes, millipedes and insects. Their common feature is the presence of breathing tubes – one pair for each segment.

South African velvet worm

Slugs and Snails

Molluscs are the second largest phylum in the animal kingdom after the arthropods. It is only natural that such a wealth of species is marked by great diversity. Their number includes species with soft bodies protected by shells of widely varied shape and colour. In others the shell is so reduced that there is no trace of it on the surface of the body. Some species move very quickly, but most travel very slowly with the aid of a muscular foot. The majority of molluscs are completely aquatic. Some pulmonates (air-breathing species) live exclusively on land.

Molluscs have a widely varied diet. Some species filter small organic particles from the surrounding water. Others are herbivores and some are predators. Molluscs include species ranging in size from several millimetres to as much as 30 metres! As you can see, this is truly a very large and diverse group and its way of life is so interesting that it is studied by many scientists the world over.

Let us now take a closer look at some of the land molluscs. Their marine relatives will be the subject of another chapter.

The majority of land and freshwater molluscs are pulmonates — that is, lung-bearing. The mantle cavity, which in aquatic molluscs contains the gills, is transformed into an air-filled sac, well supplied with blood vessels. The opening of this sac can clearly be seen on the left-hand side of a slug's body. Despite the apparent symmetry of its body, the slug shows an underlying asymmetry which demonstrates its descent from snails with the usual spirally-twisted shell.

Most land snails are herbivorous. They feed with a special organ called a radula. This lies on the floor of the mouth and gullet. It consists of a gristly rod, equipped with special muscles to move it forward and back, and surmounted by a band of rasping teeth. This band is continually being replaced from behind as the teeth wear away at the front. When a snail feeds it extends the radula over the plant surface and draws it back, scraping off particles as it does so. Molluscs such as snails are unusual in the animal kingdom in that they can actually digest cellulose — the principal component of plant matter. Most animals which utilize cellulose as a food source (such as termites and ruminants) digest it with the aid of protozoans or bacteria in their digestive tract.

Giant snail *Achatina fulica*

The giant snail, largest of the land molluscs with a shell length of up to 20 cm (8 in), has become the scourge of farmers in many parts of the world. In its native land, east Africa, it is more or less harmless. It feeds on all kinds of plants as well as on dead animals. However, as in many other instances where a plant or animal is introduced to other countries, the giant snail has become a pest in all kinds of plantations. As early as 1800 or so it was introduced to Mauritius and in the year 1847 several were released in the neighbourhood of Calcutta. Later, in 1928, it was taken to Sarawak and in 1936 to the Hawaiian Islands. It proved to be not only extremely voracious but also very prolific, laying up to 500 pea-sized eggs every 2 or 3 months, with an incubation period of one to ten days. Because a fertilized snail can lay eggs for many months without further mating, a single giant snail gives rise to

Giant snail

Roman snail

not as rapid as in the case of the giant snail. The Roman snail lays 20 to 60 eggs from which the young hatch after a period of several weeks. In the autumn the snails burrow in the ground and close their shells with a lid of dried mucus (slime) impregnated with calcium salts, wakening from their winter sleep in April.

Great grey slug *Limax maximum* **and Large black slug** *Arion rufus*
The slug is sometimes called a snail without a shell. This is not precisely accurate for slugs have a rudimentary internal shell buried in the mantle. They are divided according to various anatomical characteristics into two groups: keelback slugs

Large black slug

further whole colonies. Today this mollusc is found in practically the entire Indo-Pacific region, in Malaysia, Thailand, Indonesia, Vietnam, the Philippines and China.

In Africa the giant snail is a favourite food in some places as well as an important source of protein. In the case of mass occurrences, however, the ground is literally covered with snails — a calamity which is generally resolved by Africa's bush fires.

Roman or Edible snail
Helix pomatia
To those of you who give an inward shudder at the thought of eating the giant snail it should be pointed out that the flesh of the Roman snail is a favourite food mainly in southern Europe, one that is popular with the general public and not only epicures with a refined palate. In some places snails were even gathered in such quantities that

they had to be rigidly protected by law. That is the reason for the establishment of special snail farms where these tasty molluscs are raised. Reproduction, of course, is

and roundback slugs. The common European great grey slug is a keelback. It is found mostly near human habitations and in cultivated ground where it may sometimes cause damage to vegetables and stored foodstuffs. The large black slug is found in woodlands. It is not always the same colour — some individuals may be rufous, others black, however the foot is always orange on the edge. The large black slug is fond of mushrooms, to the great dismay of mushroom pickers.

Great grey slug

Crustaceans

Crustaceans belong to the phylum of arthropods, animals with a segmented body and segmented appendages. Like all other arthropods, crustaceans have a hard outer covering composed of a complex substance called chitin, but which in most also contains calcareous salts which make it very hard. The hard covering forms a kind of external skeleton that protects the animal but at the same time prevents its growth. For this reason the covering must be discarded from time to time. The freed body is soft at first but within a few days the skin again secretes calcareous compounds and hardens the new cover.

Crustaceans include crayfish, crabs and lobsters, shrimps and prawns. Few people, however, know that they also include a vast number of small animals which float freely in the water, that is, they are part of plankton. Many of these species are extremely important because they are the main food of certain fish. Other species, again, are parasitic on fish. The diversity of crustaceans is truly immense. Besides small microscopic species measuring only a few millimetres, their number also includes species measuring several tens of centimetres, for example the lobster may measure as much as 80 cm (32 in). Most crustaceans are marine animals but many species inhabit freshwaters and others live only on dry land.

Water flea *Daphnia pulex* and *Cyclops strenuus*

If you come across someone standing by the edge of a pond or pool sweeping a net in figures-of-eight over the surface, it is sure to be an aquarist collecting food for his fish. If you were to ask him what it was he was after, his answer would be water fleas. Few of us would realize that this is the name for minute crustaceans found in almost all still waters. The aquarist naturally catches a great many species without making any distinction between them. You might also learn, perhaps, that for smaller species of fish, cyclops are a better food.

These are also crustaceans but ones which differ from water fleas in many respects. Cyclops, like water fleas, are minute crustaceans 1 to 4 mm long. They both bear a large pair of antennae — in water fleas the second pair of antennae is longest, in cyclops it is the first. The long antennae of the water flea serve as a means of movement. With every thrust of the antennae the animal makes a spasmodic leap upward in the water, but if it stops it immediately sinks. This spasmodic

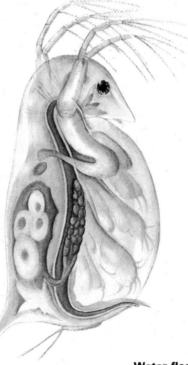

Water flea

leap gives the water flea its name. Some water fleas actively catch their food and chase after their prey, but most feed on minute water organisms which they propel to their mouths with their specially modified legs. Even though water fleas can have separate sexes, the females can lay unfertilized eggs that will develop into new individuals. This method of reproduction is called parthenogenesis. Such eggs are formed throughout the late spring and summer under favourable conditions and produce only females. Reproduction thus pro-

ceeds at a rapid rate and in suitable places the water swarms with water fleas. In the autumn the females lay larger, yolk-filled eggs which are fertilized by males, which are also produced at this time. The development of a fertilized egg halts for a time in the initial stage. The egg is enclosed by a protective cover of which the outer layer is filled with air chambers, thus enabling it to float on the water's surface where it can catch on various objects, for instance the feather of a bird. Birds then carry the egg to another pool, even thousands of miles away, thereby helping to spread water fleas to all waters.

Cyclops generally move very rapidly by means of sudden thrusts with their feet. When doing so they press their long antennae to their bodies. The female is readily identified by the pair of egg-filled sacs on each side of her abdomen. The eggs are always fertilized but under unfavourable conditions a thick chitinous cover forms to enclose them and the eggs sink to the bottom to wait out the inclement period and hatch at a later date. The larva which emerges from the egg is totally unlike the adult.

Water fleas are mainly found in freshwater, though there are a number of marine species. Within the marine plankton, the commonest animals are relatives of the freshwater cyclops, though these differ in that the females bear only a single egg sac. Other cyclops' relatives are parasitic, especially on fish — both marine and freshwater.

Freshwater opossum shrimp *Mysis relicta*

Found in the open seas of the northern Atlantic and the North Sea are vast numbers of minute crustaceans, a little over 1 cm (1/2 in), with translucent bodies. At first glance they look like tiny shrimps and that is what they are generally called. However, they belong to a separate group of crustaceans with branched (forked) swimming legs. Some species are noteworthy in that they are also found in inland freshwater lakes. Their occurrence indicates that these places were

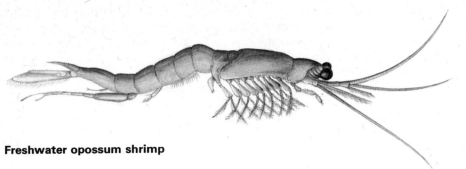

Freshwater opossum shrimp

covered by the sea or at least linked with the sea at one time, probably during the last Ice Age. There is also another means whereby the opossum shrimp made its way inland, for instance to certain lakes in northern Britain and Germany. The retreating ice sheet blocked off part of the sea thereby raising its level; in time the water from this dammed sea found an outlet in one of the freshwater lakes where the opossum shrimps survived and changed into freshwater forms.

In summer these opossum shrimps seek out deep water whereas in the autumn they are more often found near the surface. Their diet is composed of both vegetable matter and small animals.

In addition to these primitive freshwater species, there is a large number of marine and brackish water species. Many of them live in estuaries and ascend up-river into what is essentially freshwater. Many local populations are found permanently in freshwater dykes that have been recently cut off from the sea by man's activities in land reclamation.

American crayfish or Crawfish
Cambarus affinis

Crayfish include many species all with an unmistakable form. All crayfish have strong claws and a straight abdomen terminated by a three-lobed tail. Thrusts of this tail enable it to move rapidly backward and thus escape possible danger. The head is furnished with two pairs of antennae or feelers, which are important organs of touch, and a pair of stalked eyes. These enable the crayfish to see in all directions without taking the trouble of moving its body.

Crayfish mate in the autumn. The male stores his sperm or milt in special capsules called spermatophores which he attaches to the underside of the female's body. The female then crawls into a hole where she lays about 100 eggs which she attaches to the swimming legs on her abdomen. The mucus which holds the eggs fast at the same time dissolves the spermatophore thereby freeing the sperm to fertilize the eggs. The young crayfish, which are miniature replicas of the adults, emerge the following summer. At first they hang motionless on the body of the mother, but after the first moult, when they measure about 13 to 15 mm (1/2 in), they strike out for themselves.

The American crayfish is more brightly coloured than its European relatives. In 1890 it was introduced to Germany and thence spread also to France. It does not require water as clean as its European relative needs and furthermore is also resistant to certain diseases which greatly reduced the number of crayfish in many countries.

The name crawfish or crayfish is also applied to a number of marine crustaceans. These, however, belong to a different group in that they do not have pincers on the first pair of legs, only on the second.

American crayfish

Creatures with Eight Legs

Spiders, harvestmen, scorpions, false scorpions, mites and ticks, and several less important groups make up the subphylum Arachnida. There are four pairs of walking or running legs, and in front of these there are two pairs of appendages, each terminating in a pincer. The diversity of appearance in arachnids is matched by the diversity of their biology. Their numbers include predators and parasites, aquatic as well as terrestrial species, and some which are abroad by day, whilst others remain hidden. Except for regions of permanent ice they are found in all parts of the world.

Garden cross spider

Spiders

Many people find spiders repulsive, or even frightening creatures. However it cannot be denied that their webs are superb creations of breathtaking beauty. To the ancient Greeks the spider was originally a bewitched maiden who was so proud of her skill that she challenged the goddess Athena herself to a weaving contest. The girl, whose name was Arachne, won but was turned into a spider by the goddess in punishment. Here you see how its relatives, the arachnids, came by their name. A great many spiders, of course, do not weave webs but they do spin fine fibres of silk with which they line their abodes and from which they make covers for their eggs. Other species use the fibres as safety ropes. Spiders, especially the newly hatched young, also 'fly' on the delicate fibres as they are carried by the wind. We could find many more interesting things which would help us regard spiders with greater liking. Some species care for their offspring in an interesting manner, others are beautifully coloured or can even change colour. Still others have an interesting social life. In short, everyone who begins to study the life of spiders in greater detail soon loses his aversion to these interesting creatures and comes to love and protect them. There is no denying, however, that many spiders also have negative characteristics. Some, but these are few, produce a poison which is dangerous even to man. Ruthless cannibalism is another trait found in most species, but even this is not always the rule as is often stated.

Garden cross spider
Araneus diadematus

The garden cross spider is one of the few spiders for which man has some slight sympathy, perhaps because we can readily see on its vertical web how many troublesome pests it captures. There are even some people who occasionally bring their spider a fly as a special tidbit. The garden cross spider either perches in the centre of its web or else remains hidden. As soon as a victim is caught in the web the spider immediately rushes

Malmignatta spider

out in response to the signal sent out by the vibrations of the fibres and immobilizes its prey by spinning further fibres round its body. Then it kills it with poison and injects digestive juices into the body which dissolve the soft parts.

Malmignatta spider
Lathrodectus tredecimguttatus

Music composers of old, chiefly Italian and German, composed a number of pieces called 'tarantella'. This is wild dance music in which the main theme repeats itself with urgent persistence faster and faster until it ends as if completely exhausted. The origin of this musical form is linked with poisonous spiders. The Mediterranean region is the home of several truly poisonous, though not very large species. Best known is the malmignatta spider. Approximately five per cent of its bites are fatal. Only the females, about 1 cm (1/2 in) across, are dangerous. The males, which are much smaller, do not produce enough poison to harm man. As early as medieval times, however, the malmignatta's evil reputation was transferred to another species of spider — the much larger and far more dangerous-looking (but essentially innocuous) member of the genus *Lycosa* called the tarantula. It was popularly believed that its bite produced a kind of frenzy in human subjects, a wild dance which ended either in recovery or in death. In one scholarly book from the year 1643 the tarantula is even depicted with the caption 'Musica sola mei superest medicina venenni' (Only music can overcome your poison) plus three rows of music, prescribed as medicine.

The malmignatta spins a web to capture flying insects, which it kills with its poison. Its method of reproduction is of particular interest. The

Black widow spider

Bird-eating spider

delicate fibres strung out among the aquatic vegetation. During the egg-laying period the female spins a kind of platform inside the bell on which she places the eggs, keeping watch over them and later the newly-hatched young in the bottom half of the bell. Every now and then she brings down fresh air bubbles and shakes the whole bell to air it.

For the winter water spiders seek refuge in a stronger and safer place than the web bell — generally the empty shell of an aquatic snail which they fill with air and close with a fine web. The shells, which become buoyant when filled with air, rise to the surface and are carried by the wind to new places.

Bird-eating spider *Avicularia*

South and Central America are the home of the bird-eating spider, one of the largest spiders of all. The biggest live in the Amazon River region. Their bodies measure up to 15 cm (6 in) and with outspread legs they measure about 25 cm (10 in). Both body and legs are covered with hairs, sometimes so thick that they look like fur. During the day these spiders hide in a cavity or rock crevice emerging at night to hunt. They do not spin a web but capture their prey by sudden attack. Their size enables them to catch even small mammals and birds.

adult male wraps his sperms in a delicate web and carries them in his mandibles to a female which uses them to fertilize her eggs even over a period of several months. Sometimes it happens that when the sac containing the sperm cells is handed over, the female turns upon the much smaller male and eats him. As a rule, however, the male fertilizes several females. After several matings, however, he is exhausted and thus easy prey for a female. For this reason this, and other related species are called 'black widow' spiders in some regions, notably in the USA.

Water spider *Argyroneta aquatica*

The water spider is the only spider which lives permanently in water. There it spins its abode in the form of a bell-like structure filled with air. It carries bubbles of air between the hairs on the abdomen. It lurks inside the bell where it lies in wait for prey but it also captures prey in

Water spider

25

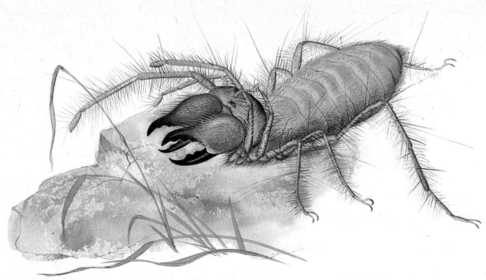

People are commonly afraid of these spiders but they are not nearly as dangerous as they look. Though their bite is unpleasant their poison is weak and not particularly harmful to man. Most people, however, have an allergic reaction to the hairs, which readily break off and catch in the skin. The female lays about 500 to 1,000 eggs which she encloses in a cocoon and carries between her front legs, keeping watch over them until the young hatch. Even then she remains close by and stands guard a few more days after which the small spiders scatter throughout the neighbourhood.

Sheep or Castor-bean tick
Ixodes ricinus

Ticks are parasites. Their mouthparts are furnished with a great many small 'teeth' with which they burrow into the skin to suck the blood of vertebrates. They are capable of sucking up enormous amounts of blood, particularly the females. By the time she has had her fill a hungry female may increase her weight as much as 230 times! The alimentary canal fills the entire abdomen which swells hugely after such a feast. The tick generally perches on a blade of grass or a twig waiting for an animal to pass by and then hops on to it. When it has found a suitable spot where the skin is thin, it pierces a hole in it and inserts its mouthparts to enlarge the opening and gain a firm hold. To keep the blood from coagulating, the tick secretes a substance called ixodin into the wound, which furthermore also dulls the pain of the puncture. Some species of ticks pass their entire life-cycle on a single host whereas others require several hosts. Ticks may also transmit certain diseases, such as encephalitis.

Sunspider
Solpuga letalis

A hairy body, huge claws, lightning-fast, soundless movements — who wouldn't be afraid of such a creature? No one would doubt that it is a poisonous animal, as is furthermore indicated by its scientific name *letalis,* meaning lethal. But this is not the case. Sunspiders are not in the least poisonous, though they can wound and even draw blood with their pincer-like first appendages which they move up and down.

Some species are nocturnal creatures, others are fond of moving about in full sunlight. The smallest measures about 8 mm but others may be as large as 6 cm (2 1/2 in) and their prey is of corresponding size — a large sunspider is capable of downing even a lizard.

Water mite *Hydrachna globosa*

Water mites look like tiny, brightly-coloured marbles crawling on aquatic vegetation or floating on the water's surface. Most are red but some are blue or yellow, often with various markings. The males generally differ in shape from the females. The eggs are laid on aquatic vegetation and when they hatch the larvae are parasitic on various aquatic insects. Adult mites generally measure about 1 mm but may be even larger. They feed on small crustaceans and other aquatic animals.

Water mites are no strangers to freshwater biologists who study their numbers with interest, for many species serve as reliable indicators of water pollution.

Harvestman
Phalangium opilio

A globose body with extremely long, thin legs is an unmistakable characteristic of these arachnids. At first glance they appear to have ten legs but a closer look reveals that

Sheep tick

Water mite

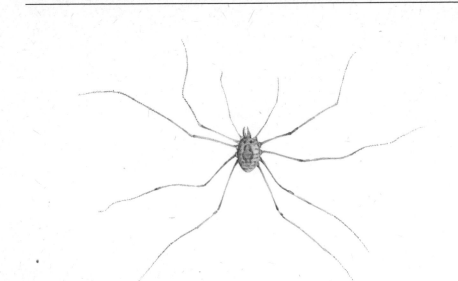

Harvestman

the first, slightly shorter pair, are elongated second appendages. Harvestmen generally perch on a wall or tree trunk, resting during the day and setting out to forage for food in the evening. Their diet is extremely varied. It includes flies, ants, caterpillars, spiders, millipedes and beetles as well as fruit.

Every small child knows that if you catch a harvestman it will shed one or even several legs which continue to jerk for a while after being detached from the body. These legs capture the attacker's attention, giving the harvestman time to escape. The legs often continue to jerk for as long as half an hour,

movements caused by impulses sent out by a group of nerve cells located in the lower part of the leg. It is interesting to note that harvestmen do not have good powers of regeneration; generally they do not grow new legs to replace the old and one often encounters specimens with an incomplete set of legs.

Carpathian scorpion
Euscorpius carpathicus
Scorpions are inhabitants of tropical and subtropical regions and only very few species are found also in the temperate zone. Nevertheless they are well known not only

by virtue of their appearance but also because of their unfavourable reputation. The dangerous-looking pincers, though quite harmless to man, are the second pair of appendages which resemble the claws of a crayfish. At the opposite end of the body, at the tip of the elongated tail, is a poison gland furnished with a sharp point — the scorpion's well-known sting. There is no need to fear the poison of the Carpathian and other related European scorpions, but that of certain tropical species is dangerous to man and the results of a sting are generally tragic. Many countries cite in their statistics that more deaths are caused by scorpions than by snakes. This is probably due to the fact that scorpions often live round human habitations — they may be found in furniture, under rugs and at night they climb into shoes and even into beds.

Although its poison is potent the scorpion uses it only rarely, generally capturing its prey — mostly insects and spiders — with its powerful claw-like pincers. Only when the intended victim is large and strong and puts up a fight does the scorpion plunge its sting into a soft part of its body, leaving it there for a while so the poison will have the desired effect on the victim before pulling it out.

Carpathian scorpion

Centipedes and Millipedes

Don't try and count them, for we can tell you that some species may have even more than a hundred legs whereas others have many fewer. Centipedes and millipedes were formerly placed in the same class but today we know that they are not closely related despite their resemblance. Both, however, have one thing in common — they breathe by means of tracheae. These are small tubules that branch throughout the body and conduct air from the exterior to all the organs and expel carbon dioxide to the atmosphere. These interesting creatures thus stand at the bottom of the important arthropod subphylum Tracheata. Centipedes exhibit a relatively close relationship to insects. Millipedes can be readily distinguished from centipedes. The body of both is segmented but millipedes have two pairs of legs on each segment whereas centipedes have only one pair. Millipedes (apart from a few exceptions) have a cylindrical body whereas that of centipedes is flattened. Millipedes are herbivorous and do not have prominent mouthparts. In centipedes, which are predaceous, the front pair of legs is modified into large poison claws. Millipedes

travel at a slow pace; the legs move in a regular sequence thus causing an undulating motion that proceeds from the front end to the rear. Centipedes generally move very rapidly, which is essential, considering their predaceous way of life. The millipede's body is very hard because its chitinous armour contains calcareous compounds as in crustaceans; in centipedes the chitin is not reinforced in this manner.

Scolopender *Scolopendra*
Scolopenders are native to tropical and subtropical regions. They include many species which look practically identical to the layman. Some are only a few centimetres long but *Scolopendra gigantea* of South America may reach a length of 27 cm (nearly 11 in)! The bite of such a large centipede may be dangerous to man, and in some cases has even been known to be fatal.
Scolopenders are noted for their care of their offspring. The female lays eggs in a mound and curls herself around them. She regularly moistens or dries the eggs, thus maintaining a favourable local environment and protecting them from being attacked by mould. When they hatch, the young, which

Scolopender

have the full number of legs and are miniature replicas of the parents, remain a while longer within the protective embrace of the mother who shields them from enemies.

Himantarium gabrielis
The order Geophilomorpha (which translated means earth-liking forms) includes centipedes with a long, thin body. They travel at a relatively slow pace and have a large number of paired legs. The Mediterranean region is the home of two species that are veritable record breakers as to number of legs and body length. Both are about 20 cm (8 in) long and as to the number of legs — *Orya barbarica* has 125 pairs and *Himantarium gabrielis* has 173 pairs!

Himantarium gabrielis

House centipede
Scutigera coleoptrata

Scutigeromorphs, or shield-covered centipedes, are a truly extraordinary group native mostly to the tropics and subtropics. Their 15 pairs of legs enable them to travel at lightning speed thus making it almost impossible to catch them. If we happen to catch sight of such a centipede in motion it looks like the wisp of a veil whisking across the wall. As yet, we know few details about their way of life. Unlike other centipedes, scutigeromorphs have the external openings of the tracheae located on their backs. Another characteristic is the compound eye which somewhat resembles that of insects. The house centipede is a skilful hunter and both its speed and type of eye indicate that even fast-moving prey poses no problem. The long slender ends to the legs serve as a lasso with which it catches its victims.

Scutigeromorphs as well as other members of the order Lithobiomorpha (centipedes living-under-stones) belong to a group of centipedes whose offspring have only seven pairs of legs when they hatch. The remainder are added only after several successive moults until they total the full 15 pairs. There are seven moults all told, but new segments and legs

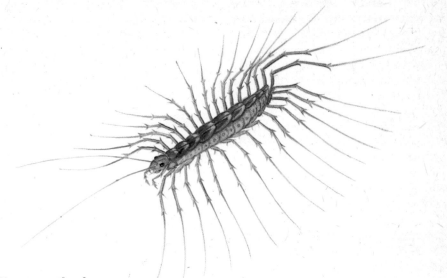

House centipede

grow only during the first four; during the last three moults the young centipede only acquires the more adult characteristics and develops sex organs.

Giant millipede
Spirobolus

Millipedes are typical nocturnal animals which hide under stones, beneath bark or in the soil during the daytime. Some species require a slightly moist environment, others do well even in dry places. They feed chiefly on decaying plant

material so that many also contribute to the formation of humus. Some species feed also on fresh leaves or fruits. Protection against enemies is provided not only by their strong armour but also by numerous glands on the sides of the body which secrete a special foul-smelling substance. In most species it is coloured brown or yellowish-brown, but it may also be red or yellow or even colourless. In some cases it is found to contain cyanic acid, in others iodine, thus furnishing its owner with an excellent chemical weapon. In Mexico the secretion of certain millipedes is used by some hunters to make poison for tipping their arrows. The secretion flows freely from the glands but some large tropical species, such as the one illustrated, are able to eject the poisonous substance a considerable distance.

Some millipedes lay eggs singly and bury them in the ground, others build nests of mud mixed with saliva and sometimes line them with silky fibres. Often the female curls her body round the nest for protection. Unlike the centipedes, the young have only three pairs of legs and the rudiments of further pairs when they hatch. However they moult several times and the number of legs as well as body segments increases with each successive moult.

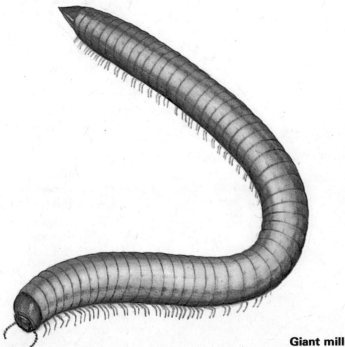

Giant millipede

The Realm of Insects

Insects are the largest existing group of animals in the world. Not only do they include a huge number of species but each species comprises a vast number of individuals. It is estimated that the number of existing species is 1,500,000 to 2,000,000.

The insect's chief protection is chitin, a tough complex molecule secreted by the outer layer of the skin, the epidermis, and forming the main bulk of the outer covering of the body. This varies in thickness. Sometimes it is only a thin membrane, for instance the body covering of a fly larva, at other times a rigid shield such as the hard wing-covers of a beetle. The chitinous cover is composed of separate hard segments joined to each other by flexible membranes so that it can bend. Because it is not elastic it does not allow for the growth of the body inside and must be cast off from time to time. Before it is shed a new cover is formed under the old but this is soft and hardens only after exposure to air. It is at this time, when the old cover has been shed and the new one has not yet hardened, that the body increases in size. During this series of moults, the growing insect usually changes shape. In many insects the young stages are called nymphs and they gradually become more adult-like at each moult. In others, however, the young stages have one characteristic form — the larva — which then, in just two moults (that is via a pupa), changes into the adult or imago. This rapid change is called metamorphosis.

Insects by the Waterside

Insects are found in all types of environments and in a wide variety of places. Water and the waterside, however, is where one will find the greatest diversity of species. The development of a great many is dependent on water. The larvae often live in water for several years whereas the imago, or adult insect, often lives only long enough to perform the duties of mating and reproduction. Some species are superbly adapted not only to life in water but also on the water's surface.

Caddis flies
Family Trichoptera

By the edge of a pool or a mountain stream you may see insects flying about which resemble small moths. These are caddis flies. The adults frequently escape our notice and so the larvae come as a surprise. They live in water and many of them build a case in which they conceal their soft, easily damaged bodies. The material used to build the case may be grains of sand, bits of wood, pine needles and even the shells of small snails. When danger threatens the larva retreats inside completely. If the larva is pulled out with care and released in the water it hurriedly builds a new case.

Alderfly

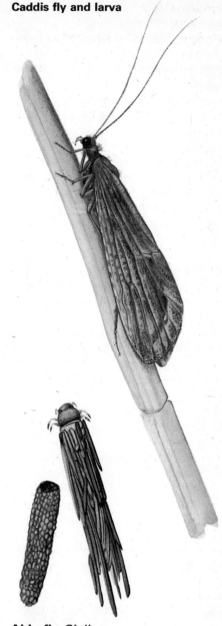

Caddis fly and larva

Alderfly *Sialis*

Late spring and early summer is the time when alderflies may be seen on blades of grass, trees and stones by the waterside. They rest with their wings folded roof-like over their body. The female lays her eggs — sometimes as many as several thousand — on aquatic plants above the water's surface. When the larvae hatch they drop into the water. They have large mandibles and are predaceous. They remain in the water until fully grown (about two years) after which they climb out and pupate under leaves, moss or simply in the soil. The pupa, which is mobile, leaves the shelter after two weeks when the adult alderfly emerges from the case.

Stonefly *Perla,*
Damselfly *Calopteryx,*
Emperor dragonfly *Anax imperator*
Stoneflies are generally found by flowing water, mostly mountain streams, because the nymphs are very sensitive to pollution. The adult fly usually rests on the underside of a leaf and may be recognized by its flattened body and the way the wings are folded flat upon the abdomen. The rear end of the abdomen is furnished with two filaments (cerci). The nymphs are generally found under stones. Some species are predaceous and feed mostly on the nymphs of mayflies.

Damselflies and dragonflies are typical predators. Damselflies are found by flowing water where they move in slow, wavering flight above the shoreline vegetation. Dragonflies, on the other hand, are more likely to be found by calm areas of water. However they may be encountered far from water in forest clearings and fields for they are excellent fliers — it is said that they may even reach speeds of 90 km per hour (56 mph). They capture their victims in flight and the large compound eye is well adapted for sighting prey. The nymphs of both dragonflies and damselflies are likewise predatory. For capturing their food (fish fry or other insects) the lower lip is formed into a formidable weapon called the mask and furnished with a pair of movable hooks.

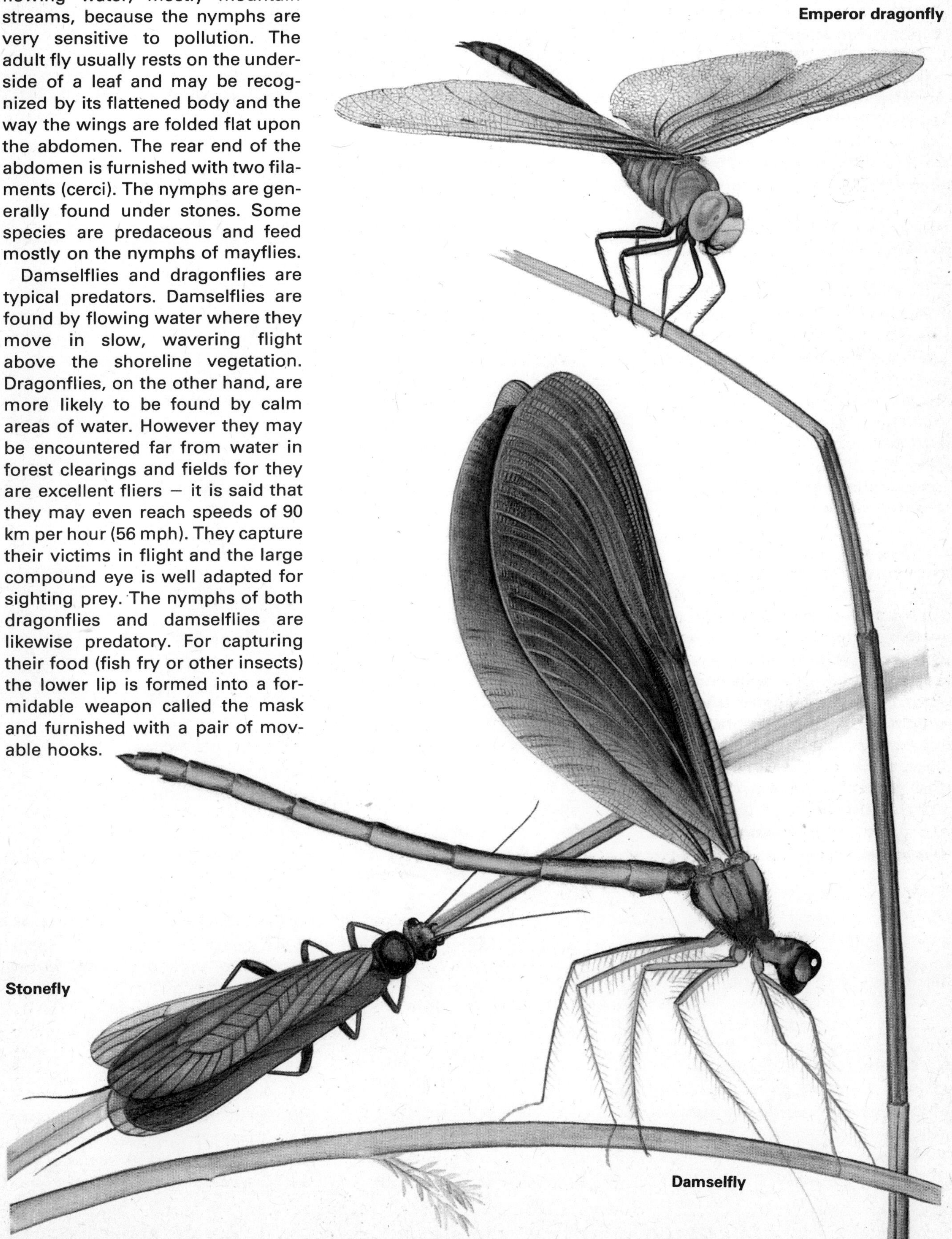

Emperor dragonfly

Stonefly

Damselfly

The Oldest Insects

Formerly insects with biting mouthparts, strong mandibles and straight wings were classed together in the same order. These insects have fore-wings which are hard, leathery and narrow whereas the hind wings are longitudinally folded, broad and membranous.

This group includes the oldest insects, ones which existed more or less as they are today as far back as the Carboniferous period. Today this group is divided into several orders. One includes the locusts and short-horned grasshoppers, a second the bush crickets (often called long-horned grasshoppers), true crickets and mole crickets, a third the cockroaches, a fourth the mantids and a fifth the stick-insects and leaf-insects. All are definitely of interest to man. Some are troublesome companions, serious pests which damage all kinds of farm crops, but there are also others that are useful in that they feed on such pests. Crickets are not of any particular importance but on the other hand their chirping is a pleasant sound and for this reason some species are kept in cages as house pets. As for stick- and leaf-insects, these are often kept as pets, too.

Great green bush cricket
Tettigonia viridissima
Giant locust *Phyllopora grandis*

Towards the end of summer a chorus of chirping comes from meadows, thickets and trees. The bush crickets are serenading in concert. Most are coloured various shades of green. The females are readily distinguished from the males by the long egg-laying organ at the end of the abdomen, used to deposit the elongate eggs deep in the ground. These however, do not hatch until the following year. The nymphs which emerge are very useful, for their favourite food is aphids. Adult crickets are also predaceous and because they may be up to 6 cm (2 1/2 in) long they naturally hunt larger insects. In the male one of the veins on the left wing is furnished with a row of tiny teeth along the edge of the right wing. The great green bush cricket is a good flier and can often be seen on a lighted window at night.

The largest bush cricket is the giant locust from New Guinea, which, measuring almost 15 cm (6 in), barely fits in the palm of the hand. The male of this species is not furnished with sound-produc-

Great green bush cricket

ing organs and so does not chirp. Its name is not quite exact, for true locusts are large migratory grasshoppers which travel in swarms. Nevertheless the name 'locust' is also used, though somewhat illogically, for certain tropical bush crickets.

Desert locust
Schistocerca gregaria

Locusts and grasshoppers can be readily distinguished from crickets by their short antennae. Unlike crickets, which are carnivorous, locusts and grasshoppers are herbivores. Their life cycles have two forms, called phases. In the first, the grasshopper or sedentary phase, the insects remain in one place. The second is the migratory or locust phase. If individuals of the first, sedentary phase occur together in one place their behaviour alters. If they live together like this for a period spanning at least one generation their appearance as well as colour changes. At first they remain sedentary but after a time they begin migrating. It is interesting to note that the change from the sedentary to migratory phase occurs only in semi-desert regions with low, sparse vegetation. Life in swarms is associated with great activity and disruption of the insect's hormonal balance. Young, as yet wingless insects set out on the march, during the course of which they gradually develop into adult insects capable of flight. The swarm's pace is now much faster and clouds of locusts are able to cover enormous distances. Swarms of locusts have been known to fly from their native Africa to Spain, Asia Minor, across Iran as far as Bangladesh and India. Records from medieval days indicate that at one time they even invaded central Europe. Such a swarm, of course, destroys and devours everything in its path. Entire crops disappear within a matter of minutes and it is no wonder that the Bible speaks of swarms of locusts as one of the ten plagues of Egypt.

Mole cricket
Gryllotalpa gryllotalpa

The mole cricket is a very curious insect. It has large shovel-like front legs, somewhat resembling those

Giant locust

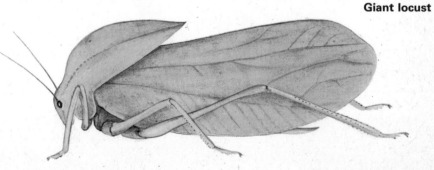

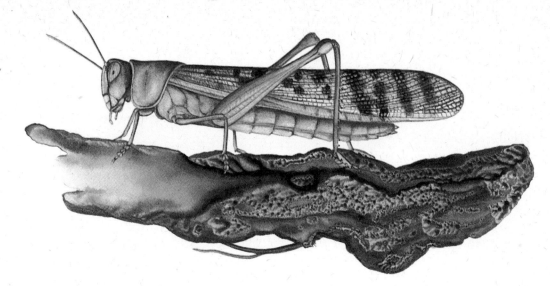

of the mole. This adaptation indicates that the mole cricket lives mainly underground where it burrows and hunts various insects. However it also feeds on the roots of plants that happen to be in its way and may thus become a troublesome pest in a flower bed or vegetable patch. The mole cricket also surfaces fairly often, particularly at night. It is capable of flight, albeit cumbersome, and may even fly to a lighted window. Those who know its 'song' will discover that the mole cricket is not nearly as uncommon as is believed. Its concealed way of life keeps it from being noticed.

The female builds her nest underground. It looks like a hollow sphere with firm smooth walls. Inside she lays as many as several hundred eggs over which she keeps watch. The nymphs remain in the nest until the second moult after which they scatter around the neighbourhood. They pass the

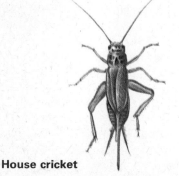

House cricket

winter deep in the ground and only when spring comes do they complete their development into adults.

House cricket
Gryllulus domesticus
This small, pale-brown cricket is found in the wild in southern Europe, North Africa and Asia. In

Mole cricket

temperate regions it occurs in man's dwellings where the temperature is the same throughout the year. Its monotonous chirping is thus most frequently heard in boiler rooms, bakeries and centrally-heated homes. Outdoors they are to be found in municipal rubbish dumps where the heat of fermenting vegetable debris provides adequate warmth. When, on a summer evening, the adults emerge to fly, they attract large numbers of bats. The chirp is made only by the male, who produces the sound by rubbing his wings together.

German cockroach or Crotonbug
Blatella germanica
The great number of common names by which this insect is known indicates how widespread and notorious it is. German cockroach, crotonbug, steamfly, shiner and many more, these are all names for this insect which may be found wherever there is a warm environment – in bakeries, hotels, laboratories and even hospitals. Man is waging a persistent but unsuccessful battle with this troublesome pest. The eggs hatch after two to three months and it takes about ten months for the larva to reach the adult, winged stage after several successive moults. At first it seemed that contact insecticides had devastating results on cockroach populations but man soon discovered that there were increasing numbers of cockroaches which were immune to these poisons. In the end he was forced to try other

33

agents, such as organic phosphates. Cockroaches are a veritable catastrophe not only because they are omnivorous but first and foremost because their faeces contaminate everything, even articles they have not damaged by feeding.

American cockroach
Periplaneta americana
In the wild cockroaches generally live under fallen leaves and decaying vegetation but they may also be found under the bark of trees as well as under stones. Some species, however, have totally adapted to life in man's dwellings where they have become troublesome pests. Man has spread these insects throughout the world and nowadays cockroaches may be said to be cosmopolitan. Cockroaches will eat anything and everything, not only foodstuffs but a wide variety of other materials as well. They will devour book bindings that contain starch or glue and even shoe polish. The American cockroach is found chiefly in the tropics but is also common in all ports. It is somewhat larger than the common cockroach and differs in that both sexes have well-developed wings. Despite its name, it is not native to America but probably to North Africa.

The female lays the eggs in a special case called an ootheca. This she carries about with her on the end of the abdomen for a time before placing it in a suitable spot. The nymphs emerge after about two months. Development is very slow — it may take even several years for the nymph to develop into a winged adult, but generally it does so in one year. During this period it moults six to twelve times.

As we have already said cockroaches may be a veritable catastrophe. Only one thing, perhaps, may be said in their favour. As far as is known they do not transmit diseases. For the natural scientist, however, cockroaches are an extremely important group of insects. Thanks to fossil finds we know that this is a very old group from the evolutionary point of view. Extinct species which were not much different from present-day forms are known to have inhabited the earth as far back as 300 million years ago. Cockroaches are thus one of the few living fossils.

Praying mantis
Mantis religiosa
Practically all the names by which this striking insect is known refer to the pose it adopts when waiting for prey. The forelegs are stout and the second and third joints are adapted to close like the blade and handle of a pocket-knife. The spines on these pierce the victim and immobilise it. A praying mantis lying in wait for prey always has its weapon ready

German cockroach

for attack. It stands motionless on four legs and holds the forelegs together as if praying. Woe betide the insect which is careless enough to come near, for the praying mantis grasps it with lightning speed and carries it to the mouth where it then proceeds to devour it alive at its own unhurried pace. Mantises feed only on insects — except certain large tropical species which occasionally capture a small bird or small lizard. Mantises, however, have many enemies in the wild, mostly birds and larger reptiles. Their only protection is their colouring — green, yellowish or brownish — which blends perfectly with the environment thus serving to conceal them from predators — this is called mimicry. Besides this some species have a body shape resembling a twig or a leaf.

In tropical rain forests one will also find brightly-coloured species which resemble flowers — their method of attracting butterflies and other insects.

Mating in mantises has its problems; because the female pounces on any moving object which approaches her, believing it to be prey, the male could readily become a victim before he has had a chance to fertilize her. For this reason he approaches the female very slowly and cautiously, sometimes taking even several hours to do so. When he has come as close as possible he rapidly seizes the female and mates. Should he happen to disturb her during the pro-

American cockroach

cess, however, she will turn and bite off his head and will then commence eating it. Nevertheless the hind end of the male's body will often complete the act of mating thereby fertilizing the female. The front end of his body is a welcome source of vitamins. After a time the female lays some 100 eggs in a spongy case which rapidly dries and hardens in the air. Each egg is placed in a separate chamber and is thus well protected from drying out

Praying mantis

derived from the Greek word *phasma* meaning apparition. There are two reasons why this order was given such an odd name. The first is the shape of the insect's body. Leaf-insects have a striking resemblance to leaves (their bodies are variously coloured in shades of green and brown) and stick-insects resemble twigs — some smooth, others spiny, so much so that it is very difficult to tell which is insect and which is plant. If we startle such a 'twig' or 'leaf' it suddenly begins to sway on its long thin legs in an attempt to scare us (the enemy) off, this is the second reason for the name phasmid. Leaf- and stick-insects are found in temperate regions. They are most plentiful in the Oriental region but one species extends as far as southern Europe.

Most leaf- and stick-insects are nocturnal; during the daytime they remain motionless, being excellently masked by their mimicry. However, some winged species are also active during the day. In recent years keeping these insects as household pets has become a popular hobby. Not only is their way of life extremely interesting but in most cases they are also very easy to rear, despite the fact that each species will feed on only one or a limited number of plants.

and from the scorching sun. The female makes approximately 20 such cases during her lifetime. The newly hatched nymphs, which emerge all at once, do not resemble the adult insects in the least. At first they hang suspended from the case by thin silky fibres produced by the abdomen but immediately after the first moult they scatter in the neighbourhood and lose the ability to produce these fibres. The young mantis may moult as often as twelve times before it develops into a winged adult. Adult mantises have well-developed wings and are good fliers, even though they use their wings over only short distances.

Leaf-insect
Phyllium siccifolium
Leaf-insects and the related walking-sticks or stick-insects belong to the order Phasmida. The name is

Leaf-insect

Bugs

Many groups of insects have the mouthparts adapted into piercing bristles enclosed in a sucking tube formed by the upper lip. With this beak they pierce the tissues of both plants and animals to suck sap or blood respectively. This is why the order in which they were formerly grouped was given the name Rhynchota, derived from the Greek word *rhynchos* meaning snout. Nowadays we know that insects with such piercing mouthparts belong to several separate orders. All, however, have one thing in common — most are parasites and pests and furthermore many transmit infectious diseases with their bites.

Two of the best known and most striking orders are the Homoptera or plant bugs, which includes the cicadas, leaf-hoppers, aphids and scale-insects, and the Heteroptera or true bugs. Both belong to the group of insects which undergo an incomplete metamorphosis, that is, a gradual development without a resting, pupal stage. The nymphs often differ markedly from the adults, not only in the shape of the body but, particularly in the true bugs, often also in colour.

I have never heard of a bug anywhere on earth being the subject of a poet's verses. Quite the contrary, many are very troublesome, for ex-

Mountain cicada

ample, they may have an upleasant odour. Like many others you have doubtless experienced eating a raspberry or blueberry which had a nasty taste because some bug had sat on it before you popped it in your mouth! Cicadas, however, are often extolled by poets for their 'song' is a typical and lovely aspect of warm, southern climes.

Seventeen-year cicada
Tibicen septemdecim
Mountain cicada
Cicadetta montana

The song of the cicada is well known in the warm temperate parts of the world, but it is only the male that sings. Cicadas have interesting structures on the underside of the abdomen with which they produce shrill sounds. Females do not have these organs and so are truly mute. The organs consist of two strong muscles attached to a plate which is vibrated to produce the well-known 'song' of the cicada. The song is heard a great distance but is very hard to pinpoint and locating the insect is a difficult task even though it remains in the same spot. Cicadas are popular insects in most of the regions where they occur. In some places people even capture them and keep them in cages as pets so they can listen to their song at home.

All cicadas are herbivorous. They pierce various parts of the plant and suck up the sap. The larvae live underground and do not resemble the adults at all. They have a large

head and stout forelegs adapted for burrowing. Sometimes their development takes a long time, the longest being that of the seventeen-year cicada from America. The female lays the eggs in the twigs of broad-leaved trees. The larvae, which hatch after about six weeks, fall from the trees and burrow in the ground, feeding by piercing plant roots. They live underground for 17 years before emerging to climb a tree and change into the adult form after the final moult. The length of this period of development apparently depends on the climate. In related species it is somewhat shorter, lasting only 13 years.

The mountain cicada lives on sunny banks throughout much of continental Europe. Its song is a long, soft, chirping sound, which cannot be compared with that of the southern European species.

Leaf-bugs
Family Miridae
If we wished to select a typical representative of the approximately 50,000 known species of bugs, the choice would be any of the moderately large ones. All are very much alike. Even though at first glance they appear to resemble beetles they have many characteristics by which they can be identified with certainty. For example the large, upper plate of the thorax, which is often large and shield-like, and the unusual formation of the wings. The hind wings are membranous

Seventeen-year cicada

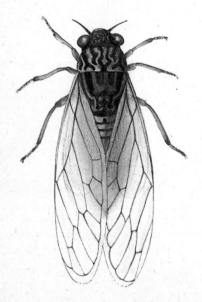

and used for flight and the front wings serve as a protective cover under which the hind wings are folded when at rest. The first two-thirds of the front wings are thickened and leathery and the terminal portion of the wings is membranous. Viewed from above the bug appears to be composed of triangles, a structure one will not find in any beetle. Bugs and beetles also differ in their form of development — bugs undergo an incomplete metamorphosis whereas beetles undergo a complete metamorphosis which includes a pupal stage.

Although some are predaceous, bugs are usually found on plants. Many are important plant pests. Not only do they damage plants by piercing them and sucking the sap, thereby weakening the plants, but sometimes they also transmit virus diseases resulting in expensive damage.

Many bugs have stink glands which secrete a fluid with a penetrating, foul smell. This fluid is released when the bug is disturbed or irritated. This is their defence against enemies and many species of birds will avoid bugs. Some bugs are furthermore coloured bright red which likewise serves to ward off enemies. A bird which has once made the mistake of capturing

Leaf-bug

a red stink-bug will avoid all red-coloured insects in the future. However, these bugs have enemies even though they are provided with such effective chemical protection. These are certain flies whose larvae are parasitic on the bugs and kill them before the bugs are fully grown.

Water stick-insect
Ranatra linearis
Bugs are found in practically all types of environments, from the regions of permanent ice to tropical rain forests. Water is likewise no exception. Some extremely interesting species make their home in water and can be easily studied. Simply look into a pool and you will see backswimmers, water-boatmen and pond-skaters. Backswimmers and water-boatmen have long, swimming legs, fringed with bristles and hunt their prey below the surface. Pond skaters float on the surface film, feeding on insects which become trapped in it. Naturally those which inhabit dense aquatic vegetation and muddy bottoms are not so readily seen. They may be seen only when they rise to the surface for air, which they take from the atmosphere through the long breathing tubes located at the tip of the abdomen.

One such species which moves slowly amidst aquatic vegetation in search of prey is the water stick-insect. Its name indicates its appearance. The water stick-insect truly resembles a thin stick and readily escapes notice amidst the vegetation. This is a predaceous insect which feeds on any animal of reasonable size. It travels slowly along the bottom on its long legs, holding the front, claw-like pair, somewhat resembling those of the mantis, in readiness for attack. Woe betide the victim which comes within its reach. The water stick-in-

Water stick-insect

sect darts forward, grasps the victim in its claws and carries it to its mouth to pierce it and slowly suck its body fluids.

The female lays the eggs on the leaves of aquatic vegetation. The eggs are suspended on two stalks and it takes about 14 days for the larvae to emerge. They resemble the adults but do not have breathing tubes and naturally no wings. They moult several times and not until late summer do they change into adults which will spend the winter in the mud.

Pests and Parasites

The relationships between insects and man are of many kinds. From this viewpoint insects may be divided roughly into four groups — those which are useful, those which are beneficial, those which are harmful, and those which are of no importance one way or the other.

The first group supplies man with food and raw materials, for example silk, honey or wax. The beneficial aspects of the second group may be judged from various points of view. Typical representatives, for example, are pollinators, in other words bees, bumblebees, certain flies, butterflies and moths.

Aphid

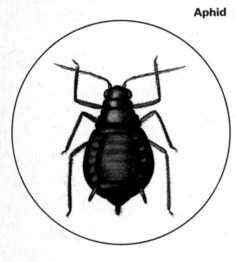

Also included in this group are those insects which destroy or at least are harmful to animals and plants which man considers to be harmful.

The third group, alas, includes a great many species. Insects cause damage to agriculture and forests and many are pests in warehouses and various storage areas. Some species live on or even in the bodies of animals, including man; these are true parasites. Some species of insects are merely troublesome, causing damage only occasionally.

Many institutions throughout the world are concerned with the control of insect pests. For this control to be effective it is first necessary to know the life-cycle of the respective pest, its natural enemies, and so on. Vast sums are spent yearly on this research.

38

Aphids or Plant lice *Aphis*

Aphids suck the juice from plants, causing curling and distortion. They are therefore one of the most feared pests in agriculture, particularly when they occur in large numbers, which often happens. The life cycle of aphids is very complex sometimes with different generations having different appearances. One is winged, the other wingless. Their requirements also differ. Some species live only on one plant whereas others have several host plants during the various stages of their development. One female can produce up to 25 daughters in a single day which are capable of further reproduction without being fertilized by a male within eight to ten days. Naturally the number of aphids can increase at a tremendous rate.

Great wood wasp *Urocerus gigas*

Perhaps you have come across a large yellow-black wasp with a long appendage on its abdomen in a forest clearing on a summer's day. This was a wood wasp and the appendage is an ovipositor or egg-laying apparatus furnished with a fine saw with which the female pierces the wood of a tree to lay her eggs. The legless larvae which emerge from the eggs bore long tunnels into the wood and pupate in a cradle at the end of the tunnels. Development takes one year, sometimes even several years.

Trees used for egg laying are not damaged by the tunnels, even though the tunnels measure about 5 mm (1/4 in) in diameter, but the wood depreciates in value and cannot be used for quality products.

Tsetse fly *Glossina palpalis*

In the late 19th century sleeping sickness depopulated vast parts of Africa. This terrible disease raged chiefly round Africa's large lakes. It is transmitted by a variety of tsetse fly, principally by the species *Glossina palpalis*. Other species inhabiting drier regions transmit similar diseases to animals. They go by various names in various parts of the continent — nagana, baleri, ghindi and suma. They are caused by trypanosomes — single-celled animals which live as parasites in the blood of diseased vertebrates. These are transferred to healthy animals as well as man by the bite of the tsetse fly. Nowadays we luckily have effective medicines which can cure sleeping sickness, at least in its initial stages. Nevertheless, the disease as well as the tsetse fly remain a serious problem in Africa and great effort is being expended to eradicate this scourge.

The tsetse fly requires a moist environment for the development of the larva as well as the adult and this is why it is most often found in waterside thickets by rivers and lakes. The fly avoids open country. Every 14 days the female produces

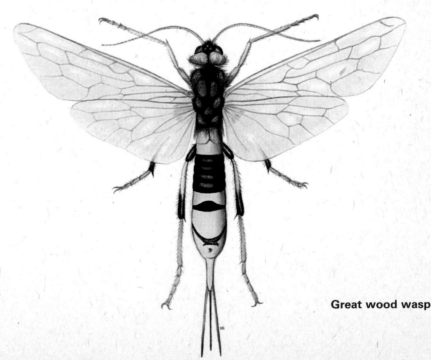

Great wood wasp

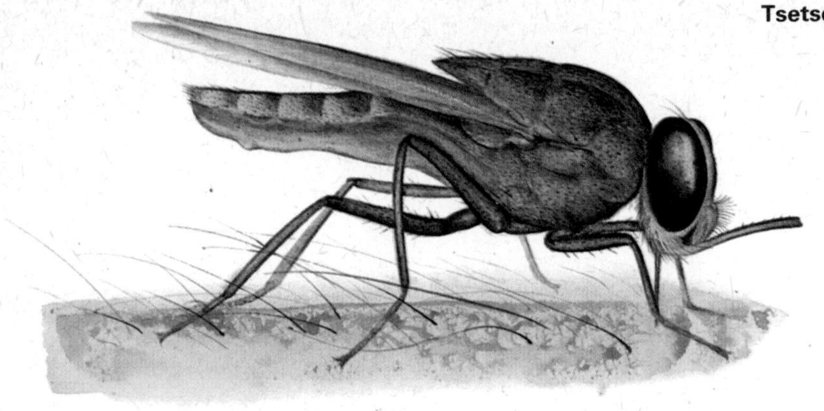

Tsetse fly

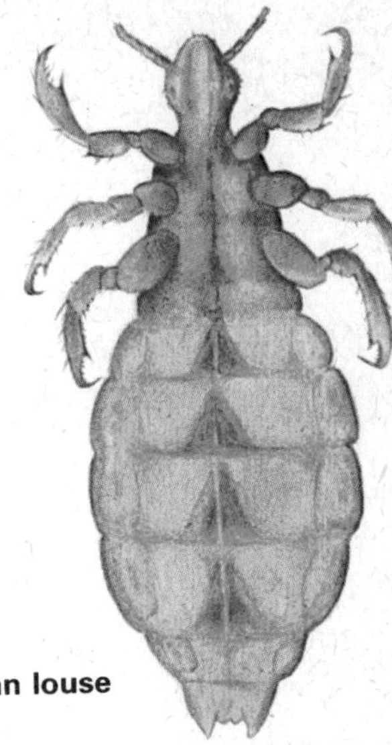

Human louse

an active larva so fully developed that it does not require any food. It burrows in the ground and after about five hours changes into a pupa. The adult fly emerges from the pupa after about a month. The tsetse fly is infected by sucking the blood of a diseased person or animal infested by trypanosomes. These undergo a rather complex development in the fly's digestive tract so that an infected fly is able to transmit the disease only after three or four weeks. Trypanosomes transferred to the blood of another host multiply rapidly. At the beginning sleeping sickness is characterized by fever and inflammation of the lymphatic glands. At this stage it can still be halted and successfully treated. In the second stage, the trypanosomes enter the nerve centres causing inflammation of the brain, and wasting of the muscles. Death occurs from absolute exhaustion of the organism.

Human flea *Pulex irritans*

The flea is one of the more famous insects with its body laterally compressed from side-to-side and well-developed legs that enable it to make very long jumps. The flattened body is a superb adaptation for moving about in the fur or feathers of the host, for fleas are parasitic on mammals and birds. It seems that the human flea became a parasite of man when, as a hunter, man tamed the first dog. It was probably then that the flea passed to man and its development to man's dwellings. Fleas are parasitic only on mammals and birds which have a permanent lair or nest – in other words, of the mammals, chiefly rodents and beasts of prey, insectivores and bats. It is interesting to note that man is the only primate attacked by fleas; lemurs and monkeys, with their less settled life-style, are not troubled by them. Of the birds, those most frequently attacked are those which nest in cavities, such as woodpeckers, titmice, sand martins, sparrows and certain kinds of pigeons. It often happens that each animal species has its special species of flea. This, however, is not a general rule for some species of fleas are to be found on a many different types of hosts.

Fleas lay their eggs in the nests of their hosts or in cracks in the floors of human dwellings. When they emerge the larvae feed on various organic remnants, which are plentiful in such places. After two weeks or so the larvae pupate. They are able to remain in this state for very long periods of time, particularly if the host has abandoned the dwelling-place in the meantime. Most species, including the human flea, are sensitive to jolts or sudden vibrations and these are an impulse for the emergence of the adults. This is an excellent arrangement which guarantees the newly-emerged flea an immediate supply of food, and the chance to reproduce, since unless it is able to suck blood it cannot multiply. However, an adult flea can also go without food for a long time. Human fleas have been known to go without feeding for almost two years.

Human louse *Pediculus humanus*

The human louse is perhaps less famous than the flea, but in fact is more prevalent in western societies. Unlike the human flea, which is the only primate flea, lice are found on almost all mammals, indeed the human louse is also found on the chimpanzee. There are two distinct races of the human louse, one specialized for living on the hair of the scalp and another which lives among clothes next to the body. In conditions where men are not able to wash (often in the past they didn't anyway) body lice may become very troublesome – they were one of the scourges of troops in the trenches in the Great War. With the advent of synthetic insecticides they can now be controlled much more easily. Nowadays there are still regular outbreaks of head lice among schoolchildren – both rich and poor!

Human flea

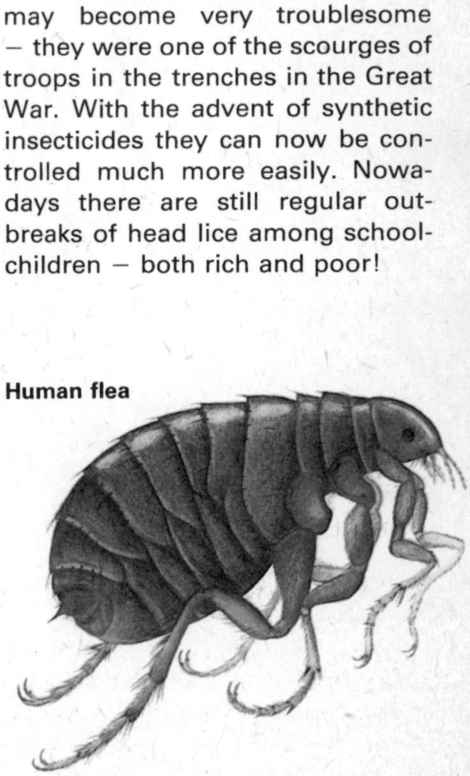

Beetles

Beetles are insects which undergo a complete metamorphosis, this means that their development includes an egg, larva, pupa and adult stage. Beetles have two pairs of wings, but only the hind wings are used for flight. The modified front wings or elytra are hard and chitinous. They serve as a protective cover for the membranous hind wings. Of course here, too, there are exceptions. In some beetles the hind wings are reduced or lost and the elytra may be fused together, in others the elytra may be soft, shorter than the body or otherwise modified. This, however, in no way alters the fact that elytra are a typical characteristic of beetles found in no other insects.

The development of beetles is as diverse as their manner of feeding. Beetles are widely distributed and are to be found throughout the world in all types of environments, both on land and in water. However, the greatest number, both as regards species and quantity, are to be found in the tropical and subtropical regions. It is estimated that there are altogether some 400,000 species but many new, as yet unknown species are discovered and described every year so that the final count will probably be much higher. The English biologist J. B. S. Haldane was once asked by a contemporary theologian what he could infer about the nature of the creator from his study of creation. Haldane replied: 'An inordinate fondness for beetles.' The size of beetles varies, ranging from minute species less than 1 mm in length to giants measuring about 20 cm (8 in.)

The unifying feature of the beetle order is the strong, external skeleton, perhaps earning them the title of the 'knights of the insect realm'. This serves them in many ways. It protects them against mechanical injury — if you tread on a beetle, it is much more likely to survive than a caterpillar! Their armour also serves as a defence against predators, but some beetle species have developed even more bizarre defences. The bombardier beetle for instance releases small quantities of a strong fluid which spontaneously explodes on contact with air! Despite their apparent unity of structure as adults, beetles have very varied larvae, ranging from active predators, through caterpillar-like forms to totally legless grubs.

Titan beetle *Titanus giganteus*

The titan beetle is the largest beetle in the world. It measures up to 20 cm (8 in) in length, and this does not include the long antennae which are typical of the whole longhorn beetle family to which it belongs. The titan beetle was first described on the basis of a single specimen found dead floating down the Amazon River. In the ensuing decades further specimens were found under like circumstances, either in the Amazon or the Oiapaque river. The titan beetle was one of the most sought-after rarities amongst beetles and German collectors paid from 200 to 400 gold marks for a single specimen, depending on its size and condition. In 1950 there were about 40 specimens of this species in world collections, 30 of which were the property of Mr Tippmann of Vienna. However, no European had ever seen a living specimen. The first to do so was Dr. Paul A. Zahl, who in October 1957 captured a handsome specimen in the offices of an oil refinery in Manaus. Later he succeeded in obtaining 15 further specimens from the natives, in other words more than found in all the entomological collections in the United States. In 1972 a Brazilian-American expedition obtained a greater number of specimens and learned something of the biology of this species. Nevertheless still very little is known about the life of this particular beetle.

Hercules beetle
Dynastes hercules

The Hercules beetle is native to tropical Central America and the northern parts of South America. The male has huge horn-like processes on the head and thorax. When he moves his head these processes open and close like pincers. If we measure this beetle including the horns we will find that it is one

Titan beetle

of the largest beetles in the world, measuring about 17 cm (nearly 7 in) in length. The females do not have these processes and are altogether much smaller. The males are said to catch hold of slender twigs with their horns and rotate round them in flight, thereby attracting females during the mating period.

Goliath beetle *Goliathus*

Goliath beetles inhabit the upper storey of the tropical rain forests of western and central Africa. They visit flowers in the treetops chiefly during the hottest part of the day, only rarely descending to the ground. Generally these are newly-emerged adults, for the larval stage is passed in decaying wood. However, these newly-emerged beetles are the handsomest. Their elytra are covered with fine velvety hairs which are extremely delicate and easily rubbed off. This is why undamaged specimens are rarely seen in collections. Goliath beetles include several species ranging in size from 7.5 to 12 cm (3 to 5 in). Some are plentiful at least during certain periods of the year, others are entomological rarities.

Stag beetle *Lucanus cervus*

Most large beetles are native to the tropics but large handsome species are to be found even in temperate regions. One such, found in Europe, is the stag beetle. The male

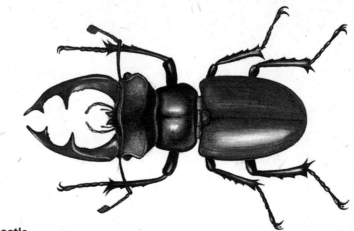

Stag beetle

has large massive mandibles and measures up to 8 cm (3 1/4 in) (including the mandibles). The female is smaller and does not have large mandibles. Stag beetles pass part of their life cycle in old oaks and so it is no wonder that currently their numbers are rapidly decreasing along with the disappearance of these large old trees. Large numbers of stag beetles are to be found only in southern Europe. There are always more adult males than females and for this reason the males engage in jousts for the females, their huge mandibles standing them in good stead for

this purpose. Males with larger mandibles are often acknowledged as victors even without a contest. The female lays eggs in the decaying wood of old oaks and usually dies immediately afterwards. It takes four to five years for the larva to reach a length of about 10 cm (4 in), after which it pupates. It is possible to tell the sex of the adult already in the pupal stage for in male pupae the contours of the large mandibles are visible on the underside. Three months later the adult beetle emerges from the pupa but is short-lived. It feeds on the sap oozing from wounded oaks.

Hercules beetle

Goliath beetle

The Ways of Life of Various Beetles

Considering the multitude of beetles and their immense diversity one is not surprised by their many differing life styles. The various environments they inhabit require varying adaptations on their part — adaptations to the shape of the body or method of reproduction. The life of beetles is studied not only by many professional scientists but also by numerous amateurs.

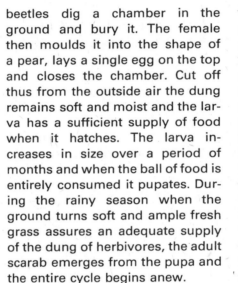

Scarab

Scarab *Scarabeus sacer*
This beetle was held to be sacred by the ancient Egyptians. It was regarded as the symbol of creation and creative power and was also symbolic of the sun god. It was copied by man in the form of seals and beads so he could constantly carry its image about with him and also carved from stone to grace his temples and burial grounds.

The scarab is one of the dung beetles feeding chiefly on the dung of ungulates. First it makes a small ball of the dung which it then rolls with its hind legs to a suitable spot where it eats it. During the breeding season both parents provide food for their offspring. First one of them brings a small particle of dung on the shovel-like edge of its head and with its front legs, furnished with finger-like appendages, shapes it into a small ball. This is often rolled about by both beetles so that further bits of dung stick to it and it increases in size. Such a ball, intended as a store of food for the offspring, is of finer dung than the ball on which the adults themselves feed. When it is big enough the

Burying beetle

beetles dig a chamber in the ground and bury it. The female then moulds it into the shape of a pear, lays a single egg on the top and closes the chamber. Cut off thus from the outside air the dung remains soft and moist and the larva has a sufficient supply of food when it hatches. The larva increases in size over a period of months and when the ball of food is entirely consumed it pupates. During the rainy season when the ground turns soft and ample fresh grass assures an adequate supply of the dung of herbivores, the adult scarab emerges from the pupa and the entire cycle begins anew.

Burying beetle *Necrophorus*
There are several species of burying beetles. Most are coloured black and yellow, some are black, but all are noted for the parental care they lavish on their offspring.

When a small animal dies its carcass is soon visited by a number of these beetles, probably attracted by the smell. The strongest female chases all the others away and then, aided by a single male, excavates the earth beneath the carcass until it sinks out of sight beneath the surface. Then, with their front legs, the burying beetles shape the carrion into a ball-like mass forming a cavity around it. After this they inject digestive juices into the decaying body to dissolve the tissues as they cannot chew but can only suck food in liquid form. Next to the carrion chamber the beetles dig a gallery on the walls of which the female lays the eggs. These are

few for she could not manage to give a greater number of larvae proper care. The larvae hatch after five days. During this waiting period the female injects further drops of digestive juice into the carrion to prepare a sufficient supply of food for her offspring. When they hatch the larvae make their way to the carrion chamber and wait for the mother. As soon as they sense her presence they raise their bodies upright and flutter their legs. One after the other they are then fed a drop of brown liquid from the mother's mouth and continue to be fed thus until they have moulted twice. The larvae shed their skin every twelve hours and after the second moult they burrow into the ball-like mass to feed by themselves. Having fulfilled her task the mother dies. The larvae remain together at first because their combined digestive juices decompose the carrion quicker thus making the food available more readily. The third, final larval stage ends after six days and the larvae then burrow into the ground to pupate. The adult beetle emerges about two weeks later.

Colorado beetle
Leptinotarsa decemlineata
The Colorado beetle is a pretty insect. It is yellow with ten longitudinal black lines on the elytra, from which it takes its scientific name *decemlineata*. The reddish-orange larvae with black head and two rows of black spots on the sides are among the prettiest of beetle larvae. Though handsome, however,

this beetle is a serious pest to man. This was not always so. At one time the Colorado beetle lived in the Rocky Mountains on buffalo burr, a plant of the nightshade family. In 1850 the pioneers who came to settle the west brought with them potatoes which are also of this family, and which they planted out in fields. Ten years later it was discovered in Nebraska that the Colorado beetle prefers the leaves of the potato to those of its original host plant. The damage was catastrophic and the potato crop completely destroyed. The pest rapidly spread from one field to another. In 1864 it appeared in Illinois, in 1869 in Ohio, and in 1874 it reached the eastern coast of the United States. The ocean was a serious obstacle to its further spread. But not for long. In 1922 the Colorado beetle occurred in large numbers on the other side of the Atlantic, in France. Individual beetles, however, had made their appearance on the European continent long before that, having made their way to the new, congenial environment probably by ship. The two world wars contributed to the rapid spread of the Colorado beetle because the world was concerned with more important and pressing matters than the control of this harmful pest. The result was that in the years 1946 to 1949 various parts of Europe were the scene of calamitous outbreaks that had to be quelled by all possible means. Nowadays this occurs only occasionally and every such outbreak is immediately localized and eradicated.

Alpine longhorn

Alpine longhorn *Rosalia alpina*

Longhorn beetles, as their name implies, have very long antennae which either curve gracefully backward alongside the body or forward in front of the head. The family to which these beetles belong contains beetles a great many species. It is presently estimated that there are some quarter of a million, found mostly in the tropics. They were, and continue to be, the most highly prized by insect collectors. The larvae of longhorns live in wood where they bore tunnels of various kinds, depending on the species. Some live only in dry wood, whereas others prefer living wood and so can become serious pests in forested areas.

One of the most handsome of the European species is the alpine longhorn, found occasionally in old beech woods in the foothills of the Alps, in the Carpathians and Beskids and in a single isolated colony in the Juras.

Great diving beetle
Dytiscus marginalis

In calm waters overgrown with vegetation you will often come across a relatively large beetle with hind legs modified to serve as oars.

It is the great diving beetle, a predaceous insect and fast swimmer which lives almost exclusively in water. However, it is not able to take oxygen from water because it breathes through tracheae and so must come up for air.

In the early spring, shortly after the last snow and ice have melted, the diving beetles mate. The male grasps the female from above, keeping a firm hold by means of suckers located on the terminal section of the front legs. The pair then swims about thus, sometimes for several days. The eggs are laid singly in aquatic plants in which the female makes openings with her short ovipositor. She lays a great many — anywhere from 500 to 1,000 eggs. The larvae hatch after about a month. They have a large head and huge mandibles, for they are predaceous and often attack prey larger than themselves. The larva feeds by thrusting the mandibles, which contain a narrow canal, into its victim, injecting digestive juices to decompose it and then sucking it dry. After the third moult the larva climbs out of the water and pupates in an earthen cocoon on the shore. The adult beetle emerges the following spring.

Colorado beetle

Great diving beetle

43

The Social Insects

There are two unrelated orders of insects (the Isoptera or Termites and the Hymenoptera — Ants, Bees and Wasps) which live in extended families — all the members are the offspring of a single queen. These may be small, as in some solitary bees, or very large, as in many termites and ant colonies, with specialized groups (called castes) performing special functions.

Leaf-cutting ant *Atta cephalotes*
The social organization of these South American ants does not differ greatly from the community organization that exists among the bees. However there is one peculiarity which marks this species out for special attention. Ants of the genus *Atta* have a curious method of feeding. They feed on fungi which they cultivate assiduously in their nests. The huge underground nest has a large crater-like entrance from which rows of workers emerge to climb trees where they bite off pieces of leaf to carry back to the nest. Other workers chew these leaf fragments into a mush for growing several species of fungi. Another caste of workers sows spores and harvests the fungus which supplies food for all. If the young queen prepares to set out on her nuptial flight she puts some fungal threads on her excrement. When she has grown sufficiently she trains gardeners to tend the fungi, feeding them eggs the while. It is not until later that workers hatch from the eggs, workers which will go out to gather leaf fragments, and the entire colony can feed on

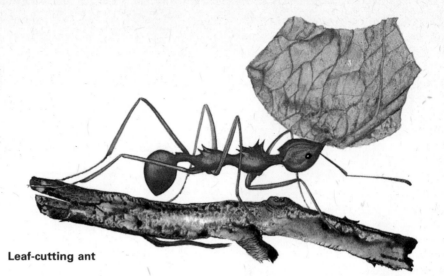

Leaf-cutting ant

the fungus growing on the prepared leaf-pulp medium.

Termites
Order Isoptera
Termites are close relatives of cockroaches and are likewise among the oldest of insects, having existed on the earth for at least 250 million years. Nowadays there are some 2,000 species, about one third of them in Africa. The termite colony is headed by a queen and king. In some species the two live together with the other termites, in others the royal pair lives in a separate chamber in the centre of the nest. The workers are the most important caste and include members of both sexes. The other large caste of soldiers likewise includes both males and females. Neither the workers nor the soldiers are able to exist by themselves. The soldiers cannot even procure food for themselves but must be fed by workers.

A further caste consists of unspecialized males and females capable of reproduction only if, for some reason, the royal pair dies or does not produce further offspring. Termite nests are complex structures composed of a system of chambers and corridors and in some species also of gardens where the termites grow fungi. Some nests are completely concealed underground, others are topped by mounds which are typical for each species.

Honeybee *Apis mellifera*
Most bees are solitary but four species are social creatures which live in colonies and produce honey, a favourite food of man. The honeybee is the most important of the four. Originally it lived freely in the wild, building its nests in tree or rock cavities in woodlands. Since time immemorial, however, man has tried to provide it with man-made cavities and eventually it became entirely domesticated, living almost entirely in hives made by man.

In every beehive there are three castes: the queen bee, who lays the eggs, the females or workers, and at a certain time of the year the males or drones, whose function is to fertilize the queen. The female workers see to the construction and maintenance of the nest from the very first. During their brief life-span, which lasts about one month, the workers perform several functions. The first three days they clean the combs and prepare them

Termite-soldier and queen

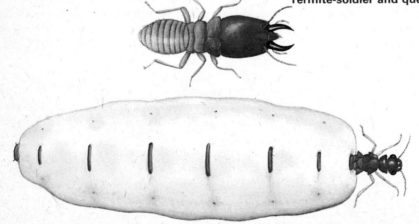

for egg laying. The next ten days or so they tend the larvae, feeding them fluid secreted by special glands. In the meantime the wax-secreting glands develop and the workers begin to build and repair the combs. They also take collected pollen and nectar for further processing and from the twelfth day on they fly out of the hive to collect the nectar, pollen, and water.

Bees start a new colony in a different manner from ants and termites. During the period when new females − young queens − are developing in special cells inside the nest, a large number of workers, led by the old queen, leave the hive. This is known as swarming. The whole swarm settles somewhere on a branch and a number of bees fly out to survey the territory and find a suitable place for a new nest. When they have found such a place they lead the whole swarm

there. The workers then set about building combs and the queen begins laying eggs. In the meantime new queens hatch in the old hive. The first queen to be born kills the other queens with her sting, aided by the workers, and then takes off on her nuptial flight together with the drones. After being fertilized she returns to the hive and the drones are all chased away or killed by the workers. The queen's lifespan is much longer than that of the workers − some four years.

The manner in which bees pass on information about sources of food is of great interest. Workers which return to the hive 'dance' at the entrance to the hive or on its walls, the curves their bodies describe, the pace and the direction of their movements revealing to the other bees the direction and distance of the food supply. The sun serves as the point of orientation.

Honeybee

cells attached to the underside of the canopy by a stalk and opening downward. Later she encloses the nest in a paper envelope. Here she lays the first eggs and raises the first larvae which she feeds by herself. When the first workers hatch they take over the queen's duties (apart from laying eggs) and build

Common European wasp and nest

Movements of a bee when 'dancing'

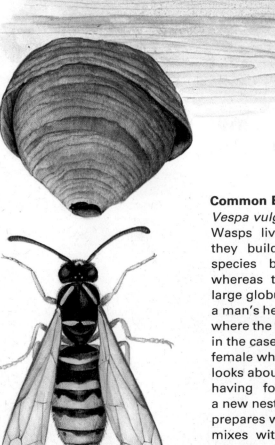

Common European wasp
Vespa vulgaris
Wasps live communally in nests they build of paper cells. Some species build only small nests whereas those of others may be large globular structures the size of a man's head. Unlike the honeybee, where the whole colony hibernates, in the case of the wasp it is only the female which does so. In spring she looks about for a suitable place and having found it begins building a new nest. With her mandibles she prepares wood shavings which she mixes with saliva and constructs a small canopy. From this she suspends the first hexagonal paper

further tiers of paper cells. The individual tiers are joined by paper stalks. Whereas the queen fed herself and the first larvae nectar, at this stage the wasps begin feeding on insects and insect larvae, mostly flies and caterpillars. When completed such a nest has six to ten tiers and measures about 23 cm (9 in) in diameter. The paper envelope provides good insulation and protects the colony from temperature fluctuations outside the nest. Fertile males and females hatch in late summer, but fertilized females are the only ones which survive the winter to found new colonies in spring.

Butterflies and Moths

Butterflies and moths belong to a large order of insects (Lepidoptera) with a characteristic body structure that is always more or less the same. No one has any trouble identifying a butterfly or moth. Butterflies and moths are insects which undergo a complete metamorphosis. Their larvae are generally referred to as caterpillars. Unlike the caterpillars, which have well-developed mouthparts adapted for chewing, the jaws of almost all adult butterflies and moths are modified into a tubular organ called the proboscis. This is always spirally coiled beneath the body when at rest. The most striking feature, however, the pride and glory of butterflies and moths, are their beautiful wings. These are membranous and in most cases covered with thin layers of minute scales. The magnificent coloration of the wings may be produced in three ways — by pigments, by the interference of light by the scales, or by a combination of both.

Butterflies and moths are found everywhere — from cold regions, where one will find mostly various species of small moths, to the tropics, which boast the greatest number and the most beautiful species. Butterflies and moths inhabit all types of environments — even water, at least in the larval stage. The number of species is much smaller than the number of beetle species — 'only' about 100,000, but the number of fanciers and collectors is not. The beauty of their wings inspires poets and composers. However, there is unfortunately another side to the picture. Butterflies and moths are not only a source of joy and beauty, their numbers also include a great many whose larvae are serious pests.

Atlas moth *Attacus atlas*
Edwards atlas moth
Attacus edwardsi

Every collector of exotic butterflies and moths longs to have specimens which are large and beautifully coloured, — and — undamaged. The latter is sometimes very hard to come by for it is not a simple matter to catch a large specimen without rubbing the scales from its wings. For this reason many collectors rear their own specimens from larvae. This is often the case with members of the family Saturnidae, which include huge moths whose large wings generally have conspicuous markings and colouring and striking glassy spots. The largest and loveliest members of this family are moths of the genus *Attacus*. The atlas moth, for example, has a wingspan of up to 24 cm (9 1/2 in) and is thus the largest of all moths.

Edwards atlas moth is slightly smaller (wingspan 17 to 19 cm (6 1/2 to 7 1/2 in) but has more conspicuous markings and colouring. The atlas moth is native to South-east Asia and its large caterpillars, more than 10 cm (4 in) long, are sometimes pests of cinchona trees, lemon trees, tea plants and coffee trees.

Monarch or Milkweed butterfly
Danaus plexippus

Some species of butterflies and moths have an instinct which leads to the formation of groups or swarms. Sometimes such swarms fly together in one direction and cover vast distances. In some cases such a journey may be an isolated event but a number of species make regular migratory journeys.

The best known such species is the monarch or milkweed butterfly of North America. It is distributed over a large area reaching from the Hudson in the east to Alaska in the west. In early autumn these butterflies begin to congregate and to fly far south in swarms. As a rule they set off from Canada in September and by the beginning of November they are already in the Gulf of Mexico. They alight by the tens of thousands on the same trees every year, remaining motionless there throughout the winter. Now and then, on a warm day, a few indi-

Atlas moth

Monarch butterfly

viduals will fly hesitantly round the tree only to settle down again to their winter sleep. These trees, covered with tens of thousands of large, brightly-coloured, motionless butterflies, are protected by law and displayed as a tourist attraction. In spring the swarm stirs and wakens, single butterflies separate from the mass and successively all wing their way north. The return migration begins in March and by the end of May or early June at the latest the butterflies are back in their northern homes again. There they lay eggs and die. How the new generation which hatches in the north finds its way to trees thousands of kilometres distant in the south remains a mystery.

Kerguelen flightless moth

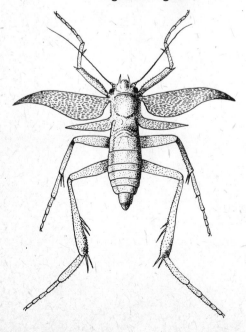

Kerguelen flightless moth
Pringleophaga kerguelensis
We have already said that the wings are the greatest attraction of butterflies and moths. However, in

some species the wings are only rudimentary or absent altogether. Such species naturally do not fit our established concept of butterflies and moths and would not even be recognized as such by the layman. In most instances the reduction, or wingless state, is confined to one sex — the female. The wingless, or rather rudimentary-winged, state is rarely found in both sexes. Far south in the Indian Ocean, nearly at 50° latitude South, lie the Kerguelen Islands. The winds there blow strongly, almost incessantly and flying insects would be swept out to sea. There are however two species of moths which have adapted themselves to

conditions on the islands. Their wings have become lost so they do not fly but merely crawl with the aid of their stout legs.

Common apollo *Parnassius apollo*
Apollos are butterflies of the mountains. They are not particularly brightly coloured, the wings generally being white with black or red markings. Nevertheless they are very lovely. Most species are relatively rare and some are even the greatest rarities of entomological collections.

The common apollo is found in the mountain regions of Europe from the Pyrenees through the Carpathians to the Balkans and north to Scandinavia. Unlike related species it occurs also at lower elevations. It is well equipped for life in the rugged conditions of its mountain habitat. The body is

Common apollo

covered with long hairs — a regular fur coat. The caterpillar feeds on rockfoil, houseleek and stonecrop. Its growth is slow and it takes two years before it is fully grown. Before pupating it spins a cocoon which protects the pupa in cold weather. The manner of flight is distinctive. It is slow and fluttering; it also glides motionless with outspread wings, borne up by the air currents. This manner of flight is common in birds but somewhat of a curiosity amongst insects.

Apollos have always been favourites of collectors and have thus become extinct in many places, particularly as they fly slowly and are readily caught.

Silkworm *Bombyx mori*

The history of silk textiles is an ancient one. The first records are from China and date from 2700 B.C. Though silk fabrics are obtained from several species of moth the most important commercially is the fibre of *Bombyx mori.* This is a domesticated species that is no longer found in the wild. The silkworm is completely dependent on man. The adult moth lives only about three days and does not consume any food. The female lays some 300 to 400 eggs from which caterpillars hatch after a quiescent period. When they emerge the caterpillars feed voraciously on the leaves of mulberry and rapidly increase in size, from a length of 3 mm to 9 cm (3 1/2 in) in about 35 days. Then they pupate in a cocoon which they spin of silky fibres. After ten days or so the cocoons are collected and graded and the pupae killed with hot air. The sticky substance joining the silk fibres is dissolved in warm water and the fibres are unwound from the cocoon. More than 3,000 cocoons are required to yield one kilogram of raw silk. Although artificial silk has replaced raw silk to some extent silk production continues to be an important branch of the textile industry in many countries.

Silkworm

Garden tiger *Arctia caja*

Tiger moths are attractively coloured moths with forewings usually a different colour from the hind wings. When such a moth rests in the grass it is practically invisible because the colour and markings on the forewings blend with the surroundings. As soon as it flies up, however, one is surprised by the striking red and yellow colouring of the hind wings. When it alights it again conceals the hind wings beneath the forewings and becomes invisible. The red colour serves as protection to surprise and deter the enemy from further pursuit.

The caterpillars are covered with long hairs, black on the back and russet on the sides. They feed on a wide variety of plants. In autumn they pupate in a loose cocoon which includes many of their hairs. The pupae hibernate but many do not survive the winter because they are often attacked by parasitic flies and wasps.

Common swallowtail
Papilio machaon

Swallowtails are among the most beautiful of butterflies. They are distributed throughout the whole world, from lowland country to high mountains. There are some 600 existing species, all very popular with butterfly collectors. Their name is derived from the fact that most species have the hind wings extended in tail-like points, quite often long and graceful.

The common swallowtail is relatively abundant and is found in Europe, Asia as well as North America. Its flight is rapid and catching it is not at all easy. One is more likely to get hold of the caterpillar, which feeds on umbelliferous plants such as caraway, fennel and wild carrot. It is black at first, later brightly coloured. Each body segment is yellow-green with a black band and red spots. If you touch the caterpillar it will stick out two orange-red horns from behind

Common swallowtail

Death's head hawk moth

pupae are spindle-shaped and encased in a smooth papery cocoon.

Because of their slow flight burnets would be easy prey for their enemies but nature has provided them with excellent protection. No bird that has once caught a burnet will ever do so again for these moths have glands behind the head that produce a yellow secretion containing histamine and prussic acid which makes it not only distasteful but poisonous. Furthermore their striking coloration likewise guarantees that a bird will not make the same mistake twice.

its head which are joined to a gland producing a secretion which has a pronounced smell, reminiscent of pineapple or fennel.

Death's head hawk moth
Acherontia atropos

The death's head hawk, though not very common, is a very popular moth, made famous by the startling pattern on the upper side of the thorax which resembles a human skull. The death's head hawk inhabits southern Europe and the southern parts of Asia but from time to time, though not at all regularly, it visits central Europe and occasionally reaches southern Britain. As a rule, these visits are made by a large number of moths at one time — they appear to be a kind of minor migration. Sometimes the moths appear in June, at other

Six-spot burnet

times not until autumn. Females which arrive in early summer usually lay eggs on potatoes and other members of the nightshade family, occasionally also on mock orange. The larvae are green with blue stripes and grow rapidly to a length of 15 cm (6 in). When fully grown they burrow in the ground where they pupate. Only rarely does an adult moth emerge from the pupa for the pupa does not normally survive the European winter.

The death's head hawk has a short proboscis and thus cannot suck nectar from flowers. It obtains sweet food by other means, often entering a hive to suck honey. As a rule, though, it is killed by the bees and because they are unable to remove such a large body from the nest they wall it in with honey. This is why bee-keepers often find these dead moths in their hives. If you capture a live death's head hawk you may hear it make a surprisingly loud sound, somewhat like a mouse. Its mouthparts are fitted to function like a whistle.

Six-spot burnet
Zygaena filipendulae

If you visit a European meadow in summer you will find striking, brightly-coloured moths with thickened antennae perching on the flowers. Usually they are coloured black and red. Their flight is slow and they can be readily caught in the hand. They are six-spot burnets. The caterpillars are generally found on plants of the pea family, the pupae on blades of grass. The

Hornet moth *Sesia apiformis*

In the insect realm we will find many instances of species resembling an inanimate object or a different animal species. This is known as mimicry and serves to protect the organism from predators. The hornet moth which resembles the hornet in form, size and colour is a good example of this. This resemblance is designed to protect the moth from bird predators which avoid hornets. Many authorities support the mimicry theory but just as many oppose it. Are birds truly unable to distinguish the hornet moth from a hor-

Hornet moth

net? In an experiment to try to solve this dispute, hornet moths were served, together with hornets, to birds which regularly feed on wasp-like insects and to birds that never capture hornets. The birds in the first group made a mistake only once, the birds in the second group carefully picked out all the hornet moths but didn't even touch the hornets. The outcome of the experiment thus showed that birds can clearly tell hornet moths from hornets and that the mimicry therefore appears to be useless. However, the problem is far from being definitely solved as yet and the dispute is not ended.

Life in the Sea

The seas and oceans cover 361 million square kilometres of the earth's surface. From the surface to its greatest depths the sea is inhabited by countless organisms. If we take the average depth as the basis for our reckoning then the available 'living space' in the seas and oceans totals approximately 1,370 million cubic kilometres. From this mind-boggling figure we might conclude that the seas contain most of the animals living on this earth. This, however, is true only as regards the number of individuals, for if we compare the number of species the result will come as a surprise. The sea is inhabited by only one-fifth or so of all known species. This is because there are far fewer barriers in the sea — barriers which not only serve to separate individual species but are also the cause of the origination of new species.

Life originated in the sea and for a long time was confined to this environment. Only after a very long

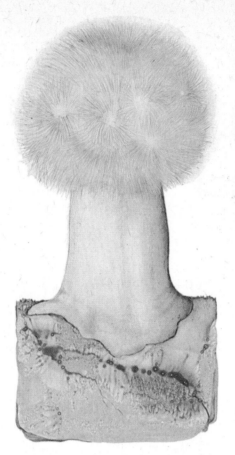

Plumose anemone

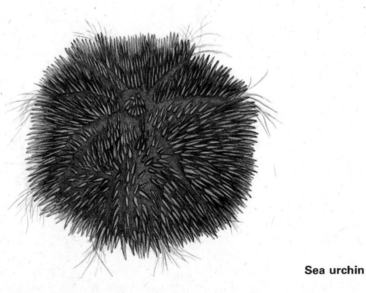

Sea urchin

time did it make its way to fresh waters and land. To this day the sea is inhabited by a great many organisms which graphically illustrate how the various forms evolved. It also contains animals regarded as living fossils.

Life in the seas and oceans has been studied by many scientists. At present it has become the focus of attention of ecologists, because

the seas and oceans appear to be the future source of food for the earth's growing population, which is slowly but surely exhausting the available sources on land.

Plumose anemone
Metridium senile
Sea anemones are found in all parts of the ocean, from the seashore to the greatest depths.

They are large, usually brightly-coloured coelenterates — sac-like animals which live attached to various firm bodies. This, however, does not mean they cannot move. Though their powers of locomotion are limited they can glide along slowly with a foot which is furnished with a suction disc. Some species can even inflate their bodies so they float freely in the water when they have released their hold on the object to which they were attached.

The body cavity is in fact one huge stomach. The mouth-opening is surrounded by numerous tentacles and the whole animal resembles a flower — which is how it came by its name. The tentacles are an ingenious mechanism for capturing prey. There may be as many as several hundred and they are furnished with minute openings at the tip. By suddenly contracting their bodies sea anemones can eject the water stored inside the digestive cavity. The tentacles are furnished with sting cells for catching food as well as for defence and they can sting like nettles when touched. These stinging arms capture and paralyze prey and carry it to the body cavity. The cavity does not contain digestive juices but the walls of the stomach penetrate the victim with 'fingers' containing digestive cells which absorb small quantities of food and dissolve it. A sea anemone which has eaten its fill and has begun digesting contracts its cylindrical body into a round ball or loaf shape and remains thus for up to several weeks.

Sea anemones are often kept in saltwater aquariums. They have no special requirements and many species tolerate even very unfavourable conditions. They are very long-lived, reaching an age of up to 100 years.

Sea urchin *Echinus*
Sea urchins resemble low or high, rounded loaves studded with spines. The spines may be scanty or thick, long or short, but they are always movable, because they have a ball-and-socket joint joining them to the sea urchin's internal

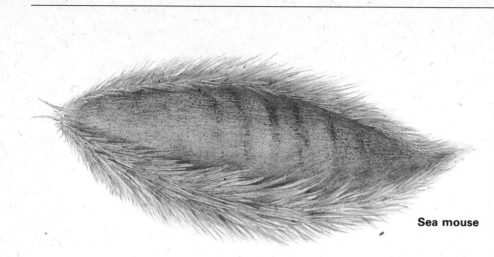

Sea mouse

calcareous shell. They are attached by means of muscles which move them. The shell of sea urchins has numerous minute openings through which the animal projects its tube feet, each of which is fitted at the end with a small suction disc. The tube feet are used in locomotion and also hold the sea urchin fast to the rock, often so firmly that it is extremely difficult to tear the animal loose. Scattered over the body between the spines, most thickly round the mouth opening on the underside of the body, are small stalks with tiny claws which are in constant motion, removing small pieces of debris caught between the spines and carrying whatever is edible to the mouth. Most sea urchins feed on vegetable matter but some also eat small animals.

The edible sea urchin *Echinus esculentus* is common off the coasts of the North Atlantic and the Mediterranean. In many regions they are considered a great delicacy. They are speared with a sharply-pointed stick, cut in two across the middle and the lower half discarded. The five ovaries are found in the upper half. These are eaten as they are or with a dash of lemon juice. In other parts of their range, these urchins are captured whole, cleaned of their spines and insides and sold as tourist souvenirs. Many species of warm-water sea urchins have poisonous spines but even the cooler water species can cause quite nasty wounds if trodden upon by the unwary.

Sea mouse *Aphrodite aculeata*
The sea mouse is considered by many to be the prettiest member of the segmented worm phylum — the Annelida. It is found in all European seas, in shallow water amidst stones. Usually it burrows into the soft sea bottom with only the hind end of the body showing. Then, of course, we can't see much of its beauty. However, if we wash the body and remove the particles adhering to it it will glow with the colours of the rainbow. Each body segment has a projection on either side bearing tufts of bristles. These gleam and glow with a play of colours when the sea mouse moves. The dorsal bristles are not as beautifully coloured, generally being grey or brown. If you should wish to capture this lovely creature you must take care, for the pretty bristles on the side break off readily and catch in the skin where they may cause inflammation. Cod and small sharks, however, do not mind the bristles and this is why the sea mouse is a regular part of their diet — a tidbit that's a regular mouthful for it may be up to 20 cm (8 in) long and 4 cm (1 1/2 in) wide. If we examine it closely we will see that,

like many other active marine worms, the head has three antennae and eyes which are relatively well-developed.

Arrow worm *Sagitta*
Floating or drifting in the ocean are enormous numbers of organisms called plankton, of which arrow worms are a part. They look like small, 14—20 cm long (5 1/2 — 8 in) transparent fish with an arrow-shaped body. Unlike fish, however, the fins are horizontal. At certain periods of the year they occur in vast numbers, floating in the water or swimming with jerking movements. In daytime they migrate from the surface to the depths and back again. In the morning and evening they stay near the surface but during the day they prefer greater depths where their transparent body is not exposed to such intense light. Arrow worms are predaceous. When prey comes near they dart out like an arrow, covering distances several times the length of their body. They grasp the prey with their jaws, which are furnished with sharp, sickle-shaped bristles. They will eat anything, including members of their own species. During the period when herring hatch arrow worms can make vast inroads into the shoals of young fish.

The distribution of the individual species of arrow worms is very interesting. Various species inhabit various parts of the ocean and the limits of their range are clearly defined. One biologist even caught one species from the prow and a different species from the stern of a ship lying at anchor. Arrow worms are somewhat of a puzzle to biologists for we do not know their relationship to other animal phyla.

Arrow worm

Molluscs

When we spoke about snails and slugs we explained more or less what molluscs are and we also mentioned that most of them are marine.

Molluscs have no external segmentation and only in the most highly developed is there any obvious division of the animal into a head, foot and 'body'. The most typical characteristic is the soft, slippery body. This is protected by the soft mantle which contains a large number of glands and secretes the hard outer cover or shell.

The shape of the shell is determined by the disposition of the mantle. If the body is drawn into a spiral then the shell secreted by the mantle will take up the spiral shape, typical of most snails. In the bivalves the mantle extends on either side and enfolds the body. Such a mantle secretes a shell in two halves with a hinge along the animal's back.

The cephalopods, that is the octopus, squids and cuttlefish, with their well-developed eyes and jet-propulsion system are considered to be the most highly developed molluscs.

Because they usually have a hard shell, molluscs are very readily fossilized and many are important in determining the age of individual rock formations.

Giant clam *Tridacna gigas*
The giant clam is veritably a giant amongst bivalve molluscs. Though the body is not extraordinarily large (the animal itself weighs about 10 kg (21 lb), the shell protecting the

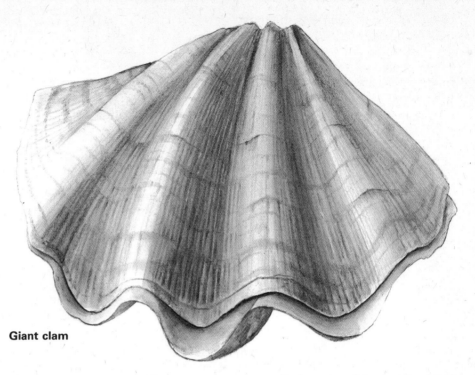

Giant clam

soft body is enormous. It measures up to 130 cm (52 in) in length and weighs 500 kg (9 cwt). The shell has prominent, often scaly ribs, the ends of which interlock like large teeth. It is found on coral reefs in the Indian and Pacific oceans. It is occasionally collected by the natives because the strong muscles which close the shell are said to be very tasty. The upper edges of the mantle of the giant clam face the light and within the tissues there are vast numbers of green and brown single-celled plants. These are an example of symbiosis — organisms living together for their mutual benefit.

The shell of the giant clam is much in demand as a museum exhibit. Formerly, however, it was used for practical purposes. In many churches, even in Europe, you will find the half-shell of this clam serving as a holy water basin or baptismal font.

European common whelk
Buccinum undatum
The European whelk is one of the commonest inhabitants of the North Sea. It is a carnivorous mollusc which burrows in the sea bottom, bores holes in the shells of other molluscs and feeds on their bodies. It is avidly hunted by fishermen first of all for its tasty meat,

secondly because it is excellent bait and last but not least because it is an unwelcome competitor for other edible molluscs. Whelks are generally caught in baskets with pieces of fish placed in the bottom. The baskets are lowered near the shore to a depth of about 20 metres and emptied daily of the whelks attracted there by the bait. Those of you who have visited the North Sea coast may have found among the objects cast up on the shore empty clumps of its egg capsules which are bean-sized, yellowish and leathery. These clumps are well known to the local inhabitants who call them 'sea soap', for boatmen use them occasionally to clean their hands.

Argonaut or Paper nautilus
Argonauta argo
Pearly nautilus
Nautilus pompilius
The argonaut is a cephalopod, in other words a mollusc with the most highly developed body structure. The male resembles a small, plump octopus. The female is much larger. She secretes a beautiful, thin, paper-like spiral shell from the inner surfaces of her broad arms. It is retained in place by the arms and serves as a repository for the eggs until they hatch. Young females do not have a shell, which they secrete

European common whelk

later. Breeding is perhaps the most interesting period in the life of the argonaut. The male gathers the sperm cells in an arm specially modified for the purpose. The arm then separates from the body and swims freely in the sea until it finds a female. Thereupon it slips inside her mantle cavity and deposits the sperms by the opening to the female's reproductive organ.

The pearly nautilus is the most primitive of the cephalopods and may rightfully be considered a living fossil for it is the only survivor of a large group which inhabited the seas during the Paleozoic and Mesozoic eras. It lives in the depths of the Indian Ocean, in places with a constant temperature and far from any turbulent waters. It is perhaps for this very reason, because of the constant conditions of its environment, that the pearly nautilus has survived to this day. The pearly nautilus is the only cephalopod which has an external shell. It is spirally coiled and divided by transverse partitions into a succession of chambers filled with nitrogen. The animal occupies the last and largest chamber and is attached to the centre of the shell by a long ligament. The mouth opening is surrounded by as many as 90 short tentacles, which, unlike those of the other cephalopods, are not furnished with suction discs.

Common European edible oyster
Ostrea edulis
Common European edible mussel
Mytilus edulis
The oyster and the mussel head the list of edible molluscs. Whereas oysters are eaten raw and alive

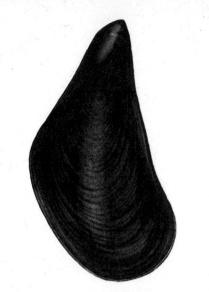

Common European edible mussel

with only a dash of lemon juice, mussels are boiled in water, or often in wine and herbs.

Oysters were eaten by prehistoric man and in no small quantities as evidenced by the great number of shells found in the refuse holes of prehistoric settlements. The ancient Romans likewise considered oysters a great delicacy, bringing them to Rome from as far away as the British Isles. Because natural oyster banks have long ceased to be able to meet the demand for oysters they are nowadays cultivated in special culture beds. The reproduction of oysters has one peculiarity. The sexes are not separate but during the breeding period the sexes alternate. In cold seas this occurs once a year but in warmer climates it may be several times. Oysters lay large numbers of small eggs; the female may lay as many as three million in one year. The larvae which hatch from the eggs

Common European edible oyster

swim about freely before attaching themselves to a solid object and changing into adult oysters.

The edible mussel is even more common in European seas than the oyster because it also lives in less salty waters. Mussels are firmly attached to the bottom of other objects by the aid of strong threads so that they must be forcibly pulled from their moorings. Mussels are also cultivated in the same way as oysters, because their consumption in Europe alone is an estimated 100,000 tonnes yearly. However, only those which have been cultivated under hygienic supervision may be served at the table. It has often happened that people have become seriously ill after eating mussels for these molluscs are fond of settling in places near the mouth of sewage systems or places where waste water seeps through. Mussels may thus transmit typhoid or paratyphoid.

Pearly nautilus

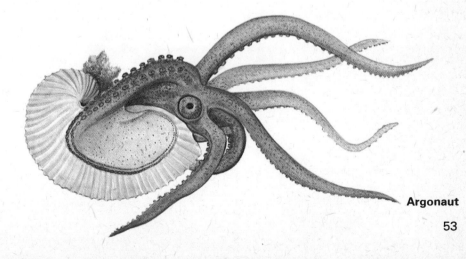

Argonaut

Crustaceans in the Sea

We have already spoken briefly about crustaceans in the chapter on water fleas and crayfish. There we saw that crustaceans inhabiting the seas include many interesting species which are often important to man. Marine crustaceans are 'gifts of the sea', sometimes as food for the poor, at other times as a tasty morsel for the wealthy. Best known of the lot are lobsters, langoustes (or crawfish) and crabs, and of the smaller species those which are generally called shrimps and prawns. Hardly anyone who has ever taken a holiday by the sea has come away without having seen at least one of these crea-

has a different function. The more slender, left pincer, furnished with regular teeth and a large number of sensory bristles, is generally used to capture molluscs. After the victim is caught it is grasped by the other, more powerful pincer furnished with several irregular but strong teeth with which the lobster crushes the mollusc's shell. After this it sets about eating the animal, snipping off bits of flesh with its left pincer. Every year the female lays a great number of eggs which she attaches to the underside of her abdomen. On hatching, which takes place after about a year, the larvae swim freely in the sea and moult

but this guarantees them only slight protection. Formerly specimens caught measured even 80 cm (32 in) on occasion but nowadays specimens measuring more than 60 cm (24 in) in length are a rarity.

Common shrimp *Crangon crangon*
'Shrimp' is the name used for all the small, slender, long-tailed crustaceans with soft, compressed bodies sold at the fish market. They are inhabitants of shallow sea-coasts. If you visit a marine aquarium you will surely see some species of shrimp there for they are the most frequently exhibited of the marine crustaceans. Their body is

Common lobster

tures. Usually, however, we know nothing about them or their way of life apart from the fact that their meat is very tasty. It will not therefore be amiss to take note of the interesting aspects of a few.

Common lobster
Homarus vulgaris
The lobster lives on the rocky continental shelf sloping from the coast of England and Norway to the depths of the Atlantic. It is far less common along the coasts of western Europe and the Mediterranean Sea. The related and equally popular *Homarus americanus* is found along the Atlantic coast from Delaware to Labrador.

A typical characteristic of the lobster are the unequal pincers. This is quite natural for each pincer

four times. Only after the last moult do they resemble the adult. They then settle on the sea bottom. Lobsters grow at a very slow rate. One weighing half a kilo is about six or seven years old. Lobsters are caught in nets or in lobster pots containing live or dried fish as bait. It is prohibited to catch lobsters from mid-July to mid-September,

generally a pale colour or semi-translucent and measures about 5 to 8 cm (2-3 in).

The common shrimp is found in the North Sea and along the coasts of Great Britain. It is caught in large numbers in drag nets or in traps in which the creatures are stranded at ebb tide. Unlike other related species the common shrimp does

Common shrimp

Common shore crab

asymmetrical, soft-bodied abdomens with appendages only on the left side. The front part of the body is likewise soft-bodied at the rear and the last two pairs of walking legs are modified to hold the crab firmly in the spiral shell by allowing the body to fit snugly inside. The shell serves not only to protect the soft abdomen but also as a shelter in which the crab can conceal itself when danger threatens. In some species one of the pincers is adapted so it can close the shell. When the shell becomes too small for the crab it finds, and then moves into a new, larger shell.

Many species carry one or more sea anemones about on the shell. In some it is a chance phenomenon, in others the crab transfers the anemone to its new shell. The sea anemone, of course, can just as well live somewhere else. The closest association between the two is to be found in the crab *Eupagurus bernhardus* which carries about one or more anemones of the species *Calliactis parasitica*. The anemone eats food remnants left by the crab and protects the crab from enemies with its stinging tentacles. The relationship between the smaller hermit crab, *Eupagurus prideuxi*, and its attendant anemone is even closer.

not change colour when cooked. Sometimes fishermen catch so many that it is impossible to process them all even in tins. Surplus shrimp are therefore dried and ground into meal which is excellent, protein-rich food for livestock.

Common shore crab
Carcinus maenas
The shore crab is found along all European coasts regardless of whether it is open shoreline with clean water or the polluted water of harbours. At ebb tide great numbers of crabs remain on the shore where they hide beneath stones or burrow into the ground with their hind legs, leaving only the stalked eyes peering above the surface. Sometimes one will find them scrambling about. They move about rapidly, not only forward but also sideways. If you try and catch such a crab it extends its pincers and makes a clapping sound by opening and closing them rapidly. Because the shore crab is very plentiful wherever it is found it is often caught and eaten as food. Especially prized are soft-shell crabs, ones that have recently moulted, which are fried whole in oil and served as a particular delicacy.

The female is readily distinguished from the male crab by looking at the underside and examining the shape of the abdomen curled under the body. The male's is narrow and triangular, the female's broad. The legs on the

second and fifth abdominal segment are modified for carrying eggs. Females with eggs are said to be 'in berry'. The larvae which emerge from the eggs do not resemble the adult crab in the least. They are called zoea and have long, curved spines that enable them to float in the water. The second larval stage, which is also planktonic, is called megalops because of the large eyes. Both larval stages serve to spread the species to new locations.

Hermit crab
Eupagurus bernhardus
Hermit crabs are an interesting group of crustaceans. They have

Hermit crab

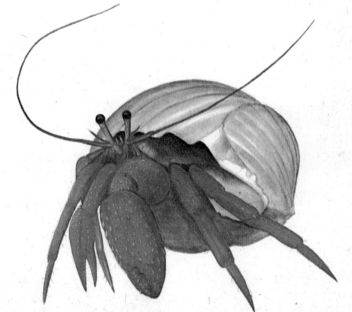

Sharks and Rays

Sharks and rays belong to the group of cartilaginous fishes, in other words fishes with a skeleton which is not bony but cartilaginous. The scales of these fishes, however, are composed of bone tissue. When you pass your hand over the skin of a ray or shark it feels like sandpaper. Many species have tiny, pointed teeth jutting from the skin which are a characteristic feature of cartilaginous fishes. When we say teeth we really mean it, for the spiny tips of these special scales are composed of dentine covered by enamel the same as in human teeth. The base of the scales, which is flat and anchored in the skin, is bony. The dreaded teeth of sharks are of the same origin — they are modified skin teeth.

Everyone knows what sharks and rays look like, if not from their own experience then from pictures. However, some sharks look much like rays and, vice versa, some rays resemble sharks. However, if we examine the gills we will have no trouble identifying them. In sharks the gill openings are at the sides behind the head, whereas in rays they are on the underside of the body, always below the large pectoral fin.

Sharks and rays are popular fish, particularly in recent years when scuba-diving has become such a popular sport. Many popular books as well as films (often quite fantastic of course) have likewise contributed to our knowledge of these fish.

Manta ray or Devil fish
Manta birostris

The manta is the largest of the rays, measuring up to seven metres across with pectoral fins outspread and weighing two tonnes. It has two horn-like processes on the head which are generally held level but are very mobile and are the reason why it is sometimes called devil fish. Like all rays the manta also has a horizontally flat body and rigid spinal column so it cannot bend to the side. The organs of locomotion are the huge, widely-expanded fins attached to the flattened sides and forming 'wings'. When swimming the manta truly resembles a bird in flight. Unlike most rays the manta inhabits the open seas, generally staying near the surface close by shoals of small fish or crustaceans, which are the mainstay of its diet. Its huge mouth thus poses no danger to man or a larger animal. The manta swims with mouth wide open, collecting food directly into the gill chamber where it is caught as in a net.

The hind section of the anal fin in males is modified into a copulating organ and so fertilization in rays as well as sharks takes place inside the body. Like many other cartilaginous fishes that inhabit the open seas the manta is viviparous but bears only a single offspring. The egg is not deposited outside the female's body but remains inside where it develops until the young fish emerges. The female's ovaries expand into a flap to which the young fish is attached by means of the yolk sac. This serves as a structure through which it receives nourishment from the mother's body. Mantas, as we see, thus grow a kind of placenta just as in mammals. Otherwise relatively little is known about this ray. Observation of its life in the open sea is impossible and its size and diet make it unfeasible to keep it in an oceanarium. We must therefore be content with chance observations or the study of freshly caught specimens.

Great blue shark *Prionace glauca*

The great blue shark is one of the species that are commonly called man-eaters, for it occasionally attacks man even though its usual food is fish and cephalopods. It naturally also attacks marine mam-

Manta ray

mals, such as dolphins. The great blue shark is a large species reaching a length of six metres though it is generally smaller, about three metres long. It is found in all seas in the tropical, subtropical and temperate zones. However, it is not equally plentiful in all places. Whereas it may be encountered frequently in the Mediterranean Sea, it is relatively rare in the North Sea. It is generally found in the vicinity of large shoals of fish, chiefly herring, mackerel and tuna. Often whole packs of these sharks travel in the wake of migrating fish. They are not loved by fishermen for they often tear nets filled with fish to get food 'free', without much effort on their part. Whaling ships likewise have had bad experiences with great blue sharks. Packs of these sharks converge round the dead whale and tear large pieces of flesh from the body. It has not been proved that a shark will attack a live whale, but attacks on dolphins are not at all uncommon. The great blue shark is a typical inhabitant of the open seas and rarely strays to the shallow waters of the shore, therefore it is relatively rare to hear of a blue shark attacking a human.

Common spotted dogfish
Scyliorhinus caniculus
The common spotted dogfish is a small spotted shark found along coastlines from Senegal to Norway. It is the commonest shark in the Mediterranean Sea and in view of its size the one most often displayed in saltwater aquariums. Only rarely does it attain a length of about one metre and 80 cm (32 in) long specimens are already considered to be large. The brownish colour of the back, the dark spots on the body and the pale belly indicate that the dogfish is an inhabitant of the sea bottom. It hunts chiefly crustaceans, molluscs and other marine invertebrates, occasionally also small fish. Dogfish are oviparous. The female lays 18 to 20 eggs each enclosed in an angular, 4 to 6 cm (1 1/2—2 1/2 in) long case with a long filament at each corner.

When laying the eggs the female swims amongst the shoreline vegetation on which the eggs are caught by these filaments. The filaments coil in a spiral about 15 cm (6 in) long so that each egg hangs in the seaweed in a soft cradle. The young dogfish emerge from the eggs after about nine months. Because the egg cases are transparent the development of the embryo can be clearly observed from beginning to end. A newly-hatched dogfish is 9 to 10 cm (3 1/2—4 in) long and unlike the parents is horizontally striped.

Dogfish are often caught and the headless bodies sold at fish markets as 'rock-eel', 'huss' or 'rock salmon', perhaps to make them sell better!

Common spotted dogfish

The Fish We Eat

Sea fishing is becoming increasingly important. The gifts of the sea are not only profitable articles of commerce but also an increasingly important item in man's diet. For maritime as well as some inland countries sea fishing is a vital part of the national economy and an important supply of food for the nation. According to statistics compiled by FAO (the Food and Agricultural Organization of the United Nations) some 50 million tonnes of sea fish and some 4.5

nets; the usual method is with bait of various kinds. Their large size, however, makes removing the hooks a difficult task. The Japanese have successfully tested equipment which sends an electric charge into the hook thereby stunning the fish and making it easy to land.

The cod is the world's second most important economic fish, after the herring. It inhabits shallow, cool waters, usually occurring at depths of 40 to 250 metres with a temperature of 0° to 16°C. The

remain in the same place. The greatest number is caught by trawling, but because they are predatory fish they are also caught with bait in the same way as the tuna.

Atlantic halibut
Hippoglossus hippoglossus
The order of flatfishes, which includes the flounders, halibuts, soles and plaices, has a remarkable body structure with both eyes on one side. These rest on the sea bottom on their broad flat side and swim with the aid of the dorsal and anal fins. The young fry, however, have a normal shape, the body becoming asymmetrical and the eyes shifting to one side only later, during growth. Halibuts are typical inhabitants of sea bottoms. All flatfish have tasty flesh and are commonly fished. Some species, including the Atlantic halibut, are important fish of commerce and the halibut is rightfully considered one of the most valuable sea fish. The flesh is oily and the liver is up to 200 times richer in vitamin A than cod liver. The halibut is a large fish reaching a length of more than four metres and weight of more than 300 kg (5 1/2 cwt). Naturally, however, the fish which are caught are much smaller, up to one metre in length. The annual catch is approximately 20,000 to 50,000 tonnes. In Europe the greatest numbers are fished off the coasts of Iceland and Norway. Shoals of halibut arrive at Iceland's shores in March and April and off the coast of Norway in May and June. They are caught by trawling as well as with bait. In America the greatest number is caught on the western coast of Canada.

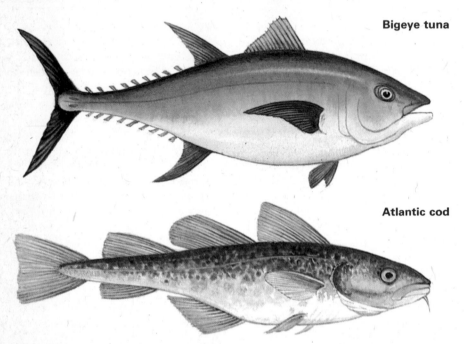

Bigeye tuna

Atlantic cod

million tonnes of other marine animals are caught yearly, mostly in the northern hemisphere.

Most fish caught commercially are taken by nets, either the trawl, for bottom-dwelling species such as the cod, or the drift-net for surface-swimming species, notably the herring.

Bigeye tuna *Thunnus obesus*
Atlantic cod *Gadus morhua*
Tunas are among the largest of fish. It is stated that they may weigh as much as 600 kg (11 cwt). Their flesh is very tasty and thus they are often fished for food. Tuna fishing, however, often presents many difficulties, for the fish, which travel in shoals, are constantly on the move and must be tracked. Because they generally swim near the surface airplanes can be used for this purpose. Tunas are rarely caught in

optimum temperature it favours is between 2° and 7°C and that is the reason for the varying success in cod fishing. A summer that is warmer than usual is enough to send the fish much farther north than expected. Fish which migrate are more than three years old. Younger, sexually immature fish are more or less resident, i. e. they

Atlantic halibut

Sailfish *Histiophorus albicans*

The sailfish has two interesting features. The first is the markedly prolonged, sword-shaped upper jaw similar to that of the related swordfish. Unlike the latter, however, the sailfish has still another remarkable characteristic — the large sail-like dorsal fin which extends practically its whole length. We all know that the sword-shaped jaw of swordfish and sailfish is used as a weapon both for defence and attack. How it is used in catching prey is not precisely known, perhaps the fish can even impale its victim with it. The sailfish is a very important game fish wherever it occurs. Catching it with a hook and line is even more exciting than tuna fishing. However it is also an important fish of commerce.

Greek moray *Muraena helena*

Round the year 90 B. C. the ancient Romans were already known to have built tanks for breeding fish which they called vivaria and sometimes also piscines. There they fed and fattened certain species of fish for feasts and banquets, usually morays, because they have very tasty and delicate flesh.

Morays are unusual predators, concealing themselves in rocks or coral reefs during the day and setting out to forage for food in the darkness of the night. Some species are very aggressive and thus a grave danger to careless divers who might want to investigate rock crevices with bare hands in search of 'treasures'.

Morays are popular specimens in all marine aquariums both for their interesting appearance and their notorious reputation. Some species are furthermore brightly and attractively coloured so that they are a very decorative element in the aquarium.

Of greater importance for sea fishing nowadays are certain Asian species. Some 50,000 tonnes are caught yearly off the coasts of Asia, more than half of that in Japan.

Redfish *Sebastes marinus*

The northern waters of the Atlantic are the home of the handsome attractively coloured redfish, found at depths of 200 to 300 metres. It is an important fish of commerce despite the fact that it has a very slow rate of growth. The redfish is viviparous. In its spawning grounds round the Lofoten Islands off the northwest coast of Norway and off the coast of Newfoundland the females bear five to eight mm long larvae in the autumn. At first these grow about 5 cm (2 in) a year but later the increase in size is less than 1 cm (1/2 in) a year. Redfish that are caught usually weigh about 2 kg (4 1/4 lb) and are approximately 20 years old.

Greek moray

Redfish

59

Fish of All Shapes and Sizes

The underwater realm is very large and marked by great diversity and so it is no wonder that its inhabitants also exhibit a multitude of forms. Fish number some 21,000 different species and include creatures of all shapes, sizes and colours as well as diverse ways of life. The largest fish is the whale shark which reaches 20 metres in length, the smallest fish and likewise the smallest vertebrate is the pygmy goby from the Philippines, which measures 7.5 to 11.5 mm when fully grown. Most fish have a highly

Seahorse

efficient, streamlined body shape but, as we have seen, some, such as the halibut, are superbly adapted to life on the sea bottom. And as for the fantastic shapes of some deep-water fish we haven't the slightest idea how and why they came about. Pages could be written about the breeding of fish and their care of the young. Some tend their offspring with care, even hatching the eggs and carrying the fry about in their mouths. Other species take no heed of the eggs and sometimes even eat them. The greatest mystery is the cannibalism exhibited by certain viviparous sharks. The two first-born devour all the other offspring while still inside the mother's body.

Let us take a look at a few interesting fish and learn something about their lives, keeping in mind, of course, that the following is but a small drop of knowledge in the sea of the unknown.

Seahorse
Hippocampus hippocampus
Have you ever seen a fish with the head of a horse, the tail of a monkey, the pouch of a kangaroo, the eyes of a chameleon and the armour of an insect? Such a fish is the seahorse. It bears a close resemblance to that of the knight in chess not only as regards the shape of the head but also because the body is covered with bony plates that make it look as if carved from wood. The seahorse always swims in an upright position. It has only small pectoral fins and a dorsal fin which is the chief instrument of locomotion. When it travels at full speed the dorsal fin moves left and right up to 35 times a second. The tail fin is absent and the tail is modified into a grasping organ with which the seahorse catches hold of seaweed and other objects in the water. The seahorse feeds on various small animals, chiefly tiny crustaceans and fish. Its mouth opening is so small it couldn't even seize larger prey. Sometimes it sucks up food through the long tube-like snout from as far away as 3 cm (1 1/4 in). When foraging for food it uses its eyes, being able to move them in-

dependently of each other — looking upward with one, for instance, and backward with the other. Not only does the seahorse have an unusual appearance, it also has an unusual method of reproduction. The male develops a brood pouch in which the female places the eggs with her ovipositor. The pouch, which is wide open at first, then closes, leaving only a small opening. Its walls swell into a spongy tissue thickly interlaced with blood capillaries. This tissue nourishes the eggs in much the same way as a placenta. The young, exact replicas of the parent, emerge after about four to five weeks. Each is ejected from the pouch by spasmodic contractions and after each 'birth' the male is clearly exhausted. This is another curiosity that can be added to the ones we've already mentioned — a male with labour pains!

Butterfly fish
Heniochus acuminatus
The tropical fishes of the family Chaetodontidae are called butterfly fish. All are beautiful and brightly coloured and thus popular specimens in marine aquariums. There are about 150 species and if we were to use their common names it would probably cause confusion over their identity. For this reason it is best to use the scientific, Latin

Butterfly fish

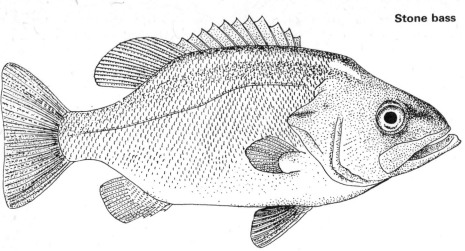

name when referring to a specific species. *Heniochus acuminatus* is a fish of the coral reefs of the Indo-Pacific region, from the Red Sea to Hawaii. In the various parts of its range it is known by different names such as wimpelfish, pennant, coral-fish, coachman and pavillon. It is the most plentiful member of the whole genus to which it belongs and also the most beautiful. As a rule it swims in pairs or small shoals round the coral reefs. When disturbed it seeks shelter in cavities and crevices, where it wedges itself in with its dorsal and anal fins so that it is almost impossible to pull it out. Young butterfly fish often approach large fish and appear to nibble them on the sides, head, fins and gills. Their behaviour is thus the same as that of cleaner wrasses which remove parasites from the bodies of other reef fish in this way.

The butterfly fish is a striking aquarium specimen and one which is quite easy to keep.

Common clownfish
Amphiprion percula
The coral reefs of the Indo-Pacific region are inhabited by small, brightly coloured fish which swim with slightly jerky, swaying movements. Both their coloration and their movements are the reason they came to be called clownfishes. There are many species but all have one interesting characteristic in common — they live in symbiosis with large sea anemones of the genus *Stoichactis*. Symbiosis is the intimate association of two kinds of

organisms which is of benefit to both. The clownfish never goes far from the anemone and when danger threatens rapidly seeks the shelter of its stinging arms. The anemone immediately stuns or kills any fish which touches its tentacles but never uses its stinging cells on the clownfish. Just to make sure, however, the clownfish's skin is protected by a special layer of mucus. The anemone, naturally, also benefits from the association; the fish provides it with stray particles of food. A single anemone may have several such 'boarders', depending, apparently, on its size. As a rule one anemone is inhabited by one large female and several smaller male fish. Because of its relationship to the anemone the clownfish is also sometimes called the anemone fish.

Clownfish are very popular fish of saltwater aquariums. They may be kept without anemones but if they are to prosper and live longer then

symbiosis with an anemone is necessary.

Stone bass *Polyprion americanus*
The perch-like fishes include small species barely 4 cm (1 1/2 in) long, as well as large species. The smaller fish are popular amongst aquarium owners. There are both freshwater and saltwater species, many of which are well known to fishermen. The largest are the groupers, or rockfish, and the sea basses. These may measure more than two metres in length and weigh more than 60 kg (127 1/2 lb). Large specimens have gained a bad reputation amongst divers since they may be aggressive and attack. Their mouth is large but the teeth are relatively small and set in broad rows, not only on the jaws but on the tongue as well. Their chief food appears to be small fish and crustaceans, mostly barnacles. Countless observations confirm that sea basses are partial to places where barnacles occur in large numbers, for example wrecked ships, the wreckage of various wooden constructions as well as drifting wood.

Groupers are inhabitants of warm seas and stay mainly round coral reefs. The stone bass also occurs in colder regions. Though it is found in the tropical parts of the Atlantic it is also found in the Mediterranean Sea and occasionally also along the coast of Norway and in the North Sea. Its flesh is tasty and so the stone bass is a popular game and food fish.

Common clownfish

Flying fish *Exocoetus volitans*

The flying fish is readily identified by the pectoral fins which are long and wing-like. However, it does not actually fly but glides with fins outspread. Nevertheless, it has a 'motor' which facilitates its flight to a degree, particularly if it wants to cover a greater distance in the air. During the take-off the fish gathers speed just below the water's surface and then leaps into the air, spreading its pectoral fins and gliding, often aided by air currents, about one metre above the surface. The length of such leaps varies, but is usually between 40 and 50 metres. If it wants to fly further the fish uses its 'motor' — the tail fin, which is asymmetrical in all flying fish — the lower lobe is larger than the upper lobe. The lower lobe of the tail fin is submerged in the water and by rapidly moving it from side to side the fish soon regains the speed necessary for further flight. The tail functions like a screw propeller — as many as 50 strokes per second have been counted. The flying fish thus attains a speed of about 50 to 55 km per hour (31—34 mph). The entire process may be repeated three or four times, thereby extending the flight to as much as 200 metres. If it travels into the wind the flying fish attains greater heights than the usual metre and may thus sometimes even land on the deck of a ship.

Flying fish lay eggs in seaweed. The young are no different from the young of other fish — their pectoral fins are not enlarged. When they are 2 to 5 cm (3/4 to 2 in) long the fins are broader but rounded and they swim near the surface. Not till they attain a length of 8 cm (3 in) do the young flying fish develop elongated fins and make their first trial leaps.

Grunion *Leuresthes tenuis*

The grunion is an inconspicuous fish about 15 cm (6 in) long and would doubtless remain unnoticed were it not for its unusual manner of reproduction. The grunion inhabits a very limited area along the flat, sandy coast of southern California. During the spawning season it congregates in huge shoals numbering as many as several thousand fish. Spawning takes place only at night during certain high tides between March and August. At high spring tides, in other words always at full moon and new moon, the grunions allow themselves to be swept up on the sandy beaches where they lay their eggs. Each female burrows into the sand up to her pectoral fins and is encircled by one or two males. With rapid movements the female deposits her eggs here and there in the sand and the males fertilize them immediately with their milt.

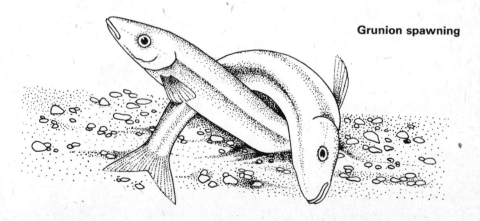

Grunion spawning

62

All this takes place on the sand, in the brief interval between two waves. The next wave carries the grunions out again to their saltwater environment. After high tide there is no evidence that the beach was visited by vast shoals of fish. The eggs develop in the sand at a depth of about 5 cm (2 in), being washed out by the next high spring tide two weeks later. This also triggers the hatching of the young fish which are washed into the sea by the waves. Young grunions grow fast and are fully grown and able to multiply within one year. However, they do not live long, dying at the age of three or four years.

Emperor snapper

Emperor snapper or Red emperor
Lutianus sebae

Snappers are a large group of predatory fish inhabiting all tropical seas. They reach a length of more than one metre and are the most important game fish in many parts of the tropics. They generally stay at greater depths near the bottom where they form small schools and hunt small fish and crustaceans. Some species are also found in lagoons, the brackish waters of mangrove stands and in tidal pools. The snapper family includes species that are coloured yellow or red, often variously striped, when fully grown. The young of some are a different colour to the adults. One such is the emperor snapper whose young are white with sooty-black stripes. The young fish stay in shallow waters, moving to the depths and changing colour during the course of their development. The black stripes turn brown, then reddish-brown and finally mahogany red in the adult. When they are excited the white ground colour changes to salmon, which is why this fish is sometimes also called red emperor.

Snappers are favourite aquarium fish. They have no special requirements and thus are one of the few species of reef fish suitable even for the novice marine aquarist.

Common porcupine fish
Diodon hystrix

The porcupine fish is a well-known species. The body, covered with spines which can be erected, is reminiscent of the porcupine and hence its name. This characteristic is quite unique in the fish realm. Besides this they are capable of expanding the body many times over by puffing themselves into a globular form. Such a porcupine fish is a comic sight — it looks like a spiny soccer ball with a pugdog head and fish's tail. Porcupine fish, like the related puffer fish, have a special sac which they can rapidly fill with water or air. The skin on the belly is loose thus enabling the body to expand enormously. As a rule the sac is filled only rarely. When the porcupine fish rises to the surface the sac is empty, and when it descends to the bottom it fills the sac to increase its weight. In this case, however, the size of the body remains the same. Only if the fish is attacked or irritated is the sac filled with such a quantity of water that the body is puffed into a globular shape.

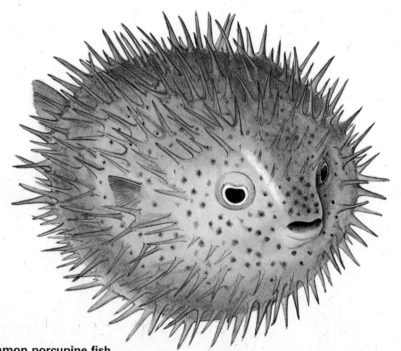

Common porcupine fish

Freshwater Fish

Bowfin

All the fish on the previous pages dwell in the sea. Many fish are found in freshwater — rivers, ponds and lakes. However, a surprising number can be found, at least for part of their life cycle, in both fresh and salt water. It is these migratory fish which are of great interest to students of fish evolution. How did these migratory patterns evolve? It is thought that the ancestral bony fish evolved in freshwater and from there they colonized the sea. Many fish, such as the salmon, breed in freshwater and then, when the eggs hatch, the young make their way to the sea. There they feed and grow, before returning, when mature, to the rivers to breed. A few fish, however, such as the North Atlantic eels and the Australian and New Zealand galaxy-fish, breed in the sea but enter freshwater to feed and grow. In between these types there are partial migrants — like the flounders, which breed in the sea but often ascend rivers to feed, and the sticklebacks, which breed in freshwaters but may overwinter in estuaries.

Strange American Fish

Arapaima *Arapaima gigas*
Freshwater fish are generally much smaller than saltwater fish and only very few species reach gigantic proportions. The arapaima of South America is often considered the largest freshwater fish of all. South American Indians say that it may reach a length of four and a half metres and weight of 200 kg (3 3/4 cwt), but in our day none which has been caught has measured more than 210 cm (84 in) or weighed more than 123 kg (2 1/4 cwt).

One of the arapaima's interesting characteristics is its large air bladder which enables it to breathe air. It is composed of cells which resemble lung tissue and is joined by a special canal with the mouth cavity and gill chamber. Arapaimas inhabit shallow waters and regularly take air into their buoyant bladder. Their chief food is fish, but they will eat almost anything — aquatic gastropods, shrimps, plants, snakes, turtles, crabs, frogs, etc. have been found in their gut. Young arapaimas, however, feed on plankton. Spawning takes place between December and May, during which time the males are readily distinguished from the females. The female is entirely brown, whereas the male has a black head and reddish tail. Each female lays altogether 180,000 eggs or so; these are deposited at intervals in several different hollows excavated by the males in the sandy bottom. The young fish, about 12 mm (1/2 in) long, hatch after five days. They are coloured black and because they remain clustered round the black head of the male they are practically indistinguishable. The female stays close by and protects the father as well as the young fish from enemies. As soon as they are able to catch small fish by themselves the young scatter, otherwise they would be in danger of being devoured by their parents.

Bowfin *Amia calva*
In the long distant periods of the earth's evolution, in the Jurassic and Cretaceous periods, there existed a large order of fish of which only one species — the bowfin — has survived to this day. It is found in calm and slow-flowing waters east of the Rocky Mountains. It is partial to shallow places with thick vegetation. Except for the spawning season it lives a concealed life and in autumn retreats to the depths where it waits out the winter. With the arrival of spring, as soon as the water in the shallows registers about 16°C, it returns again to its old habitat, the males outnumbering the females three to one. The females remain hidden in the dense vegetation, waiting until the males prepare the nest. In a selected spot, sheltered on one side by a submerged stone or roots, the male removes all vegeta-

Arapaima

tion and a layer of mud, laying bare a dense tangle of plant roots which form a mat at the bottom of the nesting hollow. Such suitable places are few and far between and so the nesting hollows are often located side by side, thus forming a kind of nesting colony. Each male guards his nest against intruders. When a female approaches a nest the male swims out with his mouth open, grasps her carefully by the snout and pulls her to the nest, where she then lays the eggs. Sometimes several females lay their eggs in a single nest, at other times one female visits males in

remarkable paddlefish, which takes its name from its long paddle-shaped snout. This is one-third to one-half the length of the body. The paddlefish inhabits open waters containing an abundance of plankton. It swims near the surface with its huge mouth open wide. The dense gill rods function as a sieve. The paddlefish consumes enormous quantities of plankton and may reach a length of up to two metres and weight of more than 80 kg (180 lb).

The paddlefish was formerly found in great numbers throughout the Mississippi River system and

all that remains of this order is a single family numbering seven species distributed in the rivers of North and Central America. The most plentiful of the species is the longnose gar, found in calm waters in the area extending from the Great Lakes southward. The gar lies practically motionless in dense masses of aquatic vegetation. It approaches its prey slowly and stealthily, grasping the victim by a sudden thrust to the side. Its chief food is fish but it will eat any animal. With its long beak-like snout it can seize several fish from a shoal at one time. However, its reputation

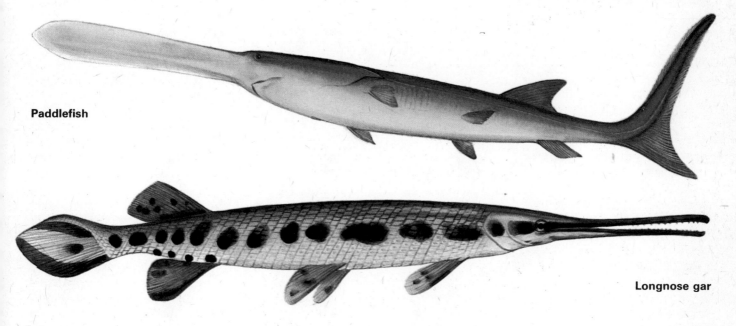

Paddlefish

Longnose gar

several nests and then disappears. The males keep watch over the eggs and ward off enemies. The young fish, coloured black, hatch after eight to ten days and hang on the edge of the nest at first. When they are about nine mm long and have practically consumed the yolk from the egg they swim away from the nest like a black cloud. About nine days later they leave the shelter and cluster beneath the body of the male, who guides the whole brood slowly through the water. He cares for the young until they begin to forage for food larger than planktonic crustaceans.

Paddlefish *Polyodon spathula*
The Mississippi River system and the Great Lakes are the home of the

was caught by the natives for its flesh as well as eggs, which were used to make quality caviar. The overfishing, poaching, water pollution and the construction of dams which prevented their migration caused a sharp decline in their number. Nowadays the paddlefish is a rare species. In the north Mississippi basin it is in danger of dying out and in the Great Lakes it has in all probability become extinct.

Longnose gar
Lepisosteus osseus
The gar is a living fossil like the paddlefish. It belongs to a very old order of fish which reached the peak of its development in the Mesozoic era, that is between 70 and 220 million years ago. Today

of being a voracious feeder is exaggerated. Gars consume relatively little food, they feed irregularly and digestion is slow. It takes 24 hours for a gar to digest its prey and that is a long time compared to other predatory fish. Nevertheless, the gar is the fastest-growing fish. The larvae that hatch from the eggs in late spring measure about seven mm. They grow in length as much as two and a half mm a day and by the end of the year the young fish measure approximately 20 cm (8 in). Then the growth rate slows, the yearly increase being roughly 2.5 cm (1 in), and stops when the fish is one and a half metres long. Because it moves about so little all the energy obtained from food can be used for growth.

Freshwater Predators

Piranha *Serrasalmus piraya*

Piranhas are considered to be the greatest predators in the fish realm and their voracity is the subject of horrifying tales. Most are greatly exaggerated but that does not mean piranhas are not dangerous. Of the 18 existing species, however, only four may be dangerous to man under certain circumstances. The most predaceous and dangerous is *Serrasalmus piraya* which inhabits the waters of the São Francisco River in eastern Brazil. It reaches a length of about 40 cm (16 in) and is one of the largest of the piranhas. The large and very sharp teeth on both the

during the spawning season are particularly aggressive. It is said that piranhas are attracted by the smell of blood in the water but it seems they are attracted by anything that is unusual. They are the only fish which hunt collectively in shoals.

In the Orinoco delta piranhas are said to play an important role in the customs of the natives. During the flood period they cannot bury their dead in the ground for many months of the year. The Indians therefore suspend the dead body in the water and the piranhas, called caribo (this means cannibals) by the natives, devour the flesh leav-

habit flowing as well as calm waters with plenty of hiding places and dense vegetation. Places where rivers rise and flood the surrounding meadows are the best breeding grounds for pike. The female lays a vast number of eggs on vegetation — large specimens may lay as many as one million. Because they are so sticky the eggs may catch in the feathers of birds, for instance ducks, and so sometimes the pike suddenly shows up in places where it has never occurred previously.

Of the six existing species the largest is the muskellunge, which inhabits the Great Lakes region of North America. The muskellunge reaches a length of two metres and specimens have been caught weighing as much as 50 kg (125 lb). This pike was always valued as a game and food fish and so its numbers are rapidly declining. It is also no easy thing to catch such large specimens, which naturally are very old. Today's 'good' catch weighs about 20 kg (42.5 lb), the record in recent years being 31.3 kg (64.5 lb). American fishermen fish for muskellunge chiefly in winter. The bait is a small, live fish attached to a float so it doesn't sink.

The European pike *Esox lucius* does not attain such proportions as the muskellunge. The record British pike is almost 25 kg (53 lb) but pike of 15 kg (31 lb) would now be considered a good catch. The European pike is also found in the brackish waters of the Baltic Sea where it breeds in shallow waters.

Piranha

lower and upper jaws interlock precisely and may cause grave wounds. One hears reports of fingers which have been bitten off and pieces of flesh cut from the body as with a sharp razor. It is interesting that the same species is extremely aggressive in some places and entirely harmless in others. The reason given in explanation is that males keeping watch over the eggs

ing only the bare skeleton. This is then dried, painted and decorated and put in the burial grounds resting on piles above the water.

Muskellunge or American pike
Esox masquinongy

Pike are popular game fish throughout the northern hemisphere (they are not found in the southern hemisphere). They in-

European perch
Perca fluviatilis

The European perch is the most colourful freshwater game fish. It is

Muskellunge

European perch

own or other species, and can cause great damage to the eggs of other fish. Sunfish were often kept in aquariums and thus found their way also to Europe. They were then introduced by aquarists to various waters in the wild and nowadays one may come across whole populations which have become acclimatized and are thriving in the new conditions. However, if sunfish occur in ponds they pose a grave threat to young fish and are therefore not popular with fishermen.

Their breeding habits are of interest; the male prepares a depression in the sand, a kind of nest, in which

Glass perch

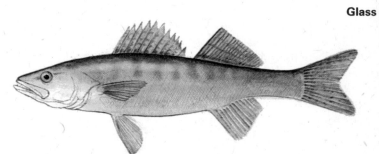

found throughout Europe and northern Asia and has been introduced in many places where it was formerly unknown. It thrives best in slow-flowing and calm waters but is also found in cold trout streams. There, of course, it does not grow as rapidly and is not as large. Though fishermen generally catch specimens weighing about 1/2 kg (1 lb), in congenial conditions one may catch specimens weighing 2 (4 1/4 lb) and occasionally even 4 kg (8 1/2 lb). Growth naturally depends on the abundance of food. It is interesting to note that until they reach a length of 15 cm (6 in) perch feed mostly on invertebrates. Only when they are larger do they begin to feed on fish, even the young of their own species.

Like all predatory fish, perch have very tasty flesh, comparable to that of trout and salmon. One drawback, however, is their small, rough scales that hold fast in the skin and cannot be scraped off easily. For

that reason it is necessary to skin them before cooking.

Pumpkinseed *Lepomis gibbosus*
Sunfish, one of which (shown in the illustration) is called the pumpkinseed, are found in North America from the Great Lakes region to Texas and Florida. They prefer clean, cold waters with sandy bottoms. Usually they occur in shoals. They feed mostly on small animals, including young fish, either of their

the female lays the eggs. The male then keeps watch over the eggs and when the young fish hatch guides them about for a while.

Glass perch
Stizostedion vitreum
The glass perch of North America is closely related to the European pike perch. It inhabits deep, clean waters from the Great Lakes and upper Mississippi region southward to Georgia and Alabama and eastward to Pennsylvania. What strikes one about this interesting fish at first glance are the eyes. They are large and extraordinarily transparent; this is how this fish came by its Latin as well as common name. The eye truly looks as if it were made of glass.

The glass perch is a very predaceous and large fish, reaching a length of one metre and weight of 9 kg (19 lb). This makes it the largest of all pike perches. It is a popular game fish and furthermore has very tasty flesh without any tiny bones. Fat is not deposited in the muscles but round the internal organs, chiefly round the gut.

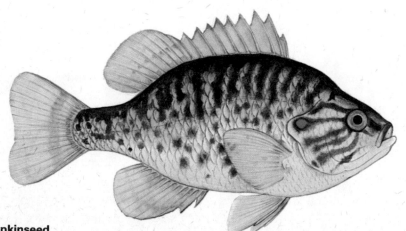

Pumpkinseed

Salmon-like Fish and Catfish

Atlantic salmon *Salmo salar*
The Atlantic salmon is one of the most famous — and delicious — fish in Europe and eastern North America. It is the classic example of a migratory fish. The adults, having grown to maturity in the sea, ascend the rivers in which they themselves were hatched. Before they begin their journey they weigh 8 to 13 kg (17 to 27 1/2 lb) (though occasionally they may reach 36 kg) and measure about 110 cm (44 in) in length. Their flesh is a rich red colour due to the presence of fat in the muscles. The up-river journeys of salmon are best known because of their abilities to jump up seemingly insurmountable waterfalls — some such waterfalls are important tourist attractions in the breeding season. The males and females mate and breed in the upper reaches of the rivers after which the males, and sometimes the females, die. The hatchlings live in the rivers for two to five years after which they return to the sea for a further period of growth, which under suitable conditions can be very rapid — up to 1 kg (2 lb) per month.

Rainbow trout *Salmo gairdneri*
Of all the trout the rainbow trout is unusual in that it has been introduced over almost all of the world and in most cases has become es-

Rainbow trout

tablished in its new home. It is native to the western United States, where one will find both resident as well as migratory forms. There are a number of reasons why the rainbow trout is so popular and introduced so often. The first is its greater resistance to water pollution. It is less affected by temperature, being found even in places where the temperature rises above 20°C in summer. Another important reason is its rapid growth. In two years the rainbow trout reaches a length of 25 cm (10 in) whereas it takes the brown trout a year longer to reach the same length. For anglers the rainbow trout has still another advantage. It is not as wary as the brown trout and will take any natural or artificial bait and at any time of the year.

Common European grayling
Thymallus thymallus
The grayling, like all the salmon family, has a small fatty (adipose) fin just in front of the tail. Its appearance otherwise is quite unlike a salmon, with its large scales and long dorsal fin. Grayling are found over most of Europe in clean, fast streams but at a lower altitude than the trout. They feed on the bottom on insect larvae but will also take insects on the surface and can thus be caught by the expert angler. His reward is the tasty flesh with the thyme-like aroma which gives the fish its Latin name.

Wels or European catfish
Silurus glanis
The wels shows the characteristic features of the catfish family — the long barbels or feelers around the mouth which it uses for finding food on the bottom. The skin is also smooth and virtually without scales. The wels is a large fish up to 30 to 50 kg (63 to 106 lb) (or even, exceptionally, 300 kg)(5 1/2 cwt) and prefers deeper water where it can shun daylight. The recent increase in such waters, owing to the damming of streams by man, has resulted in a corresponding increase

Common European grayling

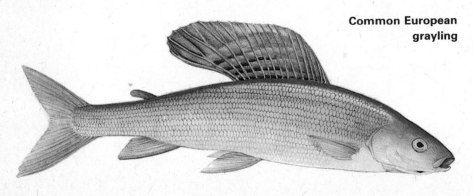

in the numbers of wels. They are found in much of the mountainous parts of central and eastern Europe.

Horned pout or Brown bullhead
Ictalurus nebulosus

The horned pout, also of the catfish family, was introduced to Europe from North America in the late 19th century in the belief that it would be useful for European fish breeding. These fish have relatively few requirements and in their native habitat reach a weight of 1.5 kg (3 lb). Today it may be said that from the commercial viewpoint the experiment was not a success. Pouts are not as long-lived in Europe as in their native land and also not as large, ones weighing 1/2 kg (1 lb) are rare. From the viewpoint of acclimatization, however, the experiment was successful. Partly by natural means and partly thanks to aquarists the horned pout has become a regular inhabitant of many European rivers. The horned pout is a bottom-dweller, generally hiding in various shelters during the day and emerging to forage for food at night. It swims close to the bottom, feeling the ground ahead of it with its bar-

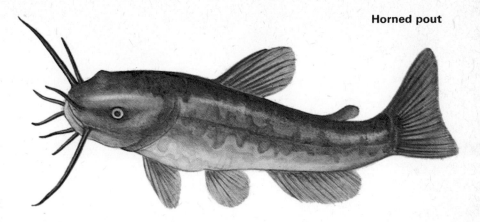

Horned pout

bels and looking for food. It is not popular with fishermen because it generally swallows the bait, making it difficult to remove the hook.

During the spawning period, usually in June, the males clear a place in the sandy bottom in which the female lays groups of eggs. These are watched over by the male who continually cleans the ground round the nest. He also keeps watch over the small black fry when they hatch, guiding the shoal about for a while before leaving them to fend for themselves.

Black-bellied squeaker

Black-bellied squeaker or Upside down catfish
Synodontis nigriventris

The whole of Africa south of the Sahara is the home of an interesting group of catfish with a markedly curved back and practically straight underside. The three pairs of short barbels round the mouth are feathery and the body is smooth, without scales. In the Zaire River basin is one member of this family that is worthy of note. It is the black-bellied squeaker, which reaches a length of only 6 cm (2 1/2 in) and for that reason was

introduced into Europe as an aquarium fish. It is very popular, particularly for its peculiar habit of swimming belly-side up. That is the reason why its coloration is reversed — the back and sides are pale grey or cream with dark spots and the belly, which in other fish is a lighter colour than the rest of the body, is black. Whereas other species of this family remain near the bottom, the black-bellied squeaker stays close to the surface where it is fond of nibbling the algae on the underside of the floating leaves of aquatic plants.

Wels

Carps and Their Allies

Common carp *Cyprinus carpio*
The carp is probably the best known and most valuable food fish in Europe. Its origin must apparently be sought in the wild Danubian carp that was caught as far back as the Neolithic period. When it was first domesticated, or at least kept in captivity, is not precisely known. The Romans, who were acquainted with the fish from cities in the Roman province of Pannonia, were probably the first to do so. Christianity was likewise doubtless responsible for the spread of the carp. It was one of the foods eaten on fast days and so large ponds were built beside monasteries to provide a supply of fresh fish for the more than 100 fast days in the year. Another centre of the carp's domestication was China, where we find references to the breeding of carp dating from the third century B.C. From these two centres the breeding of carp then spread throughout the world. It is interesting to note that in the United States the carp is not popular and is even exterminated as an undesirable fish.

Over the decades breeders have succeeded in developing several different forms, very popular being the mirror carp with a few, large scales, and the scaleless or leather carp. Carps sold at the fish market generally weigh 1 to 2 kg (2 to 4 1/4 lb), and are usually in their third year. Currently Israel is the world's greatest producer — its annual yield per hectare totalling more than 2,000 kg (40 cwt).

European bitterling
Rhodeus sericeus
The bitterling is a small, colourful fish found in almost all parts of mainland Europe. Its name in the different languages indicates that its flesh has an unpleasant, bitter taste. Its small size makes it of no particular value to man. And yet, the bitterling figures in all zoology textbooks. The reason for this is its unusual method of breeding, which is dependent on the pond mussel.

As the eggs ripen the female develops a long ovipositor with which she deposits the eggs in the mantle cavity of the mussel. The male simultaneously releases milt above the intake opening to the mantle cavity thereby fertilizing the eggs. The young fry remain inside the mussel for about 14 days until they have consumed the yolk sac. During this period they are not very mobile and thus more vulnerable than at any other time. Protection is provided by the mussel. After the yolk sac has been consumed the young bitterlings leave the shelter of the mussel and begin foraging for food. Bitterlings generally spawn twice a year — the first time in March or April, the second time in August. In congenial conditions this species often occurs in excessive numbers. Adult bitterlings are one of the several hosts of mussel larvae, which are parasitic for a time on their gills or skin.

Tench *Tinca tinca*
The tench is a fish of quiet backwaters with dense aquatic vegetation. The most conspicuous characteristic is the smooth, slippery body. This is because the scales are very small and generally embedded in the skin. For this reason some cooks skin the fish before cooking

but others don't, because the scales turn soft after cooking and do not spoil the pleasure of eating. The flesh is very tasty. In some countries the tench is even more popular than the carp and runs a close second to the rainbow trout. In pond husbandry the tench is generally not put by itself but together with carp. Unlike the latter it feeds also on molluscs and so does not seriously compete with the carp for food. Tench spawn later than carp — not until late June and July when the water is sufficiently warm. The young fish grow quite slowly. However, this is no drawback, because restaurants buy only small specimens, up to 20 cm (8 in) long — the exact size for one serving.

Tench were introduced from their native habitat to other continents. They are bred with success in South Africa, New Zealand and Java. Besides the beautiful green form, breeding ponds also occasionally contain the golden form. Tench is also popular with sportsmen. It is known to play for a long time with the bait and hooking it is not easy.

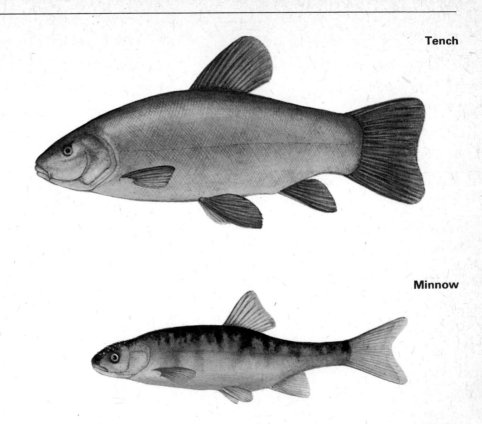

Tench

Minnow

Common bream *Abramis brama*

The deep, laterally compressed body and extensible mouth tell us a great deal about the life of the common bream. This fish is adapted to life at greater depths, where it forages for food in the mud. It feeds chiefly on red midge larvae, red tubifex and other animals which live on the bottom. Because the bream's mouth can be thrust out a long way the fish can collect food even in thick layers of mud. Breams live in shoals. In May or June, depending on the temperature of the water, they travel to shallow waters to spawn. At this time the males develop a conspicuous 'nuptial rash' on the head and sides of the body. The females lay an enormous number of eggs on plants below the surface — it is said as many as 300,000. In fine weather the fish spawn over several days. The young fry remain in shallow water for a time, feeding on plankton. Only later do the young fish forage for food on the bottom. Their great fertility and good adaptability to various conditions sometimes results in great numbers of these fish. Such a population, where the natural balance is disrupted, can contain individuals that are dwarfed or otherwise show poor growth.

Minnow *Phoxinus phoxinus*

If you go fishing in Europe in May in a clear, cool, oxygen-rich stream, even a small brook, you will most probably catch a small fish about 6 cm (2 1/2 in) long and coloured like an exotic aquarium fish. It is the minnow which in its 'nuptial dress' is the most colourful of the carp-like fish. The colours do not fade even in the non-breeding season — only the males lose their bright red undersurface and the white rash on the head. Minnows live in shoals which are continually on the move, rising to the surface only to swim down again and scattering to all sides when disturbed, after which they converge again in shoals. To sportsmen fishing the trout streams they are a welcome sight for minnows are a favourite food of trout and are often used as bait.

Common bream

Curiosities of the Fish World

Electric eel
Electrophorus electricus
Quite a few species of fish possess the ability to communicate an electric shock to animals with which they come in contact. The best known for this is the electric eel. Despite its name it is not at all related to the eel family but to the South American knife fishes. Most of its long body consists of special organs composed of 5,000 to 6,000 segments assembled in the same way as the cells in an electric battery. The electric organ is divided into three parts — two small batteries and one large battery, the main one. The large battery is not used constantly. When the electric eel is in motion the small battery in the tail sends out electric impulses at intervals of 20 to 50 per second. These impulses serve as a means of navigation in murky water. The second small battery energizes the large battery which produces a series of three to six waves of charges, each lasting five thousandths of a second. These are high voltage charges which stun or kill the eel's victim. An adult measuring about two metres in length produces charges of 600 volts and almost two amps. Such a charge will stun a horse, which may then drown. A man will survive a single charge but not repeated shocks.

Australian whitebait

Ubangi elephant fish
Gnathonemus petersi
The elephant fishes are another group of fishes capable of producing an electric charge. The electric organ is a complex one but the charges it produces are small and are not used to stun or kill prey. The fish produces an electric field around itself which serves as a means of orientation or to locate prey. The electric field also defines its territory. Elephant fish have four such organs — one pair facing upward toward the back, the other pair downward toward the belly. These are separated by insulating tissue. This system is not necessarily needed to locate food. The ubangi elephant fish is furnished with a long fleshy appendage on the lower jaw with which it forages for food in the soft bottom. It would seem, then, that the sense of touch is more important and that the elec-

tric field is used to locate food only occasionally. The name 'ubangi' derives from the African tribe whose women artificially enlarge their lower lips.

Australian whitebait
Galaxias attenuatus
The galaxy-fish, to which family the Australian whitebait belongs, are of interest for two reasons. Firstly their distribution is confined to the southern continental margins, being found in the rivers and coasts of South Africa, South America, Australia and New Zealand, indeed they are the principal native fish of New Zealand fresh waters. This distribution could well be the result of continental drift, the process by which the present continents are thought to have separated from each other some 100 million or more years ago. Secondly they include the only fishes other than the Atlantic eels which breed in the sea but which spend most of their time in fresh water. The whitebait, however, which is found in Australia, New Zealand and Chile, spawns in freshwater but spends most of its life in the sea.

Butterfly fish
Pantodon buchholzi
Certain small areas of the Zaire and

Ubangi elephant fish

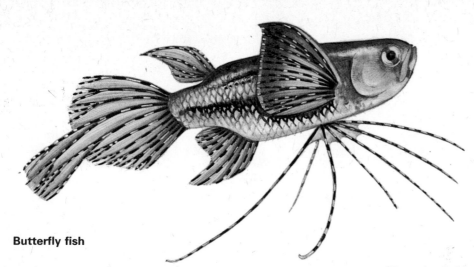

Butterfly fish

Mosquito fish

Nigeria river basins are the home of the butterfly fish. It is found in quiet waters and oxbow lakes with dense vegetation. It lies in wait for its prey just below the water's surface with its pectoral fins spread wide rather like wings. Spread out beneath its belly are the long rays of the ventral fins which serve as organs of touch. The butterfly fish feeds almost exclusively on insects. It collects them on the surface but occasionally may even leap out of the water. Its leaps may be up to two metres in length. The fish covers this distance by gliding with its wing-like fins and for this reason is sometimes known as the freshwater flying fish. The butterfly fish cannot fold the fins alongside the body like other fish but can move them up and down.

The butterfly fish's entire life is passed on or near the water's surface. Even the eggs float on the water and the tiny young fry, which hatch after three days, feed on minute insects which fall on the surface from the instant they emerge.

Mosquito fish *Gambusia affinis*
The mosquito fish is a small, viviparous tooth-carp. Though it is not particularly popular with aquarists, unlike others of its kind, it has made a name for itself in another way. This small but predaceous fish is found in the southern United States from eastern Texas to Alabama. Mosquito fish are very hardy, tolerating temperatures as low as 3° or 4°C and doing equally well at a temperature of 30°C. The quality of the water and amount of oxygen it contains are not of vital importance. These are the reasons why it was selected by man to aid him in the fight against malaria.

This disease is transmitted by mosquitoes and the mosquito fish is the sworn enemy of their larvae. This is why it was introduced into bodies of water in parts of southern Europe infested with malaria. It eats vast quantities of larvae daily and is thus very effective in the extermination of mosquitoes. Its contribution, however, is not without a price. The mosquito fish has adapted so well to its new environment that it feeds not only on mosquito larvae but on the eggs and fry of other fish as well. In some places it has even caused the total extinction of certain south European species of fish.

Australian lungfish
Neoceratodus forsteri
Lungfish are primitive fish, living fossils that graphically illustrate the course of evolution in the animal realm. Their ancestors, not dissimilar to the Australian form, lived on the earth as far back as 350 million years ago. Lungfish are found in South America, Africa and Australia. These lungfish can breathe by means of gills but have also developed a special bladder, a kind of primitive lung, with which they get oxygen from the air. The lungs of American and African lungfish have two lobes while the Australian species has one. The paired fins of the Australian lungfish are entirely different from those of almost all other fish. The framework of the fins is cartilaginous and the rays are not fan-like but arranged on either side of a central axis. The Australian lungfish can even walk with these peculiar fins! The Bennet and Mary rivers in Queensland, which are muddy, are the home of this fish. Its air bladders enable it to survive the summer months when the water contains a minimum of oxygen.

Though lungfish are not the direct predecessors of the amphibians they shed light on the path of evolutionary development. One last point of interest — the distribution of lungfish, like the galaxyfishes, can be directly attributed to continental drift.

Australian lungfish

Their Element is Both Land and Water

Existing amphibians are only the small remnant of a group which was widely distributed on the earth at one time. If we compare the existing amphibians we see that they can be divided according to their appearance into three groups, each of which apparently developed independently. Newts and salamanders have an elongated body with a long tail and two pairs of short legs. Frogs are tailless and their hind legs are longer than the forelegs. The third group includes the caecilians, which are legless and resemble worms. All amphibians have a smooth skin with a great number of glands. The skin plays an important role as an organ of respiration. Except for the caecilians, the development of amphibians usually includes a larval stage during which they breathe by means of gills and usually live in water. In the adult stage they breathe by means of lungs, though the absorption of oxygen through the skin also plays an important

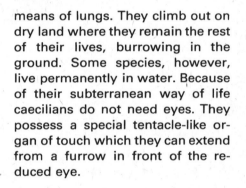

Mudpuppy

role. In some species the latter method fully satisfies the animal's needs and the lungs are reduced.

Caecilians and Salamanders
The legless body of the caecilian is reminiscent of a worm rather than an amphibian. This impression is further heightened by the fact that it is segmented. The eyes, if they are at all visible, are small and no longer serve their function for caecilians spend their whole life in the ground. The largest measure more than a metre in length, the smallest a little over 10 cm (4 in). Caecilians have no tail and the anal opening is thus located at the very end of the body.

Salamanders are another group of amphibians with a long body. Unlike caecilians, however, they have a tail and legs and at first glance resemble a lizard. Salaman-

ders include species which live on land as well as in water. In some species both sexes are alike, in others different, particularly in the newts where the males develop an ornamental fin on the back and tail during the mating season.

Caecilian *Siphonops annulatus*
Caecilians are inhabitants of tropical and subtropical regions. In America their range extends from Mexico to northern Argentina, in the Old World they are found in South-east Asia and some parts of Africa. They come to the surface

only rarely, when heavy downpours flood the ground where they live, and so it is no wonder that we know very little about their way of life. The male and female are alike. The male has a penis-like reproductive organ so that, unlike other amphibians, fertilization takes place inside the body. The female then lays

a small number of relatively large eggs. She curls her body around them, remaining thus until the larvae hatch. The newly-emerged larvae show little resemblance to the adults. They have normally developed eyes, the head of a salamander and internal gills which open to the outside at the sides of the head. They also have a tail used for swimming in water, which the larvae enter as soon as they hatch. Towards the end of the larval stage, however, the tail disappears, the gill openings close and the young caecilians begin to breathe by

Caecilian

means of lungs. They climb out on dry land where they remain the rest of their lives, burrowing in the ground. Some species, however, live permanently in water. Because of their subterranean way of life caecilians do not need eyes. They possess a special tentacle-like organ of touch which they can extend from a furrow in front of the reduced eye.

Axolotl *Ambystoma mexicanum*
The ability to breed in the larval stage is called neoteny. In amphibians this interesting phenomenon is relatively common and the axolotl is one such example. In the wild the axolotl is found only in Lake Xochimilco near the capital of Mexico. It breeds in the larval stage but in laboratory conditions it is

Axolotl

74

Alpine newt

their life is spent on land but they return to water during the breeding period. At this time the males develop dorsal fins which are often brightly coloured. Some species leave the water very early, others remain there a long time. The alpine newt of western and central Europe enters clear pools as early as April and does not abandon them until August, later than any other newt. At this time the male's fin disappears, and the skin roughens. After this the newt is rarely seen because it lives under stones, under the bark of tree stumps or in other moist places, sometimes quite far from water.

Fire salamander
Salamandra salamandra
The fire salamander is probably the most strikingly coloured European

Fire salamander

sometimes possible to raise adult specimens when hormones produced by the pituitary gland are fed to the larvae. The larva has well-developed branched gills as well as lungs and in oxygen-poor water often comes to the surface to breathe air.

Mudpuppy *Necturus maculosus*
Another example of neoteny is the mudpuppy, inhabiting the area from southern Canada to the Gulf of Mexico. If you catch this salamander it makes barking and mewing sounds which is how it came by its name. Mudpuppies are nocturnal animals. During the day they hide under stones or burrow in the mud. In the tangle of vegetation on the dim bottom mudpuppies hunt worms and the larvae of insects, crayfish and small fish even in daytime. Before settling for their winter sleep they mate, but the female does not lay eggs until spring. The clutch, which sometimes numbers only 18 but at other times more

than 800 eggs, is watched over by the female until the larvae hatch.

Alpine newt *Triturus alpestris*
Newts are distinguished from their close relatives, the salamanders, chiefly by the tail, which is compressed from side to side. Most of

amphibian. Populations in the eastern part of its range are a glossy black with irregular orange-yellow patches whereas populations in the western part of its range have the yellow patches joined in longitudinal bands. The salamander is a nocturnal creature and may be seen in

Giant salamander

daytime only in rainy weather. Though very plentiful in places it is absent elsewhere and thus has a local, discontinuous distribution. Few persons know that salamanders can make quacking sounds. Unlike newts, salamanders mate on land. The female, however, does not lay eggs but produces living larvae in water.

Giant salamander
Megalobatrachus japonicus
The giant salamander is the largest salamander of all. It reaches a length of about 150 cm (60 in) and is, not surprisingly, hunted for its tasty flesh. It lives in the mountain streams of Japan, where it hides on the stream bed. It is a nocturnal creature and its biology is interesting. In August and September the females journey to the places where they will lay eggs in holes excavated in the bottom. One female lays about 500. The males arrive after the eggs are deposited, fertilizing them and then keeping watch over them. The larvae, which float freely in the water, hatch after eight to ten weeks. Not until they have reached a length of about 20 to 25 cm (8 to 10 in) do the external gills disappear, after which they live on the bottom. It usually takes seven years before the giant salamander is fully grown. Unless captured it lives to a ripe old age — in captivity even more than 60 years

— and there is no reason to suppose its longevity is any less in the wild.

Frogs and Toads
Of all the amphibians, frogs and toads are the most highly developed. Despite its great wealth of species, numbering more than one thousand, this order has nevertheless retained a uniform body structure. In the first place there are the long hind legs which enable frogs to make their characteristically powerful leaps. Frogs and toads are also the most highly adapted of the amphibians to life on land. Only a few species pass their entire life in water. Most remain on land, apart from the breeding season, though they stay in damp places or leave their shelters only at twilight when the atmospheric moisture increases. However, their number also includes species which have abandoned water altogether and even breed on land. Frogs are found in practically all kinds of environments excepting the sea. Many live in trees and in some the large webs on the long toes are even modified into a kind of parachute which enables them to glide short distances from one tree to another. Other species live almost permanently in the ground or in termite nests, where they always have a tidbit to feast upon. The skin of frogs produces a secretion which

strongly irritates the inside of the mouth, and thus serves to protect the frog from many enemies. The secretion has a much greater effect, however, if injected into the blood. A relatively small dose may be fatal. The secretion of some frogs is thus an important component of poisons used to tip arrows.

Common or European tree frog
Hyla arborea
Tree frogs are admirably adapted for their life in the trees. The toes are furnished on the tips with cushion-like or cup-shaped adhesive discs that give them a firm grip, even enabling them to climb over smooth leaves. Most species are native to the tropics. Only one, the European tree frog, inhabits the temperate regions of the Old World. It is found chiefly in warm lowland country but in some places it is also found in mountainous areas. In spring, during the breeding period, these frogs congregate round water. There, particularly in the evening, one can hear the loud, croaking voices of the males, amplified by the vocal sac they inflate in the throat. No one would believe such a small frog can make such loud sounds. Tree frogs have the ability to change colour so that they blend perfectly with their environment, making them invisible amidst the foliage of trees and shrubs. Tree frogs breed in water,

Common frog

as 'more rum' or 'a jug o'rum'. The female lays up to 25,000 eggs which float on the water amidst aquatic vegetation. The tadpoles emerge after about a week but their metamorphosis into the adult frog is very slow and may take as long as two years.

Fire-bellied toad
Bombina bombina
Fire-bellied toads, which include four species, are found only in the Old World. They are small, incon-

Green toad

where the female lays clumps of eggs on the bottom. The tadpoles, which are active and have a golden gleam, metamorphose into frogs after about two months.

Green toad *Bufo viridis*
The word toad makes most people shudder and conjures up visions of an ugly creature. This, however, is mere prejudice. Take a look, for example, at the eye of the common toad, at the lovely golden colour of the iris. And as for the green toad, we can say without exaggeration that it is one of the most attractively coloured of all amphibians. Practically no two specimens have the same coloration.

The green toad has a wide range, extending from north Africa through much of continental Europe as far as central Asia. It is found both by water as well as in dry places such as steppes. During the breeding period, which is in spring, the toads converge round bodies of water to lay eggs. A single female can lay as many as several thousand in clumps joined by a gelatinous substance. The green toad remains in water much longer than other toads, sometimes more than a month, when you can hear the melodious sounds made by the males during this period.

The green toad forages for food

Fire-bellied toad

at night. Even though the hind legs are adapted for leaping it generally travels by crawling rapidly. The diet consists of insects, worms and snails and thus every toad may be said to be a very useful creature. However, it captures only moving prey because its eyes register only objects in motion. Smaller prey is swallowed whole without trouble. When it catches larger prey, the toad uses its forelegs to help push it into its large mouth.

In late September and October the toads hide in ground holes, rock crevices and similar shelters to hibernate far from water.

Bullfrog *Rana catesbeiana*
The bullfrog is the largest North American frog. Adult specimens may measure up to 20 cm (8 in). Only rarely is this species found outside water. It prefers ponds, marshes and slow-flowing rivers, where it rests in the water or on the shore on the lookout for food. It feeds on insects and insect larvae, worms, spiders, crustaceans and molluscs but can capture larger prey. Its stomach has even been found to contain other frogs, small turtles, young alligators and snakes, including poisonous species such as coral snakes. Exceptionally large specimens are capable of catching even a mouse or small bird.

The bullfrog is so called because of its voice — a deep loud croak heard during the breeding season. This is repeated three to four times in succession followed by an interval of about five minutes and then the 'song' continues, the words being popularly interpreted by some

spicuous toads, at least when viewed from above. The back is coloured grey or greyish-green and is a very effective camouflage when the toad lies on the water's surface or rests on the muddy banks. When alarmed, however, the toad assumes a peculiar defense posture to show the bright orange warning colours on the belly. It lifts its head, arches its back and raises and twists the forelegs outwards. Sometimes it even turns over rapidly on its back, startling the enemy with the whole expanse of its brightly-patterned belly. At the same time its warty skin produces a whitish secretion which is poisonous and causes serious inflammation of the membranes of the mouth.

In spring, when fire-bellied toads leave their winter shelters, they congregate in bodies of water. Because pairing takes place the whole summer they are found in water practically all the time. The males have a melodious, relatively soft croak. Fire-bellied toads often live in communities so that their monotonous voices seem to sound from all around, though the songsters themselves remain invisible thanks to their excellent camouflage.

The World of the Reptiles

Reptiles of the present day represent only a small fraction of what was once a large and diverse group. Millions of years ago they inhabited not only the land but all the seas as well. Reptiles first appeared on the earth in the Carboniferous period, about 200 to 250 million years ago, and their 'golden age' lasted until about 70 million years ago. Then they were surpassed by more highly developed classes — the birds and the mammals. Many of them became extinct, leaving the four groups that have survived to this day.

The first group is represented by a single species—the tuatara, the

Tuatara

second by the turtles and tortoises, the third by crocodiles and alligators, and the fourth by the lizards, chameleons and snakes.

Tuatara *Sphenodon punctatus*
The tuatara of New Zealand is the only existing rhynchocephalian, lizard-like reptiles which were widespread during the Triassic and Jurassic Periods roughly about 170 million years ago. At first glance the tuatara looks very much like a large lizard but its internal body structure exhibits a number of in-

teresting characteristics. For example, the spine is composed of vertebrae which are hollowed out at both ends. The ribs are furnished with hooked processes to which the muscles are attached. The massive skull is distinctive. On the crown is an opening above which is located a third, so-called parietal eye. It has a lens and retina but no iris. Traces of the parietal eye are to be found in other reptiles but its purpose is not known. The life processes of the tuatara are very slow. The animal takes one breath every seven seconds but at rest this number drops to as little as one breath per hour. Pairing takes place in January but not until October or December does the female lay 5 to 15 eggs in shallow depressions in the ground. Development of the embryo is very slow. The young hatch after 12 or 15 months.

At one time the tuatara was very plentiful in all of New Zealand but nowadays it inhabits only a few rocky islands along the northeastern coast of North Island and in Cook Strait between North and South Island. Thanks to the rigid protection provided by the government its numbers have been in-

creasing in recent years. The tuatara is active only during twilight and darkness. During the daytime, however, it often suns itself in front of its burrow.

Mississippi alligator
Alligator mississippiensis
Alligators, crocodiles, caimans and gavials are members of a very old group of reptiles which also included the dinosaurs. Caimans and alligators look very much like crocodiles but can be readily distinguished by the teeth. Crocodiles have the teeth of the upper and lower jaws in the same line whereas in alligators and caimans the upper teeth overlap the lower when the mouth is closed. In both groups the lower jaw is fitted with a large, prominent pair of teeth on either side. In crocodiles these teeth fit into a notch in the upper jaw and are visible when the mouth is closed, whereas in caimans and alligators they fit into pits in the upper jaw. Alligators are very well known creatures. They are found in the south-eastern United States and were at one time most plentiful in the Mississippi region. However they were killed in great numbers for their hides and so have completely disappeared in some places. In those places where they still exist they are now rigidly protected. Nowadays they are also reared and bred in alligator farms to supply leather for fancy goods.

During the breeding season the males utter deep, loud cries, whereas the females attract partners by producing a secretion from scent glands located on the throat and near the cloaca. After mating the female prepares the nest, a tall pile of rotting vegetation, in which she lays her eggs, keeping watch over them until the young hatch.

Mississippi alligator

Komodo dragon

Komodo dragon
Varanus komodensis
The komodo dragon, native to the Komodo, Rintja and Flores islands of South-east Asia, is the largest living lizard. It may reach a length of three and a half metres and weight of 150 kg (nearly 3 cwt). Europeans did not discover this giant until 1912 but natives of the Sunda Islands knew it well and apparently were well aware of its uniqueness. They are fond of carrion and if a visitor wants to observe or photograph the Komodo dragon he usually puts out a dead deer or pig as bait, thereby attracting them to the spot in great numbers.

Gila monster
Heloderma suspectum
Only two of the 3,000 existing species of lizards are poisonous — the Gila monster and the beaded lizard. They look alike and live in the desert regions of the south-western United States and neighbouring parts of Mexico. The scales of both species are small and convex and the animal looks as if it were covered with beads. The Gila monster may measure up to 60 cm (24 in) in length and the beaded lizard as much as 80 cm (32 in). Unlike snakes the poisonous glands of these lizards are located in the lower jaw. Most of its life is spent buried in the ground. It emerges only in rainy weather and then only at night. Its movements are slow and so it feeds only on the young or eggs of rodents and birds which it captures in the nest.

Jackson's chameleon
Chamaeleo jacksoni
Chameleons are unusual lizards found in Africa, Madagascar, India, Ceylon and even in southern Europe. They have several interesting characteristics which immediately attract the notice of the

Jackson's chameleon

observer. The body is deep and greatly flattened on the sides. The legs are modified into claw-like appendages which keep a firm grip on branches when it climbs. The eyes are large and bulging, and the lids are joined with only a small slit in the centre for the lens. The eyes move independently of each other and can observe two separate objects, say one in front and one behind, at the same time. The chameleon can also change the colour of its skin, much more quickly than can other land animals which have the same ability.

Its manner of catching prey is well known. The chameleon has a long, agile tongue which can be projected to a great distance to catch insects. In some species the tongue is sticky at the tip, in others it is even furnished with a kind of grasping organ at the tip. The long tongue is thrust out in 1/16th of a second and drawn back again in a quarter of a second. Some chameleons have various ornamental processes on the head, for instance helmets or horns. Jackson's chameleon, which lives in east Africa, has three such horns.

Gila monster

Lizards of All Shapes and Colours

Ocellated lizard

Ocellated lizard *Lacerta lepida*

The ocellated lizard is the largest and prettiest of the European lizards. It is native to southern France, the Iberian Peninsula and northwest Africa, where it is found in thickets and olive groves, of which it is particularly fond. It climbs skillfully on branches and when it cannot evade an enemy by hiding in its burrow it always seeks shelter in a tree. It generally reaches a length of about 60 to 70 cm (24 to 28 in), including the tail, which accounts for two-thirds of its length. On the Iberian Peninsula, however, one occasionally comes across larger specimens, the record being held by a female which measured 90 cm (36 in) from the tip of the nose to the tip of the tail. The males are generally smaller than the females but have a more robust head and tail with a thick, swollen base.

Green lizard *Lacerta viridis*

The green lizard is the second largest European lizard. It is native to southern Europe, its range extending eastward to the Ukraine and northward to the warmer regions of central Europe. In the west it is found in the Channel Islands but does not occur in Great Britain. Attempts have been made to establish it, for instance, in southern Devon, northern Wales and

Wall lizard

elsewhere, often in places where subtropical plants do well, but without success. In central Europe the green lizard was exterminated in many places by collectors who wanted it for their terrariums so that it is now protected by law in the places where it is still found.

Wall lizard *Lacerta muralis*

The wall lizard is the commonest lizard in the Mediterranean region whence its range extends to western and central Europe. The wall lizard fully merits its name. It climbs expertly with the aid of its long, sharp claws that provide it with a firm grip on rough surfaces. It can climb even vertical walls and those of the gardens and vineyards of southern Europe are its favourite habitat.

European glass lizard
Ophisaurus apodus

At first glance this legless lizard looks like a snake but examination of its skeleton reveals that it is indeed a lizard. Its close relatives are the monitors. Beneath the scales covering its body the glass lizard has bony plates which form a sort of armour.

The glass lizard, so called from the brittleness of its long tail, is native to southern Europe and South-west Asia. It belongs to the same family as the slow-worm, another legless lizard commonly found throughout Europe. In America this family is represented by many species, some of which are legless and others have well developed legs.

Green lizard

Long-legged skink
Eumeces schneideri
The skink looks like a lizard with a long tail and small, smooth scales. The legs are short, sometimes rudimentary and sometimes absent altogether. This is an adaptation to their way of life, for most skinks live buried in the ground. Within the skink family, however, we find all manners of locomotion ranging from running over the surface to burrowing underground. Skinks are found in the damp

European glass lizard

Long-legged skink

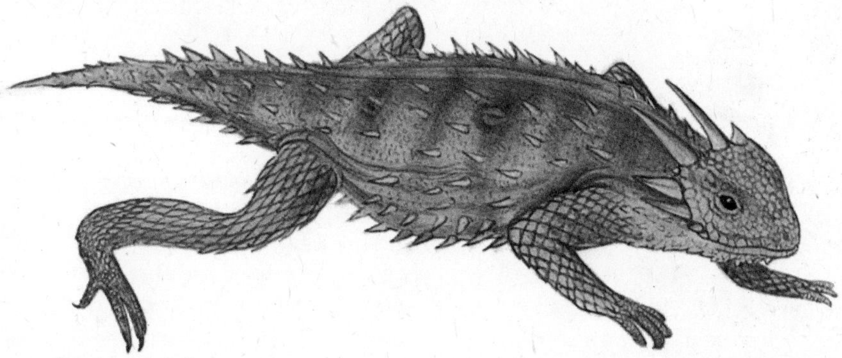

Texas horned toad

leons and so are sometimes called American chameleons even though they belong to a different family and are in no way related to the true chameleons. They are found throughout central America and the Caribbean area. Anoles are tree animals. Each male defends his territory and chases away all males which come near by rapid movements of his red throat. When this does not have the desired effect then he engages in combat. The victor remains green but the defeated lizard turns brown and gives in.

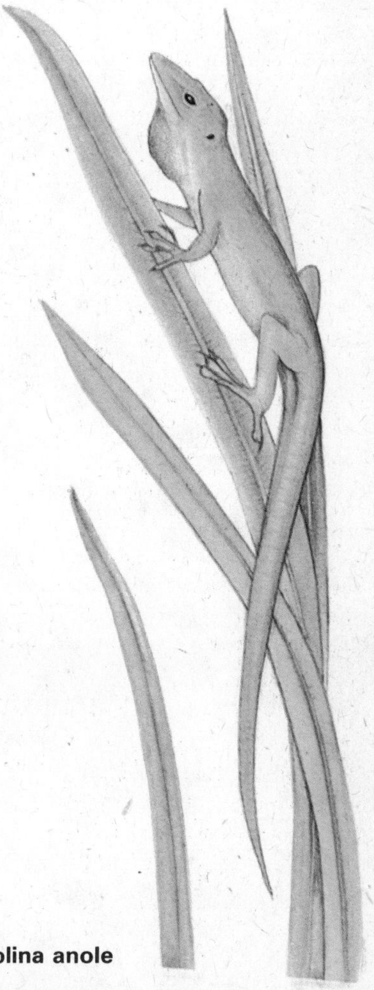

humus of rain forests as well as in deserts. Some species even climb trees, but these are exceptional. Some skinks which live in the desert are called 'sand fish'. They truly swim superbly in the sand, pressing their legs to their bodies and 'swimming' with undulating motions like a fish in water. The long-legged skink is common in an area extending from north Africa to central Asia.

Texas horned toad
or Texas horned lizard
Phrynosoma platyrhinos
The iguana family includes such diverse forms that at first glance one would not believe that they were related. Some live in trees, some on the ground and some even in water. Most iguanas feed on vegetable

matter but some are predatory. The horned toads are probably the strangest of them all. They have a short, plump, toad-like body, a round head with broad mouth, and a short tail. The back is covered with horn-like spines and this, coupled with the shape of the body, is how they came by their name. Horned toads are diurnal creatures. Their movements are relatively slow and so they feed only on desert beetles and above all ants. At dusk they burrow into the sand where they pass the night.

Carolina anole
or American chameleon
Anolis carolinensis
The anoles are the largest group of the iguana family. They can change the colour of the skin like chame-

Carolina anole

Legless Reptiles

The words 'legless reptiles' immediately call up visions of snakes, even though lizards also include several legless species. For snakes, however, this is a typical characteristic. Snakes, when it comes down to it, are a distinctively modified branch of lizards adapted for crawling and for capturing large prey. Because the limbs are absent the vertebrae are not differentiated and are divided merely into two sections — the body and tail sections. All the vertebrae in the body section are fitted with movable ribs attached at the free ends to bony plates on the underside. These are set in motion by the contraction of special muscles thereby making the snake glide. The movable connection of the jaws and certain other bones of the skull enable the snake to seize and swallow even large animals. The separate bones are joined by very elastic ligaments.

Many snakes have poisonous teeth or fangs in the upper jaw. In some these are located at the front, in others at the back. They are larger than the other teeth and are either perforated by a canal or fitted with a groove down which the poison runs into the body of the victim. Sometimes the poison fangs may be fitted with both a canal and a groove. Unlike most lizards the eardrum cavity is reduced so that snakes are deaf to airborne sounds. They are also unable to close their eyes — the lids are joined and transparent. Snakes shed their skin at intervals at which time the skin of the joined eyelids is also sloughed off.

Giants Amongst Snakes

The largest snakes are the members of the boa family. In many the skeleton bears traces of the pelvis and hind legs, also evident on the skin in the form of short appendages near the anal opening. These snakes are not poisonous and kill their prey by suffocation. They seize the victim with their teeth, crush it by coiling their bodies round it, and then swallow the dead animal whole. The best known representatives of the large snakes are the boas and pythons. Besides differing in other ways they also differ in their manner of reproduction — boas bear living young whereas pythons lay eggs.

Anaconda *Eunectes murinus*

Few snakes are the subject of so many tales and legends as the anaconda. Much has been written about its size and about whether it can or cannot swallow a human being. Travellers and Indians say that the anaconda may reach a length of twelve metres but the largest known specimen to date measured 9.40 metres. The New York Zoological Society, wishing to end the disputes about the size of anacondas, offered a reward of 5,000 dollars for a specimen ten metres long. The fact remains that no one has yet come forward to claim this reward. There is no denying, however, that the anaconda is a giant amongst snakes. Its weight is enormous — an eight-metre snake may weigh more than 150 kg (2.75 cwt). There are only two known instances of man being attacked by an anaconda. The first was a 13-year-old boy who was said to have suddenly disappeared while out swimming. An anaconda was later found at the spot. The victim's father shot it and the snake is said to have regurgitated the boy's body. In the second instance

Anaconda

an anaconda killed a grown man swimming in the river. The corpse showed traces of the snake's constriction but the anaconda released its victim without attempting to swallow it. It would seem, then, that the voracity and ability of the anaconda are greatly exaggerated.

Common boa *Boa constrictor*
The common boa inhabits wooded and scrub regions of the South and Central America. It is found mostly on the ground; only the young are fond of climbing. The adult boa reaches a length of about five metres and feeds chiefly on small vertebrates; only rarely is one of its victims a young ungulate. Boas are peaceable snakes which never attack man without provocation. Natives know this and are therefore not afraid of them. Boas will often visit chicken coops where they catch mostly rats. Smaller specimens are tamed and kept by the Indians to catch troublesome rodents. A tame snake is 'taboo' and no one is allowed to harm it. Otherwise, of course, the boa is considered to be very tasty and is avidly hunted in many parts by the natives. Two or three always set out together for they know that when attacked such a boa is not easily taken and a single hunter might readily be killed by the powerful constriction of its coils.

Reticulated python
Python reticulatus
The damp, humid regions of Southeast Asia are the home of the world's longest snake, the reticulated python. Though it never weighs as much as the anaconda it definitely holds the world record for length. The longest specimen found to date measured nearly ten metres. Such a large snake can devour even a fairly large animal and it has been known to swallow a pig or sambar-deer. No wonder, then, that there are many tales about man-eating pythons. However, there are only one or two con-

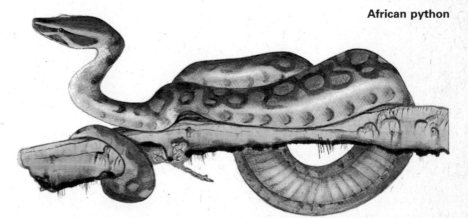

firmed cases on record of a python killing and swallowing children.

The reticulated python coils its body round eggs, remaining thus for the whole period of incubation — two to three months. Only rarely does it leave the clutch to quench its thirst. When they hatch the young pythons are about 60 to 70 cm (24 to 28 in) long. Growth is very rapid the first few years. During the first year they increase their length three-fold.

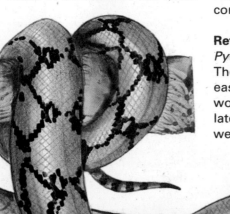

Reticulated python

83

Vipers and Rattlesnakes

Vipers and rattlesnakes have long, hollow fangs located in the front part of the upper jaw. A poison gland opens directly into the fang and the venom is injected through a tiny opening on the front side of the tooth when the snake bites. The poison contains several substances which destroy or coagulate blood.

Vipers are found only in the Old World, that is in Europe, Asia and Africa. Rattlesnakes and related species are native to the American continent, though some are found also in Asia. Most vipers and rattlesnakes bear living young, but

Common viper

both groups include species which reproduce by means of eggs.

Common viper or Adder
Vipera berus
The range of this, the commonest poisonous snake in Europe, is truly vast. It is even one of the few reptiles which occurs beyond the Arctic Circle. Eastward its range extends through the temperate regions of Asia as far as Sakhalin. Adders hibernate during the cold season. Often they congregate in certain places in the autumn and then hibernate communally in selected places called hibernacula.

Late March until May is the breeding period for adders. The males stake out and defend their territories, facing one another with raised bodies, swaying from side to side, coiling round one another, and trying to press their opponent to the ground.

The body colouring is very variable and in some places, particularly in peat moors, one may find all-black specimens. The adder is ovoviviparous, which means that the eggs remain inside the body of the female until the young are fully developed. When they emerge the young adders are encased in a thin membrane which they rupture by jerking movements of the body. They emerge in late summer.

Prairie rattlesnake
Crotalus viridis
Rattlesnakes are related to vipers, from which they differ, among other things, by possessing a special organ on either side above the nostrils — a deep pit whose lining contains a great many nerve cells. Experiments show that this serves to perceive heat some distance away. This organ is very sensitive, sensing and distinguishing gradations of one-tenth of a degree. It is probably used to locate prey. Rattlesnakes are known primarily for their rattle, also called bell, cloche, buzzer and whirrer. The rattle is composed of a series of interlocking horny rings at the end of the tail. The rings are freely joined and produce a characteristic rattling or buzzing sound when the tail

is moved back and forth. The snake does not shed its rattle, on the contrary — a new ring is formed every time it sheds its skin.

The usual number of rings is about 8, and it is interesting to note that this is the optimal number, which produces the loudest sound. Also of interest is that both the amount of poison produced and its effects may differ within the same species depending on the environment. For example, prairie rattlesnakes from the plains are about three times as poisonous as members of the same species found in the Grand Canyon.

Detail of rattle

Prairie rattlesnake

Eastern diamondback
Crotalus adamanteus

Rattlesnakes may be divided according to their colouring into several types. Three North American species have black diamond-shaped markings on the back which is how they came by their name. The eastern diamondback is the largest of the three. It may reach a length of two and a half metres and weight of 10 kg (21 lb). The poison fangs are up to 3 cm (1 in) long and the poison sacs produce

Eastern diamondback

a large quantity of poison. It is the most venomous of the North American snakes.

All rattlesnakes are superbly adapted to life in a dry environment. Practically no water evaporates through the skin. Water is lost only by breathing. The snake's consumption of water is ten times less than that of mammals of the same weight. Nevertheless, rattlesnakes do not avoid water. When they drink they suck up water with their mouths.

Fer de lance *Bothrops atrox*
South and Central America are the home of snakes related to the rattlesnakes but minus the rattle. They are very poisonous and dangerous to man. Though not aggressive, their bite kills many people every year. They are nocturnal snakes which lie coiled in the grass or thick undergrowth during the day. Because they are very

plentiful and often occur in plantations it is not surprising that they are readily encountered by man. Such a snake naturally attacks when disturbed. The most dangerous of these species is the fer de lance. This name, by which it is known throughout the world, was given to it by the natives of Martinique. Its poison acts very rapidly and when it bites a man death may occur in a matter of minutes. It has a marked decomposing effect so that the wound and surrounding

Fer de lance

area rapidly turn black, the eyes become bloodshot, and the victim bleeds from the mouth, nose and ears.

Water moccasin
Agkistrodon piscivorus

The eastern and middle United States and the eastern parts of central and South-east Asia are the home of snakes of the genus *Agkistrodon*. They are related to the rattlesnake but do not possess a rattle. When disturbed they often raise the tips of their tails and vibrate them the same way as rattlesnakes. The largest member of this genus is the water moccasin of North America which reaches a length of one and a half metres. As its name indicates it is found alongside or in rivers, lakes, pools and swamps. Its chief food is fish, which it swallows without biting them beforehand. However frogs and small mammals, which it feeds on occasionally, are first killed with its poison. The water moccasin is an extremely poisonous snake but luckily not aggressive. Nevertheless it is feared by those who work in rice fields. Like all snakes even the water moccasin will attack if unexpectedly disturbed.

Water moccasin

Colourful Snakes

Many snakes are soberly coloured and readily escape notice in the wild. Others are beautifully coloured but escape notice because they blend perfectly with their environment. Their coloration often serves as excellent protective camouflage, for example there are various species of green snakes which live in trees. Frequently, however, we come across snakes with various markings on their backs, often arranged in attractive patterns. The snake realm includes the entire spectrum of colours apart from blue. The loveliest colours are those of a snake which has

corresponds to the size of the snake, which reaches a length of 60 to 70 cm (24 to 28 in). Coral snakes are very poisonous, their bite causing death within 24 hours. Luckily they are slow and very peaceable. The poison fangs are short, so that often thick clothing or thick shoes afford sufficient protection. Danger of snakebite is thus small and most fatal bites are caused by careless persons who attempt to handle this pretty, quiet snake. One explanation of their distinctive coloration is that if an enemy is bitten, but not killed, by such a snake, then it will remember the snake's appearance

hatch after three months. They are about 5 cm (2 in) long and their coloration is the same as that of their parents, only somewhat lighter. Newly-hatched snakes have a fully developed poison gland and their bite may be dangerous. The amount of poison, of course, is small. Their first food is insects, but they soon begin hunting small lizards and snakes.

Poisonous snakes will usually attack if suddenly disturbed or irritated. The harlequin snake, however, generally will not. It reacts in a peculiar manner, perhaps intended to frighten off the intruder.

Coral snake

just shed its skin. The new skin has a velvety, often opalescent sheen. Young snakes are generally more brightly coloured than their parents — the colours fade with age.

and be wiser the next time. Even more important, many mammals and birds will learn from watching one of their number suffering after being bitten.

Coral snake *Micrurus corallinus*
There are a number of snakes with colours arranged in regular cross bands. Such snakes are among the most beautiful creatures in the world of nature. This is particularly true of coral snakes. The coral snake inhabits the drier virgin forests of northern South America. It is a nocturnal creature which conceals itself in holes and other hiding places in the daytime. Its diet consists of snakes, lizards and small rodents. The size of its prey

Harlequin snake
Micrurus fulvius
Two species of coral snakes have a range which extends to the southern United States. The harlequin snake is found in northern Mexico, Florida, the Carolinas, Texas, southern Ohio and the Mississippi valley. Its biology is no different from the other species (approximately 50) of coral snakes found chiefly in South America. Coral snakes lay eggs in the ground or various moist places. The young

Harlequin snake

King snake

When disturbed it hides its head under the body and twists from side to side.

King snake
Lampropeltis getulus

King snakes are non-poisonous North American snakes of the grass snake family. A peculiarity of these snakes is that they capture other species of snakes, often ones which are extremely poisonous. They are naturally immune to the poison and this is how they came by their name. King snakes are very colourful, some species even mimic the coloration of coral snakes and may thus acquire some immunity to attack by enemies who have learned to avoid the coral snakes. They are most active in the afternoon and evening. It cannot be said that they pursue other snakes — indeed snakes are not their chief food. King snakes kill many mice, frogs and lizards, but if they encounter a snake they grasp it by the head, kill it by crushing it in their coils, and then eat it. Because the king snake may measure up to two metres in length even rattlesnakes are common victims.

King snakes mate in spring and in summer the female lays 10 to 30 eggs. Sometimes she remains by the clutch for several days, at other times she soon abandons it. The young, which measure about 12 cm (4 3/4 in), hatch in four to six weeks. Unlike coral snakes, young king snakes are more brightly coloured than their parents.

The king snake is a very welcome sight wherever it occurs and is protected because it destroys other, poisonous snakes.

Black-banded sea-snake
Laticauda laticaudata

The peculiar family of sea-snakes fully deserves its name. All its members are excellent swimmers mostly inhabiting the warm waters of tropical seas. The tail and much of the body are flattened from side to side which greatly improves their propulsive power. Some are so completely aquatic that they never crawl out on dry land. The black-banded sea-snake is one of those which breed on land. The female lays eggs in the sand and then returns to the sea. In some places these snakes are very plentiful. They congregate in large groups sometimes even in their thousands. When travelling in the sea they often cover vast distances. They are found in the Indian and Pacific oceans from Madagascar to California. Their chief food is fish.

Sea-snakes are extremely poisonous, even more so than cobras. However they are peaceable, non-aggressive snakes which generally do not bite, even when taken up in the hand. Fishermen know them well and if a sea-snake becomes entangled in their nets they are very careful when freeing them. In some regions the natives catch sea-snakes for food, usually eating them smoked.

Black-banded sea-snake

Cobras, Mambas and Others

Poisonous snakes fall into three main groups. Firstly, there are several members of the grass-snake and racer family — the *Colubridae* which have grooved fangs at the back of the mouth and which produce poison. These are not usually dangerous but a bite from some species such as the South African Boomslang, *Dispholidus typus,* can be fatal. The vipers and rattlesnakes, as mentioned earlier, have long, hollow teeth which are folded back when the mouth is closed. The cobras and krait family have medium-length fangs with a groove along the front which is

King cobra

covered over to varying extent to form a canal. The venom of these snakes has more effect on the nervous system and less on the blood itself than that of the vipers.

King cobra *Ophiophagus hannah*
The king cobra, which reaches a length of five and a half metres, is the largest poisonous snake in the world. It is found throughout the whole of South-east Asia from India to the Philippines.

The king cobra is dangerous mainly during the breeding season. The female builds the nest, bringing the necessary material in a coil of her body. In the lower part of the nest she constructs a chamber in which she lays the eggs; the upper part serves as a resting place where she stays to watch over the clutch. The male stays close by. When guarding the nest cobras are more aggressive than at other times and this is also the period when they

are most likely to attack. However, natives nevertheless seek out the nest and kill the cobras, for the eggs are considered a very great delicacy. The flesh of the cobra is also valued. Snakes are the main food of this cobra. It is not immune to their poison, however, and so hunts chiefly nonpoisonous species. This, of course, does not mean it will not catch a poisonous snake occasionally, even attacking smaller members of its own species.

Spitting cobra
Naja nigricollis
The best known cobras are members of the genus *Naja*. This scientific name is derived from the Sanskrit word *'naga',* meaning snake. Cobras, of course, are found not only in Asia, in fact Africa is inhabited by many more species, some of which are even expert at 'spitting' poison. Best known of these dangerous snakes is the spitting cobra. The groove at the front of its poison fang closes, leaving only a tiny opening at the tip through which the cobra can eject poison for a distance of up to several metres. If such poison enters the eye it causes extreme pain and even temporary blindness. There are instances of the poison being ejected for a distance of four metres! When the snake bites, its fangs remain clamped in the body of its victim and the poison is pressed into the wound by chewing movements of the snake's jaws. The effect of the poison depends on how long the snake retains its hold and how much poison enters the wound. In human beings a further important factor is their general state of health. Those which succumb most readily are children, elderly people and those with a weak heart.

Spitting cobra

Aesculapian snake

Black mamba

Dendroaspis polylepis

The black mamba is the most feared of African snakes. Horrifying tales are told about this deadly species. The truth is that unlike other poisonous snakes the black mamba is very agile and covers short distances with great speed. If danger threatens it immediately charges with mouth wide open — and then quickly disappears. Its poison is extremely toxic and death follows quickly unless serum is administered. The black mamba is the second largest poisonous snake in the world, after the king cobra. It reaches a length of more than four metres. Unlike the closely related green mamba it prefers open country. Only rarely does it climb trees. Its favourite haunts are old termite

nests where it suns itself and lies in wait for prey. It feeds chiefly on birds, also small rodents, lizards and even snakes. Besides poison fangs the mamba has a row of long teeth on the lower jaw.

Egg-eating snake

Dasypeltis scabra

A great many snakes feed on eggs at times but some species in Africa and India feed exclusively on eggs. They are even furnished with special equipment for this unusual diet. The snake locates the eggs by its sense of smell, carefully feels them with its tongue, rejects any which are spoiled or unfertilized, and then sets about eating the good eggs. Its jaws are practically toothless but the mouth is lined with sticky tissue so that the eggs cannot slip out. The snake opens its jaws wide, enveloping the eggs with its mouth until they are swallowed. It is in the gullet that the snake has its special equipment. The processes on the underside of the vertebrae are sharp and project into the gullet where they slit the egg shell like a saw. There are 17 or 18 such sharp 'teeth' in the gullet. Behind these are several broad processes which crush the shell and six or seven more blunt processes which mould the crushed egg-shell and its contents into the shape of a cylinder pressing out the liquid

Egg-eating snake

Black mamba

inside. This is digested by the snake and the crushed shell fragments are regurgitated.

Aesculapian snake

Elaphe longissima

Aesculapius, the god of medicine in Roman mythology, was depicted with the symbol of his power — a snake twined around a staff. We are all familiar with this symbol, for it is the symbol of physicians to this day. Few, however, know that this is the Aesculapian snake. The ancients had great faith in the power of snakes. When the plague raged in Rome in Fabius' and Brutus' day people brought snakes to the city and built a temple for them on one of the islands in the River Tiber believing the snakes would dispel the deadly disease. The Aesculapian snake is native to southern Europe, its range extending far north. In some places, however, it was probably introduced. It is assumed that this may have even been done by the Romans because these snakes are often found at the sites of one-time cities, built by the Romans on their military expeditions. Some such places even included Roman baths where disciples of the God Aesculapius undoubtedly worked.

Tortoises and Turtles

Turtles and tortoises are interesting animals which lived on the earth as far back as 200 million years ago and are thus among the oldest of reptiles. They developed an unusual type of protective armour. They are furnished with a shell within which the head, legs and tail are withdrawn for protection. The shell is composed of two parts — a flat undershield or plastron and a dorsal shield or carapace which is usually convex. The two parts are either fused or joined by ligaments. The body of turtles also has a number of structural peculiarities. Unlike all other land animals the shoulder girdle is inside the ribs, which are joined with the bones of the shell. The legs do not extend below the body but to the sides. Most turtles can withdraw the legs as well as the head and neck inside the shell. These animals are noted for their longevity. The oldest was probably a tortoise given to the natives on Tonga Island by Captain Cook in 1777. It died a natural death in 1966 and was thus more than 189 years old.

Greek tortoise or Spur-tailed Mediterranean land tortoise
Testudo hermanni

In southern Europe tourists encounter two species of land tortoises which are often mistaken for each other. These are the spur-tailed or Greek tortoise which is found in the Balkans and Italy, and the spur-thighed or Iberian tortoise of the western Mediterranean. Formerly both these species were very numerous and to this day they are still frequently encountered in some places. Many, alas, die because tourists and inexperienced terrarium keepers take them home from their holiday as a souvenir or pet. Those with terrariums are chiefly interested in small tortoises and unbelievable quantities are taken from the Mediterranean region every year. According to official records the number taken just from Yugoslavia to Germany in a single year totalled two million! What with the increase in tourist travel tortoises are becoming increasingly endangered and so it will be necessary to take steps to protect them. Today the Iberian tortoise is a protected species in Spain and North Africa.

Box turtle *Terrapene carolina*

In the United States the box turtle is the equivalent of Europe's Greek tortoise in popularity. Its common name refers to the peculiar arrangement of the plastron. This has two sections which can be inclined to the carapace at the front and back so the shell is completely closed, like a box. The box turtle is found mostly in dry situations, but sometimes also in moist places. It tends to go out only at dusk, generally avoiding full sunlight by hiding in the shade or burrowing in the ground. It feeds on both plant and animal matter. It is fond of fungi and woodland berries and also carrion. Allegedly it also catches insects and so is often kept and allowed to move about freely in the household.

The breeding period begins in spring as soon as the turtles wake from their winter sleep. In early summer the female lays four to five eggs in a depression she makes in the ground. Some eggs hatch in the autumn of the same year, but some hibernate and do not hatch until the following spring. The male's eye has a red iris, whereas the female's is greyish-white.

Painted terrapin
Chrysemys picta

Southern North America is the home of two species of tortoises whose young are often sold in pet shops — the painted terrapin and red-eared terrapin. These are tortoises which live in water. The colourful young terrapins measure about 2.5 cm (1 in) on hatching and their shell is nearly circular so that they look like a coin with legs. Thousands are caught yearly, but most are fated to die a slow death in the hands of inexperienced owners. These turtles have no special food requirements but need a relatively high temperature and plenty of sun which in winter must be pro-

Box turtle

vided in substitute form with a sun-lamp. If they survive their owners have a further disappointment in store. When they are two or three years old the turtles already measure 8 cm (3 in) and their bright coloration gradually fades. At this stage they begin to eat larger prey, for instance fish, and increase in size by 1.5 to 2 cm (1/2 to 3/4 in) a year. Adult terrapins measure about 30 cm (12 in) and the original coloration has been completely replaced by a greenish-brown colour. In the wild both the painted terrapin and red-eared terrapin hibernate in the southern parts of their range only briefly, in the north longer, for up to six months, waking from their winter sleep only when the temperature of the water registers more than 10°C.

African soft-shelled turtle
Trionyx triunguis
The bony shell of turtles is generally covered with horny plates. These

Painted terrapin

turtles are predators which feed on fish, frogs, crustaceans and insects. Unlike other turtles they have sharp jaws covered by soft lips. Practically all species are found in the northern hemisphere. The only exception is the African soft-shelled turtle which is found in all of Africa south of the Sahara. It attains sizeable dimensions, specimens measuring about 100 cm (40 in) are not at all uncommon.

African soft-shelled turtle

differ in shape from the underlying bony plates so that they do not coincide, thereby increasing the strength and firmness of the shell. In some turtles, however, the bony shell is covered with soft thick skin. These are the soft-shelled turtles. Their carapace is flat and the plastron is rudimentary so that the two parts of the shell are not joined. Soft-shelled turtles are aquatic and superbly adapted to life in water. They can remain submerged as long as 10 to 15 hours. The lining of the gullet is provided with a dense network of capillaries and is able to absorb oxygen from the water, thus functioning like 'gills'. Soft-shelled

Florida soft-shelled turtle
Trionyx ferox
The genus *Trionyx* includes 14 species distributed in Asia, Africa and America. The best-known American species is the Florida soft-shelled turtle found in bodies of water from southern Canada to Mexico. It frequently climbs up on

to river banks but never goes far from water. Frequently it rests near the surface with only the tip of its pointed nose and nostrils visible above the water. It feeds on fish and frogs but large specimens may catch even small birds. The sharp jaws have a strong bite. Some old turtles are fond of molluscs and readily crush their shells. Soft-shelled turtles lay eggs by the waterside where the female buries them in the sand. The young hatch after 40 to 60 days, depending on the climate. At first they remain in shallow water close to shore. They are very predaceous and prone to bite. They differ from the adults by having more pronounced ornamental markings on the soft shell. When fully grown the Florida soft-shelled turtle measures about 40 cm (16 in) in length. Soft-shelled turtles are well known to fishermen who catch them with hooks as well as nets and think their flesh very tasty. Great care must be taken when removing these turtles for their sharp jaws can cause unpleasant wounds.

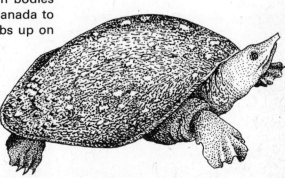

Florida soft-shelled turtle

Predatory Turtles

Common snapping turtle

Common snapping turtle
Chelydra serpentina

The family of snapping turtles has only two species — both found in America. There are undoubtedly many other species which are predaceous but none looks so much like a predator as the snapping turtle. The better-known and more widely distributed of the two is the common snapping turtle. It is found in quiet and slow-flowing waters from southern Canada to Ecuador, being partial to places with a muddy bottom. It is a large turtle, reaching a length of up to 1 metre and weight of 36 kg (76 1/2 lb). Its powerful jaws are terminated by a sharp hooked appendage that resembles a bird's beak. Its prey is anything it can overpower — fish, amphibians, birds and mammals, even quite large animals. The snapping turtle drags them underwater where it tears the flesh and devours them. Sometimes it may be encountered on land. Such specimens are usually females seeking a suitable spot to lay the eggs. Sometimes they travel far from water. After laying the 20 to 40 eggs in a depression they hollow in the ground, they return to the water. The young, which hatch about ten weeks later, instinctively take the shortest route to water. Many die along the way because their shell is still soft and does not provide them with adequate protection against enemies.

Alligator snapping turtle
Macroclemys temmincki

The other member of the snapping turtle family is the alligator snapping turtle. It greatly resembles the common snapping turtle but is much larger (it may measure up to 140 cm (56 in) and much heavier. Some specimens weigh 100 kg (212 1/2 lb). E.T. Seton, the writer and naturalist who knew the American wilderness well, tells us about this turtle in the words of Quonab, Indian friend of Rolf of the woods. 'That's Bosikado. I know her well and she knows me. We've been warring a long time; one day I'll get her. The first time I saw her was three years ago. I'd shot a duck; it was floating on the water but before I could pull it out something dragged it under and it was gone. Then a duck and brood of young ducks settled on the water. One after the other were dragged under by the turtle and finally the mother duck too. She chases every duck off and so I put out lines at night to catch her many a time. I caught a couple of smaller turtles — one weighed four kilos, another five — and they were very good. I've hooked Bosikado three times already, but each time I was pulling her in she bit through my strongest line and got away. Her back is as wide as my canoe; she clawed at the sides of my boat and set it rocking. She looked like the devil. I sure was scared!' The author doesn't exaggerate in the least. Here's one more interesting item to supplement the narrative. This turtle rests on the bottom with mouth wide open. On its tongue is a worm-like appendage that keeps twisting and turning, thereby attracting fish straight into the mouth. All the turtle needs to do then is snap its jaws shut.

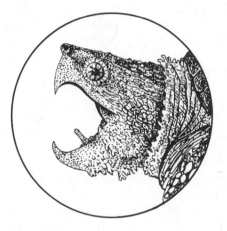

Alligator snapping turtle

Matamata

Matamata *Chelus fimbriatus*
The matamata is a most peculiar turtle. It is found in northern South America in shallow waters. It rests on the bottom extending its long neck to the surface so that only the tip of the nose and nostrils are visible above the water. The shell is generally covered with algae thus providing the animal with excellent camouflage. The long neck and limbs are furthermore covered with leathery appendages which make its camouflage even better. According to certain observations these appendages also serve to attract fish. The matamata is a small turtle measuring only about 40 cm (16 in) in length. The neck, which is only slightly shorter than the body, cannot be withdrawn completely inside the shell and so the turtle curves it sideways in an S-shape with the head pressed against the side of the body. Experiments indicate that the matamata cannot swim. In deep water where it cannot reach the surface it drowns quite easily.

This turtle uses an unusual method to capture its prey. It feeds mostly on fish. If a fish comes within range the matamata makes a sudden lunge with its mouth wide open and sucks the fish in along with a stream of water. The jaws, which are soft and different from the sharp jaws of other predatory turtles, serve in the same way as the frame of a fishing net.

Green turtle *Chelonia mydas*
The sea turtles have forelegs modified into fin-like paddles. These turtles spend their whole life in water, the females climbing out onto land only to lay eggs. The largest of all is the leathery turtle which can be as much as two and a half metres long and weigh 600 kg (11 cwt). The bony and horny protection is much reduced, hence the name. Best known of the rest is the green turtle which reaches a length of one and a half metres and weight of approximately 250 kg (4 3/4 cwt). It is an excellent swimmer, travelling through the water with a lightness one would not expect from such a huge creature. It is found in all the warm oceans — the Mediterranean, Atlantic, Indian Ocean and the Pacific. At a certain time, which differs according to the locality, the females leave the open seas and swim to the shore where they lay eggs. These are always deposited in the same place by the females of each ensuing generation. The turtles arrive at night, excavate hollows in the sand, and then lay the eggs. Each female lays about 200 and then returns to the sea. In places where these turtles are not protected by law the eggs are collected and sold in the market. The annual 'harvest' is about two million eggs.

Most sea turtles are predaceous, the green turtle being the only exception. It feeds chiefly on grass-like marine flowering plants. The jaws are not hooked but straight and fitted with small teeth.

Green turtle

Masters of the Air

Since time immemorial birds have fascinated man not only because of their brightly-coloured plumage and lovely song but above all because of their ability to fly. Besides insects, only the birds and bats can achieve true flight. Birds, however, have no rivals in the air. The bird skeleton is not only strong but also light because most of the bones are hollow. Some of the vertebrae are fused with the shoulder and hip girdles thus strengthening the skeleton. The greatest adaptation for flight is the broad breastbone or sternum with a keel to which are attached the strong muscles which power the front limbs which have been modified into wings. Some birds have lost the power of flight. Their wings and flight muscles have been reduced and the keel on the breastbone disappeared.

Great northern diver
Gavia immer

To those who live in or have visited the northern parts of the world divers or loons are a living symbol of the untamed wilderness.

Divers are the only flying birds with hind legs which are located at the very rear of the body and extend beyond it like a screw-propeller. This is excellent for swimming but when on land divers are very awkward, dragging themselves along on their bellies because they cannot use their feet for walking. When foraging for food divers submerge for only 30 to 45 seconds but when danger threatens they can remain submerged for several minutes and dive to depths of 80 metres. The great northern diver is found in North America, Iceland and Greenland.

Great northern diver

Great crested grebe
Podiceps cristatus

Grebes, which are found throughout most of the world, are quick, sleek birds which bob like corks on the water and dive expertly. They feed on small fish, insects and insect larvae, crustaceans and other aquatic animals. They also swallow large quantities of their own feathers. Most authorities believe the feathers retain the sharp bones of fish in the stomach until they have softened and can continue on through the digestive tract. Young grebes are often longitudinally striped.

Great crested grebe

Common puffin
Fratercula arctica

Its large, brightly-coloured beak and comic, swaying gait have earned the puffin the nickname 'sea-parrot'.

Puffins nest in colonies on grassy cliffs where they dig burrows. These consist of horizontal corridors ending in globular nesting chambers in which the female lays a single, white egg. Both partners feed the young nestling for about 40 days. Twice a day, in the morning and evening, they literally stuff it with small fish. Each brings as many as eight to ten in its beak at a time, generally holding them crosswise, head on one side and tail on the other. After about six weeks the parents stop bringing food and simply abandon the nestling in the burrow.

Brown pelican
Pelecanus occidentalis

Pelicans are readily recognized chiefly by the pouch on the lower beak. It can hold about 12 litres (2 to 3 galls) which is approximately triple the capacity of the stomach.

Pelicans are very social birds. They nest in colonies and remain together in flocks even outside the breeding season. When they are resting all have their heads turned in the same direction; when they are flying in a long straight line or in V-formation their movements are perfectly coordinated. Except for the brown pelican, which captures fish by plunging head-first from the air into the water, all the other five species of pelicans fish jointly. They form a long line which slowly advances towards the shore beating the water with their wings and so forcing the fish into the shallows where they are scooped up into the pelicans' pouches.

Wandering albatross
Diomedea exulans

Albatrosses are past masters at gliding — rarely will one see an albatross beating its wings. Best known of the 13 existing species is the wandering albatross, which also has the greatest wingspan of all birds — more than three metres from wingtip to wingtip!

Albatrosses nest in colonies, generally on islands. The parents take turns incubating the single white egg for a period of 77 to 82 days. Both also feed the nestling for eight months after it hatches. It takes a long time for the young albatross to fully mature and not until the age of seven years, often later, can it breed.

Brown pelican

Wandering albatross

Caspian tern *Hydroprogne caspia*

Terns and gulls are an unusual family of birds found both on seacoasts and on the shores of freshwater lakes, ponds and the like. Terns are smaller than gulls and can be recognized by their flight — they do not glide and fly with the head and beak curved downward. The Caspian tern is the largest of all terns. Its nesting colonies may be found in both the southern and northern hemispheres and on all continents. When feeding it flies about five metres above the water, hovers and then plummets into the water and seizes the fish in its beak.

Common puffin

Caspian tern

Canada goose *Branta canadensis*
The Canada goose has long been a popular game bird of American sportsmen. It inhabits a vast area extending from Labrador westward across Canada as far as Alaska and the eastern coast of Siberia. Like the wild goose in Europe, it was kept in a semi-wild state. Individuals which escaped from captivity formed fairly large populations even far south of their original area of distribution. Similar escapees are to be found in some parts of Europe, for instance England, Denmark, Sweden and even Upper Bavaria.

The Canada goose breeds on lakes and in marshes in prairie, tundra and wooded regions.

Teal *Anas crecca*
Sometimes you may have seen a flock of ducks and noticed that some of them were markedly smaller. These were teals, the smallest of the northern ducks. They are particularly fond of inland stretches of water with thick vegetation where they generally occur in groups. Teals are expert fliers. Unlike other ducks they sometimes fly a zig-zag course and can attain speeds of up to 100 km/hr (62 mph). Whereas west European teal populations are more or less resident, north and east European populations migrate south, even as far as equatorial Africa. North American populations pass the winter in Central America.

Wood duck *Aix sponsa*
The wood duck is one of the prettiest of water birds and is frequently kept by bird fanciers. The wood duck inhabits lakes and rivers

bordered by forests for it nests in tree cavities. Often it occupies cavities made by woodpeckers, at other times those abandoned by squirrels. Occasionally, however, it will also nest elsewhere, for instance in an abandoned crow's

Canada goose

nest. The nest selected by a pair of birds is used for several years. It is often high above the ground so that the ducklings, which hatch from the eggs after about 25 days, must leap from the nest to the ground.

Teal

Curlew *Numenius arquata*
If you wander in spring in damp northern meadows, moorlands or wet pastures you will hear the melodious voice of the curlew — a fairly large bird about 50 cm (20 in) in length with a long, downcurved beak. If you're lucky you may have a chance to observe its interesting courtship ritual. The male flies up with fluttering wings making his flutey call and then abruptly turns and glides down making loud trills. The nest is well concealed amidst clumps of grass. It is interesting to note that on hatching the chicks have straight bills — not

Wood duck

Curlew

until three weeks later do they begin to acquire the downcurved shape.

Purple heron *Ardea purpurea*
Shallow southern European ponds and lakes filled with reeds, the mouths of rivers and similar places are where you may come across a handsome bird of the heron family. Its long toes indicate that it can move deftly on and between reeds. When startled it freezes in a pose with head and neck outstretched so that its slender body merges with the reeds and is almost invisible.

Purple heron

Jabiru

Practically all herons nest in colonies or at least in groups. The purple heron is no exception. The nests are built where the vegetation is thickest. The birds then trample the reeds in several spots round the nest which serve as places for them to rest and later for their young when they leave the nest.

Jabiru *Jabiru mycteria*
The largest bird of the New World and one of the largest of all airborne birds is the jaribu of tropical America. This large wading stork, nearly one and a half metres (nearly 5 feet) high, inhabits bogs, swamps and branches of rivers from Mexico to Argentina. It wades in shallows where it captures fish, molluscs, reptiles, as well as small mammals. Tall trees are necessary for the purple heron during the breeding period for this is where the birds build their huge nests of branches. Such a nest is used by a pair of jabirus for a long time, being repaired and enlarged by them every year.

Raptors — Birds of Prey

Though this order includes some 300 different species the external appearance of all its members is more or less the same so that they cannot be mistaken for birds of any other order. As the name implies, raptors, or birds of prey, feed on other animals. For this their strong beak is furnished with sharp edges and their toes with strong, sharp claws. Most raptors hunt birds and mammals, but many eat only a certain kind of food. Many feed on carrion, some capture mostly insects, others feed on fish. Many smaller species catch insects. The narrowest specialist of all is probably the everglade kite, which feeds on the apple snail.

Raptors are with few exceptions expert fliers, some being able to soar for hours high in the sky; others pursue their prey amidst the thick branches of trees, and still others hover in the air in one spot and then plummet to the ground to seize their prey.

California condor
Gymnogyps californianus
150 years ago this, the largest of all North American birds was still plentiful in the area extending from

California condor

the state of Washington to lower California. Nowadays it is one of the rarest species in the bird realm. The year 1849 or so, when pioneers began settling the west, marks the beginning of the extermination of this handsome bird of prey. Hunting, however, was not the main reason for its decline in number. A far greater threat to these carrion-eaters was strychnine, used by ranchers to poison dead cattle in their fight against wolves and coyotes. Many condors were victims of this insidious poison. The

Lammergeier

California condor multiplies at a very slow rate. The female lays only one egg which she incubates for 55 to 65 days. The young take a considerable time to reach maturity and not until their sixth or seventh year are they able to breed. There was thus no way to make up for the sudden decline caused by mass poisoning. In 1965 the number of known existing condors was 38. It seems, then, that despite rigid protection, this impressive species is doomed to extinction.

Lammergeier or Bearded vulture
Gypaetus barbatus
The area of distribution of this, the largest Old World raptor extended from the Alps through the Iberian and Balkan peninsulas to Africa, and through Asia Minor as far as central Asia. In the mid-19th century, however, the bearded vulture disappeared from Europe with only a few remaining in the mountains of Spain and the Balkan peninsula. The bearded vulture, as a rule, feeds on carrion, being particularly fond of bones, which it shatters by flying up in the air and dropping them on rocks, thus enabling it to pick out the marrow inside. In the same way it shatters the shells of large tortoises.

Bald eagle
Haliaetus leucocephalus
The bald eagle is the national bird of the United States and is shown on the nation's coat-of-arms. Formerly it nested throughout the continent but is now absent in many places, occurring in moderate numbers only in Florida and Alaska. It is usually found close to water. It feeds mostly on fish but may capture small mammals and has been known to have attacked and killed the young of large ungulates.

The bald eagle nests in tall trees. The nest is used by the same pair for many years, sometimes their entire lifetime. It is a huge structure because the birds add new branches on to it every year. Though it is rigidly protected by law, the bald eagle is disappearing in some places. The main cause appears to be insecticides, chiefly DDT, which accumulates in the bodies of the fish upon which the eagle feeds.

Bald eagle

Goshawk Accipiter gentilis
The goshawk's hunting method differs from that of falcons which hunt in open country. It makes a surprise attack with lightning speed, generally in wooded country. Because it is strong it will attack prey as large as a hare.

During their first year, the plumage of young goshawks differs from that of the adults. They have longitudinal streaks on the chest and the back is coloured brown. The goshawk has an enormous range. It inhabits wooded country in practically all of Europe and temperate Asia and is also occasionally found in North America.

Goshawk

American sparrow hawk
Falco sparverius
The commonest North American falcon. It generally preys on small rodents and large insects such as locusts. Like all falcons the American sparrow hawk is readily tamed.

American sparrow hawk

Game Birds and Fowl

The domestic hen characterizes the order of fowl-like birds. This is known to everyone and has been used by man since time immemorial. The domestic fowl shows all the characteristic features of body structure and behaviour of this group. At first glance we see that all have a short, stout bill and large strong feet adapted for scraping. When a hen forages for food, she scrapes the ground away with her feet and picks up bits of food with her bill. All other members of this order feed in a like manner. There is a considerable difference between the sexes. We all know that the feathers of the hen do not begin to compare with the magnificent plumage of the cock, and so it is with most other fowl-like birds. All fowls live on the ground. This is where they forage for food and where most of them nest. Many species, however, roost only in the branches of trees, for the birds feel safer there. Domestic hens also perch on a roost at night. Fowl-like birds are not fond of flying. Given a chance they prefer to escape by running on the ground. The flight muscles, though well-developed, are composed of white muscle tissue which tires readily. This, however, is valued by man for the light meat of chicken breasts is a tasty morsel. Practically all fowl-like birds are important game birds — for instance the pheasant, partridge, grouse and wild turkey.

Common peafowl
Pavo cristatus
The jungles of southern India and Ceylon are the home of this large, striking fowl-like bird kept at royal courts in a semi-wild state since time immemorial. The cock's plumage gleams with magnificent metallic colours and the feathers overlying the base of the tail are elongated and decorated with rainbow-coloured eye-like spots. During the courtship display the bird spreads its magnificent tail feathers out like a fan. Magnificent as its plumage may be it has a loud, unpleasant wailing cry, which may be heard chiefly when the birds go to roost for the night, often also on moonlit nights.

Bobwhite

Bobwhite *Colinus virginianus*
Ask any American hunter which is the most popular game bird of the southern and eastern United States and he will probably say the bobwhite. These quails live together in pairs. The hen lays up to 15 eggs in the nest on the ground. When the young hatch, after about three weeks, they are guided about by both parents. The family remains together until the following spring. Often several families join to form a small flock in which they jointly wait out the winter. At night the flock, which may number as many as a hundred birds, forms a circle with heads facing out and tails towards the centre. This interesting grouping doubtless contributes to the safety of the whole flock for enemies cannot approach unobserved and if necessary the quails can scatter to all sides without obstructing each other.

Common peafowl

Wild turkey

Wild turkey *Meleagris gallopavo*
America has given man few domestic animals but one of the most important and most widely used is the turkey. The Spanish conquistadors were introduced to the turkey as a domestic bird of Mexico's Indians in the early 16th century. We do not know exactly when they brought the first birds to Europe but it must have been soon after, because the turkey figured as a delicacy at the court of Henry VII.

Since the reign of James I the turkey has been a common holiday dish. Its popularity, particularly in the English-speaking countries, was so great that Benjamin Franklin even wanted to have it made the national bird of the United States in place of the bald eagle.

Turkeys are woodland birds which are partial to open stands. Formerly they were very plentiful throughout their entire range from Maine to South Dakota. Indiscriminate hunting, however, was responsible for their extermination in many places.

The domesticated turkey is very similar to its wild ancestor. Selective breeding, however, has produced a white form in addition to the bronze-coloured variety. This is slightly smaller but has many characteristics which are valuable for breeding. It matures more rapidly, is easier to pluck and has broader breast muscles.

Partridge *Perdix perdix*
The partridge, which once inhabited the unforested grasslands of Europe, has readily adapted itself to cultivated land with grain fields. It is found mostly in places with lighter soil and ample places of concealment. Modern large-scale farming and mechanization, however, have made great inroads into the life of this once-important game bird. Where it formerly used to be hunted by the millions every year now its hunting is greatly restricted in many European countries and in some even prohibited. The partridge is disappearing from Europe's fields.

In early spring fields resound with the loud calls of the cocks for this is the beginning of the breeding season. The birds form pairs which remain together for the summer. In late April or early May the hen lays 10 to 15 eggs in a slight hollow in the ground and incubates them for 23 to 25 days. When the chicks hatch at the beginning of June they are guided about by both parents. The chicks grow quickly and are able to fly within three weeks. In the autumn families sometimes form groups and in winter one may often see quite large coveys.

The partridge was introduced as a game bird into other parts of the world. It has probably acclimatized best in the American Middle West, where it is called Hungarian, or simply Hun, after the country of its origin.

Partridge

Perching Birds

Blue jay *Cyanocitta cristata*

The country east of the Rocky Mountains is the home of the blue jay. This striking, well-known bird inhabits open woodlands but may even be found in the neighbourhood of man's dwellings and therefore also in large city parks. Jays are wary birds and warn of approaching danger with a loud often repeated call. A hunter may be sure that if he is spied by a jay all the forest inhabitants will be informed of his presence. During the nesting period, however, the blue jay is a very quiet creature. It hides in the treetops and takes great care not to reveal the location of its nest.

European jay *Garrulus glandarius*

Practically everything we said about the blue jay applies to its close relative, the common jay of Europe and Asia. The latter, however, is slightly larger and not as brightly coloured. At the shoulder, however, it has lovely feathers striped blue, white and black that have been used as an ornament in the hats of Europe's hunters for centuries. The European jay is omnivorous and will eat any kind of flesh but vegetable matter comprises about three-fourths of its diet. The distribution of this jay is closely linked with the distribution of the oak. It contributes greatly to the natural regeneration of oak woods for it hides acorns in various places in the ground, later finding only a few of its reserve stores. Its behaviour is a textbook example of how the balance of nature is preserved.

If young jays are taken from the nest and kept at home they are readily tamed. They are bright and quick to learn. In the wild the jay can imitate various sounds. If advantage is taken of this fact, a young jay in captivity can be taught to whistle various melodies and even imitate the human voice.

Greater bird of paradise
Paradisea apoda

Birds of paradise are without doubt the most decorative and colourful members of the bird realm. No bird can boast such diverse feather ornaments as the birds of paradise.

During the courtship display the males adopt the most fantastic positions, open their wings and beaks wide to reveal their brightly coloured throats, hop about, revolve around a branch or hang upside down.

The first news about these fantastic birds reached Europe in 1522, when two skins were brought by the ship Victoria to the Spanish emperor as a gift from the king of Batjan. In the 16th and 17th century, the era of great sea voyages, sailors brought back further skins which they obtained from the natives in New Guinea, northeastern Australia and other small islands in the neighbouring seas. When preparing the skins the natives re-moved the birds' legs which gave rise to fantastic tales, for instance that these birds continually fly towards the sun and that the female lays the eggs in a kind of hollow on the male's back. Later, in memory of these tales, the great naturalist Linnaeus named one of the first of these birds *Paradisea apoda*, meaning the footless paradise bird. In the late 19th century the feathers were popularly worn as ornament, chiefly on women's hats, and that is why in the years 1880 to 1890, 50,000 bird skins were exported from New Guinea every year! Many species were thus on the verge of extinction and only the capriciousness of fashion averted the disaster. Nowadays all species of birds-of-paradise are protected by law.

Blue jay

European jay

American robin

Crested tit *Parus cristatus*
In the woods in winter you may come across small flocks of tiny birds looking for something to eat. They are flocks of tits. As a rule these flocks include several species

Crested tit

American robin
Turdus migratorius
To Americans the American robin is the equivalent of the swallow to the Europeans. Its arrival in its nesting grounds announces the arrival of spring. The American robin nests throughout all North America, from the forest limit to the southern United States, and journeys south to the coast of the Caribbean and Central America.

of tits and often also the nuthatch and tree creeper. In the coniferous forests of Europe and western Siberia they also include the crested tit, which is readily distinguished from all the others by the crest on its head.

Greater bird of paradise

103

Serin *Serinus canaria serinus*
Canary *Serinus canaria canaria*

The original home of the serin is the region round the Mediterranean Sea. For reasons which are not quite clear it spread northward in the early 19th century and colonized all of central Europe, making its way as far as Sweden and the Gulf of Finland. When scientists observed what influenced the distribution of this small bird they discovered to their surprise that one of the important factors was – telephone wires – which is where serins usually perch when singing. A close relative of the serin is found in the Canary Islands, Madeira and the Azores. It differs from the serin not only by its larger size but also darker coloration. When the Spaniards occupied the Canary Islands in 1478 they found that this bird was kept as a pet by the natives. The Spaniards found the song of these wild canaries, as they called them, so pleasing that they began taking them to Europe in great numbers. Breeding and selection produced numerous forms differing in plumage as well as coloration and even in the quality of their song. Though the canaries we encounter are generally yellow they come in other colours too.

Cardinal
Richmondena cardinalis

The cardinal is a crested bird found throughout the warmer parts of eastern North America whence it spread southward to Mexico and Honduras. It is a universal favourite not only because of its beauty but also its song, which some observers liken to that of the nightingale. For this reason cardinals were caught and kept in cages. Americans even introduced this species to the Hawaiian Islands. The cardinal's breeding territory is at present shifting slightly northward. Thus, for example, it is now quite common even in the state of New York where it was previously unknown.

In summer the birds live in pairs.

Canary

Serin

The female does not have the bright red plumage of the male – she is coloured brownish. In the autumn and winter cardinals form small flocks which roam the countryside in search of food. They gather mostly seeds of various plants, their strong beak being able to crack even the hard grains of corn. During the nesting season the cardinal, like other seed-eating birds, collects insects which it feeds to the nestlings.

European swallow
Hirundo rustica

No other bird in the world is as popular as the swallow. Perhaps it is because they are plentiful, delightful birds which aren't afraid of man and destroy numerous flies and other troublesome insects. Many species have become constant associates of man and their nests may be found under eaves, in farm sheds and even in living quarters indoors. Their nest is made of mud cemented with saliva. Swallows alight on the ground only when gathering material for the nest, otherwise they spend most of their time in the air and perch on slender branches, the ridges of roofs and telephone wires. They catch insects, which are their sole food, on the wing. They are skilful and rapid fliers, often capturing their prey close above the ground or water. They also drink by dipping their beak along stretches of water as they skim over the surface.

Cardinal

104

Before leaving for their autumn migration they form large flocks often numbering several thousand birds. Their journey is a long one for their winter quarters are located in South Africa. At the end of March or in early April the birds return to their nesting sites. They are heralds of spring and the symbol of faithfulness to the home. Old swallows use the previous year's nest, merely repairing it somewhat. The young which hatched the preceding year construct new nests to which they return throughout their whole lifetime.

Blackcap *Sylvia atricapilla*

Warblers are small, inconspicuous Old World birds which live concealed in thickets. The various species look very much alike and are easily mistaken for one another by the casual observer. Though uniform in appearance their song differs markedly. It is by the song that one can identify with certainty the various warblers without even seeing them. The blackcap fully deserves the title of champion warbler. Its song is melodious, loud, flutey and slightly reminiscent of the nightingale's. Not in vain is it ranked second to the king of songsters by bird lovers.

Short-toed treecreeper
Certhia brachydactyla

The inconspicuous, small bird with long, curved beak that climbs jerkily in a spiral round a tree, working its way upward from the base and investigating every crack with the tip of its bill, is a treecreeper. Its colouring blends so well with the bark that it is practically indistinguishable. After systematically examining the whole trunk it flies to the bottom of another tree and begins over again, all the while uttering its soft, high, peeping note.

Both parents build the nest and later both tend the young. In this European treecreepers differ from the related species in America, where only the female incubates the eggs.

European swallow

Blackcap

Short-toed treecreeper

Interesting Relatives

Our knowledge about the evolution and evolutionary relationships of various vertebrates is often based on fossil finds. Bird fossils, however, are far fewer than those of other vertebrates. This is because the bird skeleton is fragile and the body rapidly decomposes after death. Only few birds die in such a manner that their remains are deposited in a suitable sediment, the richest source of fossils. These are the reasons why we know so little about the evolutionary links in the bird world and why authorities are sometimes uncertain as to the exact place of certain bird orders. Often the only information they can use are similarities in the external body structure of existing species without knowing if the features they consider important are not merely the result of adaptation to the environment instead of an indication of an evolutionary link.

We therefore assume, for example, that parrots are at least distantly related to pigeons and cuckoos, trogons have certain characteristics similar to those of parrots and nightjars but others that are reminiscent of woodpeckers. Toucans show a slight resemblance to hornbills but other features again clearly indicate that they belong to the same order as the woodpeckers. Deciding where certain orders of birds belong in the natural system of classification is not always a simple matter and we have yet to learn more about the evolution of these interesting relatives.

Wood pigeon *Columba palumbus*
The wood pigeon, as its name implies, is a woodland inhabitant, found in both broad-leaved and coniferous forests, in lowlands and mountains, near human habitations as well as in places where there is no trace of man. Sometimes it may be seen even in large city parks. Its area of distribution is large — all of Europe, western and southern Asia and northwestern Africa. In the northern parts of its range the wood pigeon is migratory, leaving for its winter quarters in the south and south-west in September and October and returning to its nesting grounds again as early as March.

Like all pigeons it builds a very simple nest of several twigs in which it usually lays two white eggs. This small number is offset somewhat by the fact that the wood pigeon has two and sometimes three broods a year. In the autumn the birds form large flocks which roam the countryside wherever nature has provided a rich harvest of seeds. Their favourite food is acorns and the seeds of conifers, but grain fields are a rich source of food and so their crops are often stuffed with all kinds of grains.

African grey parrot
Psittacus erithacus
Allegedly the oldest stuffed bird is an African grey parrot from the days of Charles II and which is said to have been the lifetime companion of the Duke of Lennox and Richmond. One thing is certain — that the parrot was kept as a household pet as early as the days of the Tudors and that Henry VIII had one too. It gained further popularity thanks to Robert Louis Stevenson and his book Treasure Island. Long John Silver always carried a talking parrot about on his shoulder. It is this remarkable ability to mimic human speech (of all parrots probably best developed in this species) that made it so popular with bird-fanciers and the parrot that people want above all others. If old birds are taken into captivity they never become fully tamed, but no bird is more delightful than a tamed young African grey parrot. Experienced bird-fanciers can tell if the parrot offered for sale is an old or young bird. The eyes of young parrots are dark, whereas in mature birds the iris is an ivory colour.

Budgerigar
Melopsittacus undulatus
The budgerigar is just as much a symbol of Australia as the kangaroo and if we were to count all the budgerigars kept as pets it would certainly be the most numerous parrot in the world. It first became a cage bird in 1840 when

Wood pigeon

African grey parrot

John Gold brought several budgerigars to England from his trip to Australia. Wild budgerigars are coloured green, though sometimes one may encounter a colour variety, chiefly yellow and blue. Bird-keepers purposely established these colour varieties by breeding and selection and developed many others as well. The number of budgerigars kept as cage birds increases every year. They make delightful pets. Their song is much more pleasant than that of other parrots and lacks the grating cries and screeches common in other species. Though it is one of the smallest of the parrots, the budgerigar learns to imitate human speech quite readily but of course only if it is raised and fed by the keeper instead of by its natural parents. It must also be kept in a separate cage so it can't see or even hear any of its kind. Patience and repetition are needed to teach the bird to talk. In this there are great differences between birds — some learn barely a few words whereas others soon repeat whole sentences. Males are generally more talented than females but this is not a hard and fast rule.

Budgerigar

Like all trogons, to which group it belongs, the quetzal nests in cavities. Because the male and female take turns incubating the eggs they must take care not to damage their greatest ornament — the long, fine tail feathers — in the confined space. Observations have revealed that the male turns around inside the cavity so that his beak juts out of the opening. He does not pull the tail feathers inside but throws them over his back so the tips jut out of the opening too.

Sulphur-breasted toucan
Ramphastos sulfuratus

The huge, brightly coloured bill is the main distinguishing feature of the toucan, a bird which once seen can never be forgotten. Despite its size, the bill, like the body, is very light for under the horny layer it is hollow and reinforced with thin, horny plates. We do not know the real purpose of this curious bill. Perhaps it plays a certain role in the courtship display and in visual orientation, because the various species of toucans, which all have similar plumage, differ markedly in the coloration of the bill. Most brightly coloured is the bill of the sulphur-breasted toucan. The intelligence of toucans may be compared to that of crow-like birds. In South America, their native land, the local boys often take the young from their nests in tree cavities and tame them. On hatching the young show no resemblance to the parents. Their bills are short and broad and the lower bill extends beyond the tip of the upper one. In time it acquires the same proportions as those of the adult bird. The young can be distinguished even several months after they have fledged for their bill is still slightly shorter than that of their parents.

In the wild toucans feed chiefly on fruit, but they are not averse to eating meat on occasion, perhaps a small bird, lizard or tree-frog.

Hyacinthine macaw
Anodorhynchus hyacinthinus

The macaws of South America are definitely the most magnificient of all parrots. They are the largest and

Quetzal *Pharomachrus mocinno*

'After a short while we saw emerging from the main thoroughfare an imperial procession. In front marched three dignitaries bearing staffs of gold, behind them, surrounded by nobles, came the emperor, borne on a gleaming gold sedan-chair. His head was shaded by four kaziks holding a baldachin of green feathers bordered with silver and adorned with emeralds and pearls.' This is how Hernando Cortés, the Spanish conqueror, describes his meeting with Emperor Montezuma on his arrival in Mexico. The green feathers on the baldachin and the emperor's robe were from the quetzal — one of the most beautiful birds in the world. The Aztecs and Mayans believed it to be the god of the air and never killed it. They captured the birds alive, plucked out the long tail feathers and then released them. Quetzalcoatl, the chief deity of the Aztecs, was represented as a feathered snake, the feathers being those of the quetzal. Today the quetzal is the symbol of liberty and freedom and is the national bird of Guatemala. It is depicted on stamps and seals, a province and Guatemala's second largest city bear its name, and it is also the monetary unit of that country.

Quetzal

Sulphur-breasted toucan

Hyacinthine macaw

their massive beak, half the size of a head, and long tail, often longer than the body, are characteristics by which they can be identified. Though it is not as brightly coloured as other macaws, the hyacinthine macaw is still one of the handsomest. Since time immemorial it has been kept by the Indians of the Xavanti tribe for its feathers, which are plucked regularly and used to decorate their ceremonial robes. This is simpler than hunting the birds and also ensures a regular supply of the precious feathers.

Very little is known about the life of this macaw in the wild. Most of our knowledge is based on observations of birds kept in captivity. We do not even know what it feeds upon in the wild, presumably mostly seeds and the fruits of various trees. Its strong beak can crack the hard shells of nuts. However, it can handle small particles of food equally well, being aided in this by its fleshy tongue. Frequently the macaw adroitly holds the food with one foot.

Nocturnal Birds — Owls

Owls are a very distinctive group of birds and cannot be mistaken for any other birds. Their most striking characteristic are their large eyes which face forward and look out of a facial disc of fine feathers arranged fan-like around them. The large head rests on a short but very mobile neck, enabling the owl to twist its head around a full 270°. The plumage is soft and fluffy and that is why owls fly so silently, practically without a sound.

Owls generally hunt live prey. For this they are well equipped with a strong, downcurved beak and long sharp claws. The feet are furthermore furnished with a reversible outer toe — an excellent adaptation enabling them to grasp prey between opposite-facing claws. When hunting, owls are aided by their keen senses of sight and hearing. Their daylight vision is just like that of any other bird, but because the retina of the owl's eye contains many more light-sensitive cells than that of other birds, their night vision is much better. The owl's ears are not visible at first glance for they are concealed by the feathers. Furthermore the ear openings and ear flaps are often of different shape and size, apparently so that the owl can pinpoint sounds more precisely. Many owls depend more on their hearing than on their sight when hunting prey, for at night they are more likely to hear a small animal than to see it.

Barn owl

Barn owl *Tyto alba*
The barn owl originally inhabited tree and rock cavities but nowadays it is associated with man more than any other owl. It may be found near human dwellings, in little-used or abandoned buildings such as churches, old mansions and ruins.

The barn owl is widespread throughout most of the world. It forms pairs, probably for life, which always use the same nest. Adult birds are resident, but young owls often roam great distances. The barn owl is very useful because it destroys a great many rodents, chiefly voles, which make up 70 percent of its diet. In years when the vole population is plentiful the barn owl may have several broods, whereas in years when the vole population is small it may not nest at all.

Burrowing owl
Speotyto cunicularia
The long-legged burrowing owl is a very interesting bird. It inhabits the prairie lands from Florida to the western United States whence it extends southward to Central America and on as far as Tierra del Fuego. Sometimes it digs its own burrows but usually it occupies the burrows of various rodents, of North American prairie dogs for instance. It is interesting to note that such a burrow often contains three inhabitants — a prairie dog, rattlesnake and burrowing owl, all

Burrowing owl

Tawny owl

seeming to live in peace. If you drive through prairie land where the burrowing owl occurs in large numbers you can see these delightful, comical birds resting in front of their burrows.

At dusk the burrowing owl hunts small mammals, amphibians and insects, mostly grasshoppers. It

Great horned owl

flies close to the ground on the lookout for prey. Every now and then it flies several metres lower and remains hovering in one spot, attacking silently and swiftly as soon as it spies a victim. Small animals are swallowed on the spot, with larger prey it flies off to some suitable, usually elevated spot.

Tawny owl *Strix aluco*
The tawny owl is probably the commonest owl in Europe. Its range extends as far as southern Asia and northwestern Africa. It is a woodland bird but may also be found in city parks and gardens. During the day it usually perches pressed close to the trunk of a tree or hides in a cavity. It hunts in the evening, flying silently about its territory and visiting favourite trees or other elevated places from which it looks about for prey. It chiefly hunts rodents but sometimes also captures birds, amphibians and insects.

The tawny owl generally nests in cavities. The eggs are incubated only by the female, but her mate remains close by and brings her food to the nest. Because the female usually begins incubating as soon as the first egg is laid, the young hatch successively and thus one will find newly-hatched balls of white young and grey-feathered juvenile birds together in one nest. The owlets take up to seven weeks before they begin to become independent of their parents, but after about four weeks they will leave the nest and sit near it where they can often be seen and heard, begging for food.

Great horned owl
Bubo virginianus
The great horned owl is found in all types of woodlands in North America. It flies at night in forest clearings and at the edges of forests on the lookout for prey. In view of its large size it can catch even a duck, hare or other game animal but it usually feeds on small rodents.

Snowy owl *Nyctea scandiaca*
The snowy owl is the most striking and probably the most handsome owl of all. Its lifestyle differs somewhat from that of other species. The reason for this is that the snowy owl inhabits the barren arctic tundras surrounding the North Pole where the sun does not set for months on end. The snowy owl must thus hunt in the daytime as well as at night. The barren treeless tundra affords no places for nesting and for this reason the snowy owl nests on the ground, usually on a hummock which gives it a good view of the surrounding countryside. Another unusual characteristic is the pronounced difference between the males and females. The male is almost all white whereas the female has dark markings on her breast, back and wings.

Snowy owl

111

Flightless Birds — and the Aerial Acrobats

Kiwi

The word 'bird' immediately evokes the image of flight. And rightly so, for flight is the most typical form of locomotion in the bird realm. To this day we do not know exactly how birds came to fly. Most probably their ancestors ran about and climbed with the aid of the hind limbs and used their feathered forelimbs, which could be spread wide, for gliding. Gradually their powers of flight increased until birds were capable of active, powered flight. However, even today there exist birds which do not fly. These may be divided into two groups: those which do not have a keel on the breastbone for the attachment of the powerful flight muscles (the Ratites) and those which have lost the power of flight relatively recently, chiefly certain island forms of rails, cormorants, ducks and parrots.

Birds without a keeled breastbone are sometimes classified together in a separate subclass, but more detailed study reveals that they are probably not as closely related as was formerly believed. Most of them are large birds, such as the ostrich.
Birds that fly include both good and poor fliers. Some are able to glide and soar for hours by making use of the air currents. Others travel only from one branch to another and only rarely brave a larger space. There are those which literally make their home in the air, even sleeping in flight. Some birds are veritable aerial acrobats, for not only are they able to halt in the air but even to fly sideways or backward. Whereas the flightless species include the largest existing birds, the acrobats are usually the smallest ones.

Ostrich *Struthio camelus*
The ostrich is the largest living bird. Adult males measure about 2 1/2 metres (8 feet) in height and weigh more than 100 kg (2 cwt). An ostrich egg weighs about 1 1/2 kg (3 lb).
The ostrich inhabits Africa's dry, open country and was previously found in Asia Minor and Arabia as well, where it has since become extinct. The male and female differ in coloration. The male is black with white wings and tail. His neck and legs are red, a much brighter hue during the breeding period than in the non-breeding season. The female is greyish-brown. The ostrich is exceptional amongst ratites in one way. In other species only the male incubates and tends the young, whereas in the ostrich both parents share the duties of caring for the offspring. The dark-coloured male sits on the nest at night, the paler female during the daytime.

Emu *Dromiceius novaeholandiae*
The open, semi-arid plains of Australia are the home of the world's second largest bird, which along with the kangaroo has become the symbol of this continent. During the breeding period it can be encountered only occasionally for the male, who incubates the eggs by himself, does not leave the nest for the whole eight-week period of incubation. Outside the breeding season, however, the emus form small flocks which roam the countryside. The diet consists of various fruits, seeds, leaves, grass and insects. A favourite food is caterpillars, which the emu destroys in large numbers. When you hear its voice you would not believe it was the

Ostrich

from the time spent on the nest swifts are constantly in the air, even mating as well as sleeping in flight. A sleeping flock circles at great heights. Most swifts are found in the tropics.

Ruby-throated hummingbird
Archilochus colubrius
The western hemisphere is the home of those expert aerial acrobats — the hummingbirds. Most people view them as midgets of the bird realm. And so they are, in most cases. But some may be larger than a sparrow. In view of their remark-

Emu

Alpine swift

voice of a bird. Each sex utters different sounds. The male makes repeated guttural sounds whereas the female utters a deep bellow.

Kiwi *Apteryx australis*
The home of the kiwi, one of the most interesting of all birds, is the damp New Zealand forests. There it passes the day in holes under the roots of trees until night falls, when it sets out to forage for food. Unlike all other birds the kiwi locates its prey, mostly worms and insect lar-

Ruby-throated hummingbird

vae, with its keen sense of smell. During the dry season when the ground is very hard it feeds on various forest fruits and leaves.

The kiwi lays one or two eggs which are huge in relation to the size of the bird — they weigh about 0.5 kg (1 lb) — roughly one-fifth to one-quarter the female's weight.

Alpine swift *Apus melba*
If we wished to demonstrate the perfect aerodynamic body structure we would choose swifts as an example. They are expert fliers, achieving speeds of 160 km an hour (100 mph). The spine-tail swift of Asia, the fastest of them all, allegedly can fly even 300 km/hr (187 1/2 mph). Swifts feed exclusively on insects, which they capture on the wing, often at great altitudes. They never alight on the ground, for their legs are weak and unfit for running. All four toes point forward and serve for clinging to the steep cliffs on which most swifts nest. Apart

able flying ability and their food requirements, hummingbirds have occupied a part of nature used by no other birds — the world of blossoms. These they share with insects, which have similar flight characteristics. With the tips of their wings they describe figures-of-eight in the air, moving them so rapidly that the human eye sees only an indistinct blur. The colours of an air-borne hummingbird are bright and change with the bird's every movement for they are produced by the refraction of light in special transparent cells overlying a black pigment in the feathers.

Nature Always Comes up with a Surprise

Edible-nest swiftlet
Collocalia inexpectata

The Indo-Australian region is inhabited by several species of swiftlets which cement their cup-like nests to steep cliff faces or tree trunks. These nests are cemented with a sticky saliva produced by their salivary glands, in a far greater quantity than in other birds that use the same technique. The nests of two species are even constructed solely of saliva which hardens in the air. The nests of swiftlets have been used in the oriental cuisine for centuries. They were used in gourmet soups and dishes, which were furthermore very nourishing. Most valuable are the nests from the limestone caves on the coasts of Indochina. The swiftlets find their way about in the darkness of the caves by using sonar — not unlike that of bats. To this day some 20,000 nests are gathered from the huge swiftlet colonies. If their nest is destroyed the birds build another of similar quality. The third, however, is inferior because the swiftlets no longer have enough saliva and use various plant remnants in the construction of the nest. The first two nests are delivered to the market as they are. The third is first cleaned of all impurities and the hardened protein-rich substance is supplied to the market by Chinese merchants as 'dragon's teeth'.

Brushturkey *Alectura lathami*

Australia is a continent of many surprises and so no wonder that it is here that we find birds which do not sit on their eggs to incubate them. What they do is the same as their evolutionary predecessors — the reptiles. They lay the eggs in places where they are sure to be provided with the required heat for incubation and then leave them. These birds are called megapodes or mound builders. The first name is derived from the Greek word meaning 'large foot' and this is truly a typical characteristic of them all. The second name should rightfully be applied only to those birds which build mounds of earth and vegetation in woodlands in which the females lay their white eggs. The rotting vegetation generates sufficient heat for incubation. One such bird is the brush turkey, so called although it bears little resemblance to the turkey — just the colourful, bare neck. Some species of megapodes from the Solomon Islands lay their eggs in the ground heated by steam escaping from active volcanoes. Another species lays eggs in rock crevices filled with earth, the mass of rock, warmed by the heat of the tropical sun, maintaining the required temperature.

Brush turkey

Edible-nest swiftlet

Megapodes which live in non-tropical regions have to work harder than their tropical counterparts. They must keep a constant check on the temperature inside the mound, regulating it either by scratching the vegetation away or adding more rotting material, depending upon the weather.

Lyrebird *Menura superba*

Found in the mountain forests of eastern Australia is a legendary bird whose striking courtship display has been observed by only a few people. This is because the lyrebird is an extremely shy bird. In the autumn each male stakes out a certain territory in which he prepares several display grounds, small spaces on top of mounds of earth and vegetation which he scrapes together. First he sings a while, perched on a branch nearby, then he descends to the ground, to the top of a mound, continuing his loud, penetrating song. Several minutes later he spreads his long lyre-shaped tail wide and tips it forward over his back so that it almost covers him completely. Thus veiled he continues to dance accompanied by the bubbling notes of his song, which rise higher and higher until they suddenly cease.

Roadrunner
Geococcyx californianus
Common European cuckoo
Cuculus canorus

The word cuckoo immediately calls up the image of a bird laying its eggs in the nests of other birds. This habit, however, is not limited to cuckoos but may be found in other birds as well. On the other hand, many cuckoos raise their offspring in their own nests. One example is the roadrunner, a common ground cuckoo of the open deserts of the southwestern United States. Its nest containing two eggs may often be found in stands of cactus.

The common European cuckoo is a classic example of those cuckoos which are brood parasites. The eggs of different females are variously coloured and so each seeks the nests of birds with eggs which resemble hers. The young cuckoos hatch after 12 days and eject the eggs and young of the host.

Roadrunner

Common European cuckoo

Lyrebird

Monotremes and Marsupials

The most peculiar and at the same time most primitive of mammals are the monotremes or egg-laying mammals. They are found in Australia, Tasmania and New Guinea, where they survived apparently due only to Australia's isolation from the other land masses for many millions of years. Monotremes show an interesting mixture of mammal-like and reptile-like characteristics. Their most obvious mammalian feature is, of course, their coat of hair and their ability to maintain their body temperature above that of their environment. There are also numerous internal features, especially in the blood and nervous systems which are completely mammalian. However the skeleton of the limbs is more reptilian — the shoulder girdle being just like that of a lizard. Their mode of reproduction is in fact rather bird-like, but their eggs differ in that the developing embryo cannot grow as rapidly as that of a bird, and when hatched, it is much less well-developed. Nevertheless, when the young hatch the mother feeds them on milk from glands on her underside. These glands are scattered about and are not concentrated with distinct teats. At the present time there exist only two families, each very different from the other — the platypus and the echidnas.

Duckbilled platypus
Ornithorhynchus anatinus

The duckbilled platypus fully deserves its scientific name *Ornithorhynchus*, meaning bird's bill, because it has a bill superficially like that of a duck. All the body characteristics indicate that it is superbly adapted to life in water. The tail is flat, like that of a beaver but covered with hairs, the toes are webbed, and the ear openings and eyes can be covered with flaps of skin when the animal submerges. The female platypus digs holes in the banks of rivers where she usually lays two eggs in a nest of leaves after stopping up the entrance. The young, which hatch after about two weeks, lick the milk formed in the glands on the mother's belly. Not till they are eleven weeks old do they open their eyes; six weeks later they leave the nest — only then does the mother emerge from her burrow.

Australian echidna or Spiny ant-eater
Tachyglossus aculeatus

Echidnas or spiny ant-eaters are the duckbill's closest relatives even though they look quite different. Their backs are covered with thick, sharp spines, the head and underside of the body with fur. The strong claws on the forelegs are used for scraping and the long claw on the second toe of the hind leg for cleaning the spiny coat. The long, slender snout is terminated by a small mouth opening. The

Echidna

long, worm-like tongue is the instrument with which echidnas hunt food — ants and termites. Because they have no teeth they swallow a large amount of sand and small stones which help to grind hard food in the stomach, just as in birds. The one or two eggs are carried about in a pouch on the abdomen, which the females develop during the breeding season.

In northern Queensland and New Guinea there is a rather different kind of echidna. It is larger with a longer, curved snout, the spines are shorter and on the mid back they are replaced by thick, black hair.

Duckbilled platypus

Marsupials

At one time marsupials were distributed throughout much more of the world. Nowadays, however, they are found only in the Americas and, more commonly, in Australia and neighbouring islands. Here the marsupials evolved in isolation, separated from the other continents and without the competition of higher mammals, developing a vast number of different forms adapted to the widely diverse environments they live in. Marsupials may be encountered in the rain forest as well as in open semi-desert, some live underground, others in the treetops, their number even in-

cluding species which can glide from tree to tree with the aid of a fold of skin between the body and the limbs. Marsupials include carnivores as well as herbivores, and even those which feed on the nectar of flowers. The largest marsupials are up to 2 metres (6 feet) high, the smallest barely the size of a mouse. The name marsupial derives from the characteristic pouch, or *marsupium* as it is called in Latin, on the abdomen of the females, in which the young are carried about until old enough to fend for themselves. Not all marsupials, however, have a well-developed pouch and in some species it is absent altogether. Marsupials bear varying numbers of young and thus

also have varying numbers of teats — from 2 to 27. The females of some species even bear more young than they have teats for and so only the firstborn get fed. This ensures that all of the teats are occupied, since some young will inevitably be unable to find one.

Opossum *Didelphis virginiana*

The opossum is a common marsupial of North America's forests. It is about the size of a small cat and moves about slowly in the trees. In this it is greatly aided by its bare, scaly, prehensile tail which is used like a fifth limb. The opossum can even hang by this tail in order to

Opossum

use all four limbs. Opossums are omnivorous but they prefer meat. The forelimb is remarkably like the human hand and is skilfully used to capture and grasp small animals. The opossum has a peculiar manner of passive defence — it feigns

death. This behaviour is quite common in insects, for example, but is quite exceptional amongst mammals. If you were to see a limp opossum with partly open mouth that showed no signs of breathing you'd swear it was dead. Many an enemy has been so deceived thus enabling the opossum to escape at the earliest opportunity.

Opossums bear a great many young — as many as 18. Some, however, die because they cannot get to the teats. The mother generally rears about six. The young leave the pouch after about two weeks.

Northern populations have a thick underfur under the long guard hairs which is much in demand and for which opossums are avidly hunted.

Bilby or Rabbit bandicoot
Macrotis lagotis

The rabbit bandicoot is probably the handsomest member of the bandicoot family. In the wild the presence of these omnivorous marsupials is revealed by the conical holes they dig in the ground when foraging for food. Not only are their limbs adapted for moving about underground but so are their pouches, which open backwards so that no dirt can get inside as the animal burrows. Nowadays the rabbit bandicoot has practically disappeared in many parts of Australia. It was mercilessly hunted for its lovely fur both by the whites and the native aborigines, its highly-prized silvery-black tail being worn by them as an ornament.

Rabbit bandicoot

Koala

Koala *Phascolarctos cinereus*

The koala is undoubtedly the most popular animal in Australia. The eyes are like beads and its nose like a button — rather like a children's teddy bear. It is rarely encountered on the ground, for it dwells in trees. During the day it sleeps curled into a ball in the fork of a branch, at night it climbs about in search of food. It feeds exclusively on the leaves and buds of the eucalyptus tree; each genus feeds only on certain species of eucalyptus. Koalas of the eastern coast, for example, visit only spotted gum or tallow wood, those found in Victoria visit only red gum. Furthermore, they feed only on leaves of a certain age, for the foliage may be poisonous during certain stages of its growth.

Koalas live alone or in small groups. During the breeding period the male gathers a small harem of females over which he keeps careful watch. After 25 to 35 days of gestation the females give birth to a single offspring. The mother car-ries the youngster in her pouch for about six months and after that on her back, where it keeps a firm grip, for another six months. Because koalas are not fully grown and able to bear young until they are four years old the natural rate of increase of this species is relatively slow. Though there were millions of koalas in Australia as recently as a hundred years ago, nowadays their number is counted only in the thousands. This catastrophic decline was caused first and foremost by two epidemics that in 1887—1889 and 1900—1903 killed off a large part of the population, and secondly, alas, by man. These delightful and defenceless creatures were mercilessly hunted. In the 1920s some 200,000 koala skins were sold in Australia and about two million were exported every year. Nowadays the koala is recognized as an endangered species and is rigidly protected by law. Because of its great popularity attempts have been made to raise the koala in many zoos outside Australia.

Tree kangaroo
Dendrolagus goodfellowi

Many people shake their heads in wonder when they hear of tree kangaroos, for the idea of a kangaroo on a branch strikes them as absurd. And yet there are such kangaroos which abandoned open country for life in virgin forests. Their home is New Guinea and neighbouring islands and part of North Queensland. Tree kangaroos spend much of their time on the ground but always sleep in trees. They sleep in a sitting position with head between the legs. They are mostly nocturnal creatures which feed on leaves and forest fruits but they supplement their diet with animal matter such as insects and insect larvae. If they are not disturbed tree kangaroos climb down the tree tail first. If alarmed, however, they may leap from a considerable height and flee into the thickets. Eyewitnesses affirm that the tree kangaroo can jump from a height of 15 metres (50 feet). Tree kangaroos live in small groups composed of a single male and several females. Very little is known about their biology as yet.

Red kangaroo *Megaleia rufa*

It was in 1629 when Captain Pelsat visited the Abrolhos Islands off the southwestern coast of Australia with his ship *Batavia*. There he saw an odd creature — a wallaby — which he described quite accurately in his report.

Nowadays we know of 55 species of kangaroos, wallabies and wallaroos, all members of the family Macropodidae. The name *Macropus* means 'large foot' and this is a typical characteristic of the whole family. Kangaroos have short forelegs and large, powerful hind legs. When grazing they move slowly on all fours by sliding their long hind legs in front of their forelimbs, but when travelling rapidly they make long leaps only with their hind legs; they maintain their balance with the aid of the long, heavy tail. The speed at which they travel may be as much as 40 km/hr (25 mph) and the length of a single leap as much as 8 metres (26 feet).

There is very little difference between the wallaby and kangaroo. As a rule wallabies are those species with hind feet less than 25 cm (10 in) long. Wallaroo is the name given to a species of kangaroo living in rocky areas.

How kangaroos multiply, and above all how the young get to the mother's pouch, remained a puzzle for nearly a century. And no wonder, for the offspring of the largest kangaroos are about the size of a bean — scarcely 2 cm (3/4 in) long at birth. Nowadays the puzzle has been solved, for the entire process of birth, including the youngster's

Tree kangaroo

Red kangaroo

transfer to the pouch, has been observed many times; it has even been filmed.

Kangaroos exhibit a phenomenon called latent pregnancy. After giving birth, the female will mate again, but the fertilized egg will remain undeveloped until the young in the pouch ceases suckling — through death or through becoming independent. The female thus always has an embryo ready to start development.

One of the largest of the kangaroos is the red kangaroo inhabiting plains and dry regions throughout the whole of Australia. Apart from the subspecies found in west Australia it is readily recognized — the male by his red and the female by her greyish blue colouring. In old males the throat and chest are coloured a very deep red, often bordering on purple. This is a glandular secretion which can be wiped off the animal's coat, at least in part. The red kangaroo lives in groups called mobs, composed of about a dozen individuals. Though it has a large area of distribution, larger than that of other kangaroos, it cannot be said to be particularly abundant.

Insectivores

The order of insectivores is composed of an assortment of remarkable small mammals. Insectivores include very conservative forms as well as modern, highly specialized forms. Except for Australia and most of South America they are found throughout the world in all types of environments — they live on land, in water, as well as underground. Most are inconspicuous creatures and rarely seen for they are active mainly at night. Their concealed way of life often keeps us from being aware that insectivores are very numerous. This is particularly true of shrews. Many people have never seen a shrew and many have never even heard of them. And yet shrews are everywhere — in every garden, every park, every field and every woodland. In warm regions where conditions are particularly congenial to their way of life their number runs into the millions.

All insectivores are predatory animals. Their mouth is furnished with numerous small sharp teeth, all often very much alike. What interests the zoologist as well as the layman about this group is the way the various species have adapted to the conditions of widely varied environments.

European hedgehog

European hedgehog
Erinaceus europaeus
Hedgehogs are popular creatures for their spiny coat and the ability to curl up into a ball, which for this otherwise slow and defenceless creature serves as an excellent form of defence against enemies. The spiny ball puts many an enemy off but some beasts of prey and birds such as the eagle owl know how to deal with a prickly morsel such as this. Hedgehogs include several species found in Europe, Africa (excepting the rain forest belt) and Asia (excepting the southern regions). Broad-leaved forests mark the northern limit of their range. All species look very much the same. The European hedgehog is well known to those living within its range. It is frequently encountered at dusk — especially in spring and autumn as it rambles about hunting for slugs, snails, worms and insects. Besides these the hedgehog also eats birds' eggs and young birds in the nest as well as other small vertebrates. It is gener-

ally known that it will likewise catch and eat snakes, even a poisonous viper. It is not immune to the poison, as is sometimes stated, but wins out over the snake with its tactics. It counters the snake's thrusts with the stiff bristles on its head and when the snake is worn out and wounded it quickly puts an end to its life.

Few persons have had the luck to find the nest of a hedgehog. Made of leaves, grass and moss by the female it is concealed in thickets or undergrowth. The newly-born young are pinkish-white and the spines are hidden under the skin. Before long, however, while they are still blind and helpless, they grow a soft, white coat which is already spiny.

Star-nosed mole
Condylura cristata
Whatever group of animals we choose there will always be at least one member that makes us shake our heads in wonder and moles are no exception. The eastern United States is inhabited by a species with a body and strong, shovel-like forefeet like those of the European mole but with a rough fur and tail practically the same length as the body. And as for the head, well! Around the nose is a ring of 22 fleshy tentacles which make the nose look like a flower — or a sea anemone. The tentacles are mobile and can even be retracted somewhat. This peculiar appendage undoubtedly serves as a sensory organ used by the star-nosed mole to locate food in its underground tunnels. This interesting creature bur-

Detail of nose

Star-nosed mole

Otter shrew

rows in damp ground alongside water. The rough fur indicates that it even enters water and the shovel-like forefeet serve as excellent paddles.

Otter shrew or Giant water shrew
Potamogale velox
The rain and gallery forests of tropical Africa from Nigeria to Angola are inhabited by an insectivore which resembles a small otter – the otter shrew. It is one of the largest insectivores for it is about half a metre (18 in) long. The tail is flattened from side to side and serves as both rudder and paddle. The otter shrew does not have webbed toes like other aquatic mammals and swims with snake-like movements. The head is slightly reminiscent of a shark's because the mouth is located on the underside. The otter shrew is a nocturnal creature. In many places it appears to be more plentiful than was previ-

ously assumed and has even been discovered in places where for years no one had an inkling that it inhabited. It is extremely shy and wary and encountering it in the wild is most unlikely. In daytime it hides in thick vegetation or in holes in the banks with underwater entrances. The female bears her young in such burrows. As a rule there are only two, which is a very small brood for an insectivore.

Like all insectivores the otter shrew is an agile predator, feeding chiefly on crustaceans, particularly crabs. However, it also captures many insects, molluscs, amphibians and fish.

Savi's pygmy shrew or Etruscan shrew
Suncus etruscus
Today there are almost 200 existing species of shrews and it may be said that without exception they are important members of the animal

world, because they help to maintain the balance of nature. We cannot estimate, let alone count the vast number of insects, snails and worms devoured by shrews.

The Etruscan shrew is one of the white-toothed shrews, which has representatives in practically all the not-too-cold parts of the Old World, being absent only Australia. Some are large species, for instance the house shrew common in the warm regions of Asia and Africa measures up to 26 cm (10 1/2 in), but most members of this family are small. They even include the smallest living mammal – the Etruscan shrew, which measures less than 4 cm (1 1/2 in) and weighs only 2 grams (one twelfth of an ounce).

All white-toothed shrews exhibit interesting population dynamics which have lately become the subject of intensive research. At regular intervals there is a sudden, marked increase in the number of white-toothed shrews. More specific reasons for this phenomenon are not yet known.

The red-toothed shrews are more widespread, extending into North America and also further north as well – there are even species living in Alaska!

Savi's pygmy shrew

Flying Mammals — the Bats

Vertebrates mastered the air three times — firstly the flying reptiles, secondly the birds and thirdly the bats. The methods with which they mastered it differed, but the result was always the same — they all travelled by active, powered flight. Bats fly by means of their forelimbs which are modified for the purpose quite differently from those of birds. The forearm and all the fingers, excepting the thumb, are greatly elongated. Stretched between them, the body and the hind limbs, sometimes also between the hind limbs and the tail, is a thin, sensitive membrane. Strong muscles are naturally needed to power such a flying apparatus. These are attached to the keel on the breastbone, as in the birds. Bats are a very old group of mammals. Fossil finds indicate that they appeared on the earth not long after the dinosaurs had disappeared, about 60 million years ago. They apparently evolved from the same ancestor as the insectivores. The original forms were probably tree insectivores that captured flying insects.

Bats are divided into two groups.

The first includes mostly large species found in the tropics and subtropics of the Old World and feeding chiefly on fruit. They are the flying foxes or fruit bats. The second group includes mostly small predaceous species but also fruit-eaters and carnivores. Bats are a large order of mammals. Despite a certain uniformity in external appearance they have many interesting characteristics which evolved as adaptations to widely varied conditions.

Indian fruit bat
Pteropus giganteus

Fruit bats could be divided into several subgroups. They inhabit the warmer regions of the Old World, chiefly Africa, Asia and Australia. The northernmost area is the Mediterranean, where one species lives in Cyprus. The best known and most frequently displayed in zoos is the Indian fruit bat. It is a species that forms large colonies. Fruit bats sleep in trees. Such trees are readily recognized even if there are no hanging bats in the branches for the faeces and urine of these creatures strip the trees of all their leaves and on the ground below are huge amounts of guano — bat manure.

Fruit bats hang on branches shrouded in their wing membranes. They do not awaken until dusk when they climb here and there, utter cries, and spread their wings in readiness for their flight in search of food. Then the whole colony rises and wings its way to places where various fruits are ripening. Such a group of bats may cause marked damage in fruit plantations and orchards. The number of bats in a group is sometimes enormous and sometimes several groups join, veiling the sky like a dense cloud. Towards morning the bats return to their trees, where for a while they utter loud cries as each seeks its place before settling down. When feeding, or at the roost, fruit bats always maintain a certain distance between each other. Apart from the mating season and the period when the young are nursed by the mother one will never see two bats with bodies touching. The arrangement of a sleeping colony of fruit bats is of interest. All animals which live in colonies have a certain social order and respect the social distinctions that exist amongst its members. In the case of fruit bats this order is manifested by the level at which the bat sleeps. The strongest individuals, which rank at the top of the social ladder, sleep in the uppermost branches. The lower an individual sleeps the lower its rank in the society. The reason is simple. Fruit bats do not excrete their faeces and urine until they are back in their tree. They grasp the branch with their forelimbs, let go with their hind limbs, and excrete. Those lower down are naturally worse off than those on top. A female which is tending her young, usually twins, rises in the social order and makes her way to the topmost branches, directly beneath the leader.

The flesh of fruit bats is a favourite food in some regions. However, it is permeated with a peculiar, sharp odour and must be cooked with various herbs and spices which will disguise the odour.

Indian fruit bat

Detail of head

Common vampire bat

Detail of head

Common vampire bat
Desmodus rotundus

The name itself is enough to make the blood curdle because it is subconsciously linked with the vampires of fairy-tales — dead persons returning to suck the blood of the living. Most bats which belong to the leaf-nosed bat group feed on insects and do not even consider blood. However there are three species which feed only on blood and these have earned a bad reputation undeserved by their innocent relatives. The most widely distributed and most plentiful is the common vampire bat, found from northern Mexico to Argentina, in Chile and in Uruguay. They are readily identified by the pointed ears and the bare membrane stretched between the hind limbs. The white-winged vampire is a much rarer bat. It is found chiefly in the tropical regions of South America from Venezuela to Peru and Brazil, but also occurs in Trinidad and Mexico. As its name

indicates, part of the wing membrane is white. The third and last species is the hairy-legged vampire, a very rare bat found in Mexico, Central America and Brazil. The membrane between the hind limbs is hairy and the bat's hairy coat is long and soft. Besides that this vampire has saw-like teeth.

During the day vampires rest in caves or dark places. Their colonies generally number about a hundred individuals, but they may consist of as many as two thousand. Shortly after dusk the vampires leave their hiding places to fly about close to the ground. Sometimes they land on the ground and run about with agility, in a way few other bats can, at other times they land directly on the body of their victim. They do not catch hold with claws as other

bats do but land softly on the cushioned pads on their fore and hind limbs, so softly that even a sleeping person will not be wakened by the impact. With their sharp front teeth they remove a small piece of skin, thus making a flat wound from which they *lick* the blood; they do *not* suck blood. Their saliva contains substances which prevent the coagulation of blood. The common vampire bat attacks mammals and on occasion even man. The white-winged vampire only attacks birds, as far as is known. The hairy-legged vampire feeds chiefly on birds but will sometimes attack mammals.

The chief danger is not their blood-letting but the possibility of their transmitting various diseases, mainly rabies.

Greater horseshoe bat
Rhinolophus ferrumequinum

Horseshoe bats form a separate group of bats with membranous appendages on the nose. Sleeping bats hang freely, shrouded in their wing membranes without touching each other. Numerous species are found chiefly in the tropical regions of Africa and Asia, their range extending as far as northeastern Australia and the Solomon Islands. Three species, however, have a range which extends all the way to the temperate regions of central

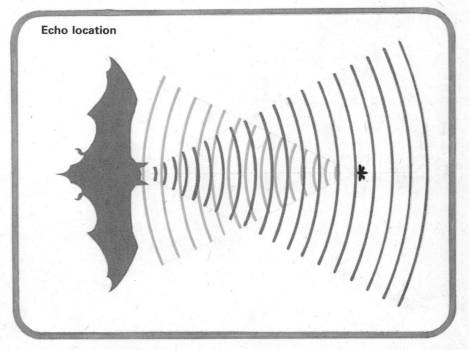

Echo location

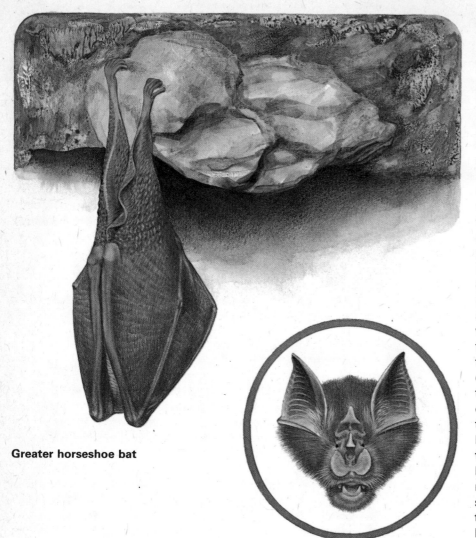

Europe; two are found even in England.

In recent years horseshoe bats have been a subject of great interest to scientists, namely the manner in which they determine the position of an object by means of ultrasound waves. In itself this is not unusual, for echolocation is a common phenomenon amongst bats. Horseshoe bats, however, use a method which has not been devised even by man's most modern technology. Unlike other bats they do not emit sound waves from the mouth but from the nose. Each signal, with a frequency of about 80-100 kHz (one kHz is 1000 cycles per second), lasts one-tenth of a second, which means that the waves reflected from objects less than 15 metres (50 feet) away reach the bat again before the signal has ended. This method of echolocation is quite different from that of other bats, and is more comparable to hunting with a searchlight! Horseshoe bats perceive the reflected signals with their sense of hearing. Their orientation by means of the constant emission of signals is perfect, much more so than the systems used by other bats.

Greater horseshoe bat

Detail of head

Mouse-eared bat
Myotis myotis
The location of objects in space by other bats is based on the principle of the reflection of sound waves. It does not matter if the object is the trunk of a tree or an insect in flight. With mouth half-open bats emit a series of extremely short sounds lasting only two- to three-thousandths of a second and having a frequency of approximately 30 to 70 kHz. The sound waves are reflected back to the sender and in this manner the bat is able to deter-

Mouse-eared bat

mine the position of an object. The sounds are inaudible to the human ear, which is only able to register tones up to a frequency of approximately 20 kHz.

The mouse-eared bat is one of the commonest mainland European bats and has been the subject of many experiments. It is found throughout central and a large part of eastern Europe, where it often forms large colonies numbering as many as a thousand individuals. In summer the females form groups by themselves, without any males, and tend their offspring. Such female colonies are most often found in attics and towers. In winter males and females pass the cold season in caves, cellars and mine shafts, often squeezed tightly together. The winter and summer quarters are often quite far from each other, sometimes many hundreds of kilometres apart. Ringing of bats by various countries working in close collaboration has revealed many interesting facts about their way of life and movements.

Long-eared bat *Plecotus auritus*

If you were to encounter the long-eared bat it would strike you as large. This is because of the ears, which are as long as the body. At rest they are folded alongside the body with only the small inner lobe (the tragus) showing. In flight the ears always point forward. Even though this is a common species its distribution is not definitely known. We know that it is found in Europe

Detail of head

Long-eared bat

from England and Scandinavia to northern Italy and the mountains of Bulgaria and from there through central Asia to Japan. It, too, forms colonies composed of females and their young in summer, during which time the males live a solitary life. In winter the bats retreat to caves, mine shafts and other sheltered places. The long-eared bat is not an expert flier. Generally it flies around trees and collects insects on the leaves. It may also be seen pursuing insects round street lamps.

Bats are very useful creatures because they catch vast quantities of insects. In many places man has disrupted the balance of nature and many species of insects which were formerly harmless have multiplied in such numbers that they have become dangerous pests of commercially important plants grown in monocultures. Bats are thus one of man's most effective helpers. Their way of life is unobtrusive and that, perhaps, is why we tend to forget them. We destroy old trees in which bats live, wall-up mine shafts and others of their winter quarters. Bats are declining in numbers but there are ways and means to prevent this. As a substitute for hollow trees we can hang out special boxes for bats; the entrances to their winter quarters may be fitted with grilles — not walled up, and first and foremost we must take care not to disturb them in their winter abodes. Bats are on the list of protected species in practically all countries.

Mouse-eared bat: detail of head

Man's Nearest Relatives — the Primates

If we were to trace man's ancestors amongst the mammals we would have to start with the insectivores. This is where the branch which finally led to the development of man had its beginnings. Today there still exist on the earth animals which indicate the path taken by that development. Found in the jungles of South-east Asia are small, squirrel-like animals with the mouth of an insectivore and the behaviour of primates. Their appearance and the environment they inhabit has earned them the name tree shrews. That was 70 to 75 million years ago. It was another five or ten million years before the appearance of creatures that could already be classed as true primates and it was roughly about 50 million years ago that the apes and monkeys branched off from these. Right from the start there were two branches of monkeys, which evolved separately and independently of each other; they differed in many ways such as in the shape of the nose and the position of the nostrils. The first are the platyrrhine monkeys, which have a broad, flat nose with a wide, flat space between the nostrils. The second branch is represented by the catarrhine monkeys, which have a slender nose with the nostrils spaced close together as in man. The evolution of the platyrrhine monkeys took place on the American continent. Their ancestors, which lived in North America, arrived in South America between 40 and 30 million years ago, when South America had already been isolated for ten or more million years. Subsequently these monkeys developed in isolation, independent of the evolution of monkeys in the Old World.

If we were to trace the path that leads to the ancestors of the second branch — the catarrhine monkeys — it would take us to Fayûm in Egypt. That is where remains of the earliest Old World primates were found. They date from the Oligocene and scientists determined their age as being 28 to 32 million years. Present-day Fayûm is located in semi-desert but at that time the appearance of the landscape was entirely different. It was located on the coast of an Oligocene sea and was covered with dense tropical forests broken here and there by tongues of shrubby grassland or savanah. Living here already at that time were two different lines of primates — one which evolved into the apes and a second which evolved into the baboons and guenons. About 30 to 15 million years ago a further group branched off from the second which produced the leaf-eating monkeys such as the colobus monkeys and langurs.

Geoffroy's spider monkey
Ateles geoffroyi

If we wished to select a typical representative of the platyrrhine monkeys the best choice would be the spider monkey. It has all the characteristics typical of American monkeys. All monkeys were originally arboreal (tree-dwelling) creatures and only the most highly developed abandoned their original habitat. The platyrrhine monkeys remained in the treetops and many developed certain very useful adaptations to their environment. One, for instance, is the prehensile tail which can be used as a 'fifth hand'.

Not only can the animal hang by its tail but it can also use it to grasp objects. The prehensile tail is most highly developed in the spider monkeys. The end of the tail, the last third, is bare and has lines and whorls like fingerprints in man. The tail is also furnished with strong muscles which enable the monkey to take a firm grasp on any object, large or small. This tail is just as dexterous as the human hand — spider monkeys can even throw objects with it.

Geoffroy's spider monkey is one of the species which abandoned their South American habitat and travelled north. Of all the platyrrhine monkeys it is the one which spread farthest northward.

Geoffroy's spider monkey

Patas monkey

Mandrill *Mandrillus sphinx*

Baboons, like guenon monkeys, are inhabitants of Africa. However they are adapted more to life on the ground and most species are found in savannahs or the rocky country of semideserts and mountains. Only two have remained faithful to the forest. One is the mandrill. It cannot be mistaken for any other monkey for none has such a fantastic, colourful face. The adult male is one of the most colourful of all monkeys for not only his face but his rump is brightly coloured too. In all baboons there are marked differences between the sexes. The males are much larger than the females, sometimes twice as large. The difference is sometimes further accentuated by the long hairs on the forward part of the body. In female baboons the area round the genitals is greatly swollen and often coloured a bright red during the mating season. The mandrill is an exception – the hind quarters of the females are coloured black.

The mandrill is found in rain forests from the Sanga River in Cameroon to Gabon. This baboon remains mostly on the ground, climbing trees only at night or when danger threatens.

Patas or Red monkey
Erythrocebus patas

The patas monkey is a guenon monkey but its way of life makes it an exception. Guenons are forest or at least arboreal creatures found only in Africa south of the Sahara. Practically all live in the forest. Only the vervet avoids the jungle forest, making its home in the bush, and as for the patas monkey – it has even gone a step further and inhabits open savannahs and semi-desert regions. Its range extends even deep into the Sahara. It is excellently adapted for life on the ground. It is long-limbed and when fleeing from danger can reach speeds of 56 km/hr (35 mph). Patas monkeys are very social animals and form bands of as many as 30 individuals. Bands have been observed in West Africa numbering more than 100. Each band is actually a harem which is headed by a single adult male and otherwise consists of females and young.

Mandrill

Apes

The word apes immediately evokes the image of large, tailless monkeys which walk in an almost erect position even though they generally lean on their forelimbs. If we compare the body structure of the apes with the anatomy of man we will be surprised by the striking resemblance. Apes have exactly the same organs as man, the only difference being in their size and sometimes their shape. The close relationship between the two is borne out also by the fact that apes are subject to many of the diseases of mankind to which other primates are immune. Despite this man cannot be said to have descended from the apes. They form a separate line which branched off from the common ancestral line some time in the Miocene or early Pliocene period, that is about 10 to 15 million years ago. It is generally believed that apes are much more intelligent than monkeys but it is difficult to determine what criterion should be used to judge the level of intelligence in animals. Experiments with so-called 'lower' monkeys showed they are able to solve problems put before them just as well as apes. There is even the instance of a capuchin monkey (which is sometimes classed as a 'lower' monkey), which obtained food beyond its reach by engaging the help of the harnessed live rat used in the experiment. Such behaviour surely indicates intelligence of a high order. Nowadays we still encounter the general belief that one of the criteria of intelligence is the size of the brain, or rather the weight of the brain in relation to the weight of the body. (This is usually done by measuring the skull capacity.) Basically this is true, but we must not forget that, for example, Neanderthal man had a far greater 'skull capacity' than modern man.

Chimpanzee *Pan troglodytes*

The chimpanzee is probably the best known of the apes. It inhabits the rain forests of the Guinea and Congo block from Guinea to Uganda to Lake Tanganyika. Its range, however, does not extend south of the river Zaire. It is not found only in the rain forest but in some places in woodland savannahs and mountain forests up to elevations of 3,000 metres (10,000 ft).

The organization of the chimpanzee band is — exceptionally amongst primates — a loose one. One may encounter single, solitary chimpanzees as well as bands numbering eighty members. Individual bands join and break up again and their make-up changes. The only constant factor in the chimpanzee community is the bond between the mother and her offspring. The individual members of a band do not have specific territories — the territory is an area occupied by the whole band. In chimpanzee society there is a certain cooperation which we do not find in other primates. These apes help one another hunt food, for example, and some individuals have been known to put some aside for their companions. The chimpanzee is the most omnivorous of the apes; some bands seem to specialize in meat-eating whereas others are strictly herbivorous.

Gorilla *Gorilla gorilla*

The gorilla is the largest, grandest and surely the handsomest of the apes. Its home is the jungle forests of Africa from the Cross River in southern Nigeria to the mouth of the river Zaire, east to the Itombwe mountains and the volcanoes of the Kahuzi and Virunga mountains. An isolated gorilla population is found also in northwestern Uganda in Kayonza forest, deservedly dubbed 'impenetrable' to this day. The gorilla reaches an average height of 175 cm (5 1/2 ft) and weight of about 180 kg (30 stones). However, some specimens are known to be much larger. The largest male, captured in the wild in present day Zaire near the village of Tschibinda, measured

Orangutan

195 cm (6 ft 4 in) and weighed 241 kg (nearly 38 stones). Gorillas in captivity, for instance in zoos, are generally fat due to the unsuitable diet and limited activity.

Gorillas live in bands of three to 30 individuals. Unlike chimpanzees they occupy far larger territories with, as a result, a smaller population density. Each band includes at least one silver-backed male and several females. The silverback is superior to all the other males and all the females. If the group includes two or more silverbacks the order of rank is determined by their size and thus probably by their age also.

Gorillas are largely herbivorous, feeding on fruit, leaves and bark.

Orangutan *Pongo pygmaeus*
The jungles of Sumatra and Borneo have their 'man of the forest'. The Malay word for this is, of course, orangutan. The Dyaks of Borneo, however, call it 'mias' or also 'monyet merah besar', which literally means large red monkey. The orangutan is probably the most greatly endangered ape at present because man is continually clearing the jungles which are vital to its existence. According to the statis-

Gorilla

Chimpanzee

tics of the International Union for Conservation of Nature and Natural Resources there is now an estimated population of 3,700 orangutans in Borneo and only about 1,000 in Sumatra.

Orangutans are solitary creatures and only rarely do they form groups of more than three or four. Most often such a group includes a male and one or two young apes of varying ages, or a pair of adult orangutans. In orangutans there is a pronounced difference between the sexes — the males are up to twice as large as the females and may even weigh three times as much. The male has an elevated crown, which makes the head look very high, and wide cheek flanges. On the throat is a large pouch which serves as a resonator and increases the sound of the animal's voice. The female has only a small throat pouch and there are no cheek flanges. Orangutans are active only in daytime.

Strange Teeth, Strange Mammals

Let us now take a look at some animals which do not even seem to belong to our day and age. They are an old group of mammals of the order Edentata, which in Latin means 'without teeth'. The name itself is strange and as a matter of fact misleading, for most edentates have teeth and one, the giant armadillo, has more teeth than any other land mammal. It also strikes one as surprising that this group of related animals includes mammals which not only differ markedly in appearance but also in their way of life. The group is of very old origin, but the primitive characteristics are intermingled with traits which indicate a high degree of specialization. The edentates consist of only a few forms. All are found in South and Central America. They compose the anteaters, the sloths and the armadillos.

With such a bizarre collection of animals one may wonder just what it is that they have in common. One feature may be the teeth, if indeed they have any. The teeth of sloths and armadillos differ from those of other mammals in that they lack the hard outer layer of enamel. The roots of the teeth are open so that the teeth grow continuously; this, in view of the lack of enamel to withstand wear, would seem to be a necessity. The other peculiarity which they have in common is an extraordinarily complicated set of joints between the vertebrae, virtually locking the whole spine into a solid unit.

It is not so long ago (from the viewpoint of evolution), that the edentates had their 'golden age' on earth. Some species were so large that they were as big as an elephant. These giants were contemporaries of Ice Age Man. Fossil finds in Ultima Esperanza clearly indicate that the inhabitants of Patagonia at that time hunted huge sloths which, unlike present-day species, did not live in trees but roamed the pampas. Discovered in Ultima Esperanza was a cave which had been inhabited by these giant sloths as indicated by the vast accumulation of manure. Remains of bones and skin, traces of fire and mounds of dry grass suggest that they were surprised here by man and killed by suffocation with smoke. The weapons used by man at that time were apparently inadequate against such large animals. Carbon dating determined that the accumulations of sloth manure are about 10,000 years old.

Giant anteater
Myrmecophaga jubata
Anteaters are the only edentates which deserve the name. The jaws are completely toothless and because there are no traces of any teeth even in the anteater embryo it is assumed that this loss had already occurred long ago. The anteater's snout is strikingly narrow and the jaws are joined and cannot be opened. The only mouth opening is the small one at the tip of the snout. With remarkably strong claws on the muscular forelegs the anteater breaks apart the hard nests of termites or rolls away stones to get at the ants and other insects hiding underneath. It then extends its long tongue with which it gathers the insects. The tongue measures up to 50 cm (18 in) in length and is moved by strong muscles attached to the breastbone. It is coated with a sticky substance with which the insects are picked up. Naturally the anteater also picks up a fair amount of

Giant anteater

Two-toed sloth

sand, pieces of wood and other impurities. One part of the stomach, however, is furnished with strong muscles for crushing and grinding such particles.

Two-toed sloth
Choloepus didactylus
The sloth spends its entire life suspended upside down. Its legs are modified into hooks by means of which it hangs and moves from one branch to another. If disturbed it can cover a short distance to attack the intruder with great speed. The permanently suspended position has caused its hair to grow in the reverse direction — from the belly towards the back. Only thus can it guarantee that the water of tropical downpours will run off to the ground. Sloths do not merit the name edentate, for they are not toothless. Their teeth, though few (only 20), include several which are strong and sharp and can inflict deep wounds. Sloths are herbivores and the sharp edges of the four front teeth, which close tightly together, function like secateurs. The other teeth are more or less uniform, the 'secateurs' of the two-toed sloth being the only exception.

The life of sloths is still in many ways a mystery. They are nocturnal creatures and this makes observation of them difficult. The two-toed sloth differs from its three-toed relative in that it has only two toes on its forefoot. All sloths have three toes on their hind feet. The two-toed sloth, when disturbed, makes a soft bleating sound — giving rise to its native name of Unau. The three-toed sloth makes a more high pitched noise, earning it the native name of Ai.

Nine-banded armadillo
Dasypus novemcinctus
Armadillos are literally mail-clad mammals. The back is covered with a shield composed of bony plates topped by a horny layer. The bony plates are formed by the underskin, the horny layer formed by the outerskin. This armour is composed of bands joined by soft skin so that it is flexible. The number of bands varies and in most cases determines the name given to the respective animal — for example the armour of the nine-banded armadillo is composed of nine movable bands.

Armadillos are found from the pampas of Argentina to Central America, the range of the nine-banded armadillo extending even to the southern USA. Most armadillos bear two young as rule. In the nine-banded armadillo, however, four embryos are formed from a single fertilized egg and these can develop into identical quadruplets. Females of the related sevenbanded armadillo bear eight or even twelve identical offspring.

In recent years it was discovered that armadillos often suffer from a disease resembling leprosy and so have become important experimental animals in the fight against this dreaded disease.

Nine-banded armadillo

131

Flesh-eating Mammals — Carnivores

The basis of all life on the earth is carbon. In complex and yet readily explained ways it is cycled through all the living organisms on the earth and we may even say that the carbon in existing plants and animals is the same carbon which was used by plants and animals many millions of years ago. Only green plants are capable of transforming non-living, inorganic matter into living, organic matter. To do this they need water and sunlight. In the presence of light, green plants are able to transform carbon dioxide, water and the minerals dissolved in it into organic substances. These basic organic substances are sugars, fats and proteins. Green plants are thus called primary producers. Part of the energy they produce is used by them for growth, development and reproduction, and part goes to waste. The energy which remains in the plant body represents a supply on which all the remaining members of the living community are dependent. The herbivores which eat parts of the plants are the first in the line of consumers and are called primary consumers. They, too, use up a certain amount of energy because they must seek food, grow and live. However, the amount they take is far greater than the amount they use. All this

time it is the same solar energy which was used by the plants and that is now stored in the tissues of the herbivore, for example a red deer. Some animals, however, are unable to make use of plant food and are thus dependent on the herbivores. These are the carnivores. When a tiger captures a deer it gets its energy 'second hand' so to speak, it is thus a secondary consumer. During their lifetimes both primary and secondary consumers exhale carbon dioxide, returning it to the atmosphere whence it is taken up again by plants. That is but one part of the cycle. Every plant and animal has a certain limited life span. When the time comes it dies. Part of the energy, transformed into body tissues, passes on. Vultures feed on a dead tiger, for example. Most of the captive energy, however, is returned to the soil where the remains of the dead bodies are decomposed by bacteria, fungi, mites, worms and other organisms and broken down into simple mineral substances which are once again used by green plants. Here the cycle begins anew. Though these examples have been simplified in the interest of making the process clear to the reader, it is evident that stored energy is being continually depleted throughout the food cycle.

This energy, of which there is increasingly less the closer we get to the top of the pyramid, is lost in the form of heat and the activity of animals in the course of their growth and reproduction. At the base of the pyramid are the green plants, at the top are the final consumers, the carnivores. And the most typical of these are the beasts of prey or flesh-eating mammals.

Canines

Canines are the most ancient of the beasts of prey, their history can be traced back some 40 million years. Among their number there are none today that differ markedly from the more-or-less uniform appearance of all. The snout is long and furnished with a great many teeth. The structure of the teeth indicates that besides meat, their chief food, canines occasionally eat plant matter as well. The sense of smell is always well developed, more so than any other sense; hearing runs a close second.

The distribution of the dog family is wide. Canines are found in all types of environments from deserts to virgin forests, from the remote Falkland Islands to high mountains. Only certain isolated parts of the earth — New Zealand, Australia, New Guinea and Madagascar — have no native canines. These were introduced there by man.

Red fox *Vulpes vulpes*
Everyone is acquainted with the fox, if not in real life then at least from tales where it is always depicted as a sly, crafty creature. Few, however, have seen it in the wild for it is a timid animal and avoids man. Nevertheless it is found in many places where it would not be expected such as in cities. The red fox is native to virtually the whole northern temperate zone, and has been introduced into Australia and New Zealand. The fox has a reputation for stealing chickens and other domestic fowl. True, it does so on occasion but statistics evaluating the stomach contents of dead foxes clearly reveal what foxes generally eat. Heading the list are mice and voles, followed by frogs, snails,

Red fox

Coyote

slugs and insects. Very occasionally the stomach contents were found to contain parts of hare, pheasant or partridge. At certain times of the year the fox supplements its diet with plant food. If the opportunity is offered foxes will feed on carrion, particularly in winter when food is scarce. Foxes which live in cities often visit waste dumps and trash bins, where besides feasting on garbage they often catch quite a number of rats. The mating season is from late December till February. After 52 days of gestation the female gives birth to the young (usually four) in the den. They are blind for about ten days, during which time the mother stays with them. Food is supplied by the male. It takes almost a month before the young emerge from the den; during this time both parents go out to hunt food. The pups leave their parents when they are two months old, at the age of six months they are already as large as adult foxes, by the time winter comes they are fully-developed and able to bear young. The fur of foxes is a popular

article. Besides normally coloured foxes their number often includes diverse variations (mutations) which are highly prized. One, for instance, is the silver for.

Coyote *Canis latrans*
Practically every book on the American 'wild west' makes some mention of the coyote. This canine is rightfully considered a typical animal of these parts. Formerly the coyote inhabited the western plains and rocky plateaux of northern Mexico. During the past 400 years, however, its range has expanded enormously to the north and east of its native habitat. It appears to be one of the most adaptable of the American mammals. In the 16th century it spread throughout all Mexico and in the 19th century as far as Canada. At the turn of the century it had already arrived in Alaska. In the east the first coyote was shot in the state of New York in the year 1912 and now it is found throughout the eastern seaboard excepting Delaware and Rhode Island. Nowadays, most states offer

a reward for a dead coyote. Even though some 90,000 are killed every year, their number is not declining. The enormous adaptability of this species has led to a stable population. The coyote will eat anything. It captures rodents, birds, as well as reptiles, it feeds on insects as well as carrion. It is not averse to plant food, gathering acorns and nibbling opuntias (prickly pears). It is even able to capture larger animals. Sometimes two or more coyotes will join together for the chase. Small animals, however, form the greater part of their diet. Mating takes place between January and March and then the long, wailing cries of coyotes may be heard even more frequently than at other times. About two months later the female bears the young in the den or some other hiding place. The reproductive faculties of coyotes are enormous. One litter comprises six to ten pups. When they leave the den about two months later the young roam the countryside together with their parents.

Wolf *Canis lupus*

The wolf is undoubtedly the best-known representative of the dog family. It is or was found in most of Europe, temperate Asia and North America. Within such a vast range there are naturally differences between the wolves of the various regions. Those found in the south are usually smaller than those which inhabit the northern regions. There are, of course, exceptions. Some of the islands of the Canadian arctic group are the home of small white wolves. However this is not the first such exception we have encountered in nature. The largest wolves are found in the wooded regions of Siberia and Canada. They weigh about 80 kg (180 lb) and occasionally even more than 100 kg (225 lb). Wolves also differ markedly in colour. The greatest diversity is found amongst the North American wolves, where a single pack may include specimens of all hues ranging from black

to all white. Originally the wolf inhabited all types of country within its range. Man, however, always viewed him as the mortal enemy of wild game and livestock and generally as a pest which must be killed on sight. Even though we now know that this is not totally true and that wolves play an important role in maintaining the balance of nature, this opinion is still quite prevalent. Even many of the national parks in the United States believe that they are no place for the wolf. Nowadays the problem of protecting the wolf is being dealt with by a special committee of the International Union of Conservation of Nature. Too late, in some instances. In many places the wolf is already extinct.

The basic unit of the wolf community is the family. Late winter, usually February, is the mating season. Two months later the female bears her four to six young in the

den. Their further development is the same as that of other canines. After about ten days they open their eyes but remain in the den with the mother considerably longer, during which time they are brought food by the male. When they emerge from the den they are fed by both parents, first of all regurgitated food; later the parents bring whole animals to the pups to play with in preparation for life on their own. The young remain with the parents two years. Only during the brief period when the mother has a new litter do the older offspring keep their distance. The family forms a pack which hunts together. At certain times, particularly in cruel winters, they may team up with several other packs to hunt deer or other, larger prey. In the far north, reindeer or caribou are the favoured prey and further south elk (wapiti) or a young or ailing moose may be attacked.

The Dog

Scientists agree that the dog was probably one of the first animals to be domesticated. This may have been about 12,000 years ago, when Ice Age Man, hunter of mammoths, wild horses, aurochs and bison teamed up with the wolf, which similarly followed the movements of animal herds. The magnificent pictures in the caves where late Ice Age Man sought shelter depict not only the animals hunted at that time but also wolves. The waste heaps of Stone Age Man undoubtedly attracted wolves and the smell of fresh meat even more so. Perhaps it all began when a wolf took a morsel from man's hand or when a hunter brought a young wolf home and raised it in his community. We will never know. The process of domestication undoubtedly took place in a number of separate places at about the same time. Thus, for instance, remnants of the skull of a relatively small species of dog that probably lived as a partly-tamed animal together with man were found by archaeologists in the waste heaps of settlements on the east coast of Denmark. The age of these settlements is estimated to be 10,000 to 12,000 years. Dating from the same period are similar finds at Bologoye near Moscow. The foregoing lines refer to dogs and not wolves. The dividing line between the two is sometimes difficult to determine. Even today, if a scientist is shown the skinned body of a wolf and that of a dog of the same size he will not find it easy to identify them.

The dog was used by man to help him hunt and as a watchdog to guard his camp. What this teamwork was like can best be surmised from the customs of Australia's aborigines, which today still live like Stone Age Man in some places. Their life was studied by the anthropologist M. J. Meggit who writes: 'When out hunting one day a group of men of the Walbiri tribe came upon the tracks of a dingo which was pursuing a kangaroo. They followed its trail the whole day ready to be in at the kill. However they came upon the exhausted kangaroo before the dingo and killed it with their spears and boomerangs. They gutted the animal and carried it back to camp. As was the custom they threw some of the entrails to the dingo.'

Man, however, did not use the dog only to hunt and guard his camp. It appears that already long ago he selected certain varieties and kept them for his own pleasure. Both in the Egyptian as well as the Inca and Mayan civilizations we find depictions of dogs that resemble some of the breeds of today. The variability of the dog and breeding and selection by man have produced a vast number of forms.

Domestic dogs

Cats

Of all the beasts of prey the cats are the best adapted for the life of a predator. The cat's most characteristic weapon are its claws. In the normal position the claws are enclosed in a sheath on the terminal bone (phalanx) of the toe. The whole phalanx is then pulled back by a special elastic ligament and the cat travels practically without a sound using only the soft pads of the toes. When the toes are stretched a tendon attached to a muscle overcomes the resistance of the ligament, the terminal phalanx straightens and the claw slides out from the sheath. The

Wild cat

sharp weapon is thus ready to capture prey or to wage battle. All cats have a relatively small, round head. The jaws are shortened and the teeth few in number. However, they are superbly adapted for grasping and eating flesh. The canine teeth are large and can readily hold or even kill prey. The cheek teeth of cats are adapted for slicing or shearing meat. Their sharp edges slice meat like scissor blades. The tongue is furnished with sharp, horny protuberances which function like a rasp to remove the last bits of flesh from bones.

Wild cat *Felis silvestris*

The wild cat has a wide distribution. According to the opinion of some authorities the term wild cat embraces several groups of cats consisting of similar forms that live in different environments. The first group includes the woodland cats of Europe, ranging from Scotland to Transcaucasia and found also in Asia Minor. They look like our ordinary striped domestic cat. The second group includes the Kaffir cats of Africa. Unlike the former they do not avoid man but often occur in the vicinity of his dwellings. As we shall see later it was this charac-

teristic which was an important factor in their domestication. The third group includes the wild cats of the steppe and desert regions of Asia.

The European wild cat is already extinct in many places. In Britain it has survived only in Scotland, in Germany in the Harz and Eifel mountains. The Carpathians contain a fairly large number. The habitat of the European wild cat is dense woodland with thick undergrowth. Cats are solitary creatures. Only during the mating period, generally in February, do they come together in pairs. However, the female soon leaves her mate to find a place to bear the young. It is always in a dry and well concealed spot. The young are like the kittens of our domestic cats. They are just as playful and just as sweet. European hunters often asserted that wild cats are serious pests to wild game but this is not so.

Domestic cat

The ancestor of the domestic cat was probably the Kaffir cat of Egypt. Individual references to and fossil finds of cats kept in households date from the time of the Badari civilization, that is about 4,000 B.C. However only as of 2,000 B.C. can we speak of the cat as a domestic animal with certainty. Africa's wild cats were prime subjects for domestication. They are undaunted by human civilization, on the contrary — life in the vicinity of man has many assets, first and foremost being the large number of rodents found in his granaries.

Domestic cat

Ocelot

Ocelot *Leopardus pardalis*

The ocelot is the commonest wild cat of South and Central America. Its range extends to the southern United States. Some ocelots are scarcely larger then a good-sized domestic cat, others which live in the tropical rain forests of Guayana and the Amazon region are almost as big as a small leopard. The ocelots of the rain forests are generally a darker colour with large ring-spots. Ocelots of open steppe regions are light with small spots. All are expert climbers and are often found in trees. Their prey is anything they can overpower — ranging from mice and lizards to the young of ungulates. Thousands are shot for their fur every year and this has brought about the danger of their extermination in some regions. It is no wonder that the International Union of Conservation of Nature and Natural Resources (IUCN) appointed a committee of authorities from all over the world to deal with the problem of spotted cats and try to persuade the governments of the respective countries to abstain from further killing.

Bobcat *Lynx rufus*

The bobcat inhabits the area extending from Canada southward deep into Mexico. It is found up in the mountains as well as in semidesert regions, in mountain forests as well as in subtropical swamp forests and spreads of cactus. Naturally in such widely diverse habitats we can expect to find forms that differ in colour as well as size. The largest, the Canadian bobcat, is the same size as the Canadian lynx and more or less the same colour as well but the two can be readily distinguished by the tail. That of the lynx is always black at the tip, that of the bobcat a light colour. Besides this the bobcat has a white patch on the outside of each ear which the lynx does not. The smallest is the Sinaloa bobcat which is only slightly larger than the domestic cat.

Puma *Puma concolor*

Puma, mountain lion, cougar, catamount — those are all names for a cat which was and still is popular chiefly in North America. This is the largest species of the cat family now found in North America. The puma inhabits, or at least inhabited, the territory from southwestern Alaska and central Canada through all of North and South America to Tierra del Fuego. In the northern part of its range it is now generally extinct. It is found only in

Bobcat

Puma

(47 mph) and it is estimated that it may even attain a speed of more than 100 km/hr (63 mph) during the final sprint. Cheetahs are definitely savannah creatures which hunt during the day and rest at night. They live singly or in pairs. Where several cheetahs form a group it is sure to be a female and her offspring, which roam with her even when they are nearly a year old. The cheetah hunts chiefly small species of antelope. It selects its victim as it moves stealthily towards the herd, then leaps swiftly out of its last cover, overtakes the fleeing antelope and knocks it to the ground with its front paws. It then grasps it by the neck and drags it to the thickets or tall grass to partake of its meal undisturbed. Cheetahs are short-distance runners. If they do not capture their prey within a distance of about 200 metres they abandon the chase.

Formerly cheetahs were found throughout all Africa, excepting the forest regions, and in Asia from Arabia to Transcaucasia and on to India. Asian cheetahs are now almost extinct.

inaccessible mountains along the Pacific coast. Remnants of populations, now under rigid protection, are still to be found in Florida and perhaps also at the mouth of the Mississippi. From the Mexican border southward the puma is a relatively common animal.

Like all other animals with a wide range of distribution, pumas from different parts of their range differ in size. The largest reach a weight of 110 kg (245 lb), the smallest only 25 kg (56 lb). Very occasionally one may come across a totally black puma or an albino.

Cheetah *Acinonyx jubatus*
The cheetah is somewhat of an exception amongst cats. It has all the typical characteristics of the cat family but quite different claws. These are not retractile. The cheetah's method of hunting is entirely different from that of other cats. It pursues prey by running swiftly, in which it is aided by its long legs found in no other cat. The cheetah is one of the fleetest of the mammals, if not the fleetest. Its speed has been clocked at 75 km/hr

Cheetah

138

Large Cats

The large cats — the lion, tiger, leopard and jaguar — differ from all the preceding members of the cat family. Small cats are only able to mew whereas large cats have a much louder voice and roar. There are also a great many aspects of behaviour which differentiate the two types. Large cats, for example, feed lying down, whereas small cats merely lower the front part of the body between the loosened shoulders. There are also many more smaller details evident only to the expert.

The large cats are native to the Old World. Only a single species, the jaguar, is found in South America whence it spread north to the southernmost parts of North America.

Lion *Panthera leo*
The adult male lion with his handsome shaggy mane leaves us in no doubt as to why he is called the king of beasts. He appears in many coats-of-arms and royal insignias as the symbol of strength, courage and beauty. No other beast of prey has won such admiration. The lion is also sometimes called 'king of the jungle'. This, of course, is quite misleading because lions do not live in the jungle and never have. The lion is found in open country covered with thickets and scattered trees. It is the only member of the cat family which lives in groups, called prides, usually consisting of families. These number 20 and sometimes 30 members. Each pride includes one or more adult males and several females with cubs of various ages. The lion is capable of a speed of 60 km/hr (38 mph) for a short distance and his leaps may measure up to 12 metres (40 ft) in length. There are records of a lion leaping over a fence 3 1/2 metres (11 1/2 ft) high. Lions generally stay on the ground and do not climb trees. In some regions, for instance Kigezi and Ishasha in Uganda, however, lions commonly rest in trees. The theories explaining this unusual behaviour are many but no one has yet come up with an explanation which is fully satisfactory.

The life of the pride is of interest. When hunting, the prey is killed by the lioness but the first to begin feasting is the lion, followed by the lioness and last of all by the cubs. Lions are fully grown at the age of two but do not reach maturity until the age of five. As the male cubs mature they are driven out of the pride by the males. When they are strong enough, a group of young males (usually brothers from the same pride) will take over an established pride by driving out the current old males.

The period of gestation is from 105 to 112 days, after which the female bears from two to five young. The cubs are nursed by the mother for almost 3 months but are not fully independent until the age of one year. The mortality of the cubs is great. They often suffer from malnutrition because they are the last to get at the food. In those places where man has affected the balance of nature by shooting male lions for trophies, food for the young was hunted by the lioness, who started eating right away and allowed her offspring to do so too. The result was that many more cubs survived than had previously been the rule and there was an enormous increase in the number of lions. They hunted much more game than usual and man, who had himself disrupted the balance of nature, sought the cause somewhere else entirely.

Leopard *Panthera pardus*
Even today we still occasionally hear the question: 'Please, what is the difference between a leopard and a panther?' The answer to this is — none. Both names are used for the same species of cat. The distinction was introduced at one time by hunters who called larger specimens with a larger head, panther and smaller specimens with a smaller head, leopard. Today we know that these differences are determined by the sex — panthers were thus mostly males, and leopards females. It is interesting to note that the name panther continues to be

Lion

used for black leopards which are called black panthers. This brings us to another question: 'Is the black leopard a different species from the spotted leopard?' It isn't. There are only colour variations within the same species. Black leopards, of course, are more plentiful in some places, for example in high mountains and damp rain forests. They were hunted most in Ethiopia, Sikkim, Thailand and Malaysia. The leopard's area of distribution is extraordinarily large. It embraces all of Africa except the Sahara, extending through the Sinai to Asia Minor and thence across all of southwest Asia to the Far East and south to Java.

The leopard is considered to be the most skilful of the large beasts of prey and rightfully so. It is expert at climbing and is strong and nimble. It drags its prey, mostly moderately-large ungulates, into a tree so it can feast undisturbed by scavengers. More than once a captured young bullock has been seen hanging in the fork of a branch 6 metres (20 ft) above the ground. This is sufficient proof of the leopard's remarkable strength. Another time a leopard carrying an adult wild ram in its jaws was seen as it leaped onto a rock ledge 3 metres (10 ft) above the ground.

In many regions this handsome beast has already been exterminated. Let us hope that it will be saved by man in those places where it is still found.

Jaguar *Panthera onca*

The jaguar is the only large cat found on the American continent. It is more thickset than the leopard and has shorter legs and a shorter tail. In weight, though not in size, it is the equal of the tiger. The average weight is more than 100 kg (225 lb), much more than that of the leopard. The largest specimens inhabit the Mato Grosso region in Brazil. The jaguar is found in dense forest as well as in sparse bush and the semi-deserts of southern Argentina and the American southwest. In some parts of South America's rain forests which are flooded with water several months of the year, the jaguar stays in the trees all the time the ground is underwater. It is expert at climbing, being even more expert than the leopard. It is fond of swimming and often enters the water where it preys on caimans, turtles and fish. However it is just as skilful at capturing large animals, chiefly capyb-

Jaguar

140

ara and tapirs. It lies in wait either by the waterside or in a tree from which it lunges at its prey.

Jaguars are solitary in habit. Only during the breeding period do they form pairs. Should two males desire the same female for a mate they engage in combat which may often end in the death of one or the other. The gestation period is thirteen weeks and a single litter usually includes one to four young. When they are about three months old the female ceases to nurse the cubs but they remain with her until she has another litter, sometimes longer. Like the leopard, the jaguar also has colour variations but these are much rarer. Black jaguars continue to be prized in zoos.

Tiger *Panthera tigris*
The tiger and lion though quite different in appearance are very much alike in their internal structure. Only the skull exhibits any marked differences. It is evident that the species are closely related. Their biology, however, is completely different. The lion, as we have already said, lives in groups whereas the tiger is of solitary habit. The lion inhabits open country whereas the tiger is found in forests and jungles. In the minds of most people the tiger is linked with hot, damp jungles but they forget that the tiger is also found in the Siberian taiga which is noted for its freezing cold weather.

The tiger is found in Asia, its range extending west to the southern slopes of the eastern Caucasus. From these parts it was not at all unusual for the tiger to make its way as far as Turkey. Recent years have brought comforting news about new occurrences of tigers in these areas. Eastward the tiger's range extends to the Amur River region in the Far East and thence to the south. From India to Sumatra, Java and Bali runs a continuous belt of its distribution in southern Asia. Naturally the Indian tiger differs from the Siberian and Manchurian tiger. The Manchurian tiger, which must survive the rigours of a cold climate, has much longer and thicker fur. It is also larger because its greater size means that the body surface is relatively smaller and it does not have to expend so much precious energy. The smallest and darkest are the tigers from the Sunda Islands, Sumatra, Java and Bali. There the tiger is already very rare and numbers only a few surviving specimens. The number of tigers is generally declining throughout their whole area of distribution. This is not only because they are hunted by man as killer of domestic animals but also because the felling of forests is depriving the tiger of its habitat.

Tiger

The Weasel Tribe

The weasel tribe is large. All its members have a long, cylindrical body, short legs, and generally pronounced scent glands, which play an important role in communication and often also in defence. Otherwise it is relatively difficult to find other external characteristics common to all at first glance. The family includes slender martens, agile weasels as well as thickset badgers and web-footed otters adapted to life in water. However, there is still one more characteristic they have in common — the fur of practically all is highly prized and much in demand. Many species have paid dearly for this.

Members of the weasel family are found throughout the world except for Australia and New Zealand. But even there they have been introduced by man, sometimes with the intention of establishing a precious fur-bearing animal in the new environment, at other times to help control pests, such as the rabbit in Australia. In most cases such attempts failed.

Stoat or Ermine *Mustela erminea*
The stoat inhabits all of Europe (except the Mediterranean region), northern Asia, Japan and North America. It is found in areas from lowland country to high mountains. Since time immemorial it has been hunted for its precious winter fur (ermine) which was used to decorate the robes of sovereigns. The summer coat is brown above, white below, with a black tip to the tail.

Stoat (ermine)

The winter coat is pure white with the exception of the black tip to the tail. And it was this white winter fur ornamented with black-tipped tails which trimmed the robes of state. The most beautiful ermine is from the northern regions or from the mountains. In more southerly regions the stoat does not change into a white winter coat or else the change is only partial.

Though the stoat is mostly active at night it often hunts also in daytime. It eats whatever it can capture — mostly rodents and often birds. The stoat has two mating periods a year but produces only one brood. This is because the fertilized eggs of females which mate in spring develop immediately whereas those of females that mate in summer remain latent and development does not begin until the following spring. This phenomenon is quite common in other species of the weasel tribe.

American mink *Mustela vison*
The American mink is probably the most widely raised fur-bearing animal nowadays. Originally it inhabited the forests from Alaska to California and was avidly hunted by man. Later, however, the demand for its fur became so great and capturing it in the wild too difficult and so the first mink farms came to be established. Besides the wild minks coloured dark brown, many different colour mutations are also raised nowadays, many of which are highly valued commercially. Sun and rain damage the fur of minks and so the cages at mink farms are roofed.

Sea otter *Enhydra lutris*
The fur of the sea otter was the most valuable of all and that is why the animal became the victim of indiscriminate hunting. Only when this species was on the verge of extinction did man begin to do something for its protection. There are two geographical races of sea otters. The smaller and darker coloured is found from Kamchatka to the Aleutian Islands and from there southward along the north Pacific coast to British Columbia. The larger and browner form has a local

American mink

distribution along the Pacific coast from Washington to Baja California. The sea otter is the largest member of the weasel tribe. It spends practically its whole life in water. Its diet consists of sea urchins, molluscs, crabs and only in small part of fish. When feeding the sea otter lies on its back and holds the food with its front paws. It is very resourceful and can even solve the problem of how to get at mussels with a very hard shell by smashing them against a stone.

Sea otters generally form groups numbering several tens and sometimes even hundreds of individuals. The young are not blind at birth but are able to see and are already furnished with teeth. The female usually bears only a single offspring, on the rare occasion twins. On dry land she carries the young otter in her jaws, in water between her front legs while swimming on her back. Sea otters are very gentle with one another. They stroke each other with their paws and play with the youngsters, tossing them into the water and then immediately taking them into their arms again.

American badger *Taxidea taxus*
The territory from British Columbia and Saskatchewan in Canada to Texas, California and Mexico is the home of an interesting member of the weasel family — the American badger. Its favourite food is prairie dogs and pocket gophers, which it digs out of their holes with its strong claws. Its strength is truly enormous. In a typical American experiment the American badger proved capable of lifting a platform

Sea otter

with a horse and rider on top. Today this flesh-eating mammal has adapted to life in cultivated grassland and feeds on various small animals as well as plant food. The burrows it digs are extensive and reach to depths of 10 metres (32 ft). The is where the female bears the young, usually five, in late summer.

In the northern parts of its range the American badger falls into a torpid state in winter when all its body functions slow down. This, however, is not true winter sleep. Badgers in the south wait out periods of extreme drought in similar manner.

Striped skunk *Mephitis mephitis*
Even those who have never seen a skunk have surely heard about its formidable weapon. Like all members of the weasel family the skunk

American badger

Striped skunk

has scent glands at its anal opening. The secretion serves to mark the animal's territory. In the skunk, however, these glands are enlarged and transformed into an efficient weapon. The animal is well aware of the force of its weapon and thus generally moves about slowly and exhibits no fear when confronted by an enemy. It simply turns its backside towards the intruder, lifts its tail and waits. If it continues to be harassed it squirts the foul-smelling liquid at the eyes of the enemy. Its aim is excellent and it hits the target precisely up to a distance of 3.5 metres (11 ft).

Bears

The evolution of bears is closely linked with that of dogs and it may be said they are modern relatives of dogs. With tongue in cheek we could even say they were huge, thickset dogs with a very short tail. Bears walk flat on the soles of their feet unlike most carnivores which walk on their toes. They have small incisors and wide molars for the bear eats not only flesh but also large quantities of plant food which it grinds with the flattened cheek teeth. The history of bears is relatively recent, the first known bears appeared only about 15 million years ago. For most of their history, bears were northern animals, confined to the cooler regions. In the past million years or so they have spread south into South America and South-east Asia.

Bears are a small, but yet versatile group of flesh-eating mammals. All in all there are now only seven existing species of bears. Compared with the number of species in other families of carnivores this is very few. Only the hyenas and aardwolves have fewer. Because there are so few species of bears we can name them all here. They are the brown bear, the American black bear or baribal (more about these later), the Himalayan black bear from the forested regions of central, eastern and southern Asia, the sloth bear from India and Ceylon, the sun bear from the rain forests of South-east Asia, the spectacled bear from the Andes, and the polar bear which inhabits the arctic regions of the northern hemisphere.

Most people think bears have long shaggy fur and most of the seven species do. On some parts of the body the fur may even be prolonged to form a sort of collar or tuft. Only the sun bear, which is found in temperate climates, has very short, smooth and flat fur. The coloration of bears is strikingly uniform. There may be white or yellowish markings on the chest and in the spectacled bear on the head, but otherwise the body is a single colour. In other animals the back and belly are a different colour but in bears they are the same.

The most highly developed sense in bears is the sense of smell, followed by hearing. Their eyesight is relatively poor and bears are able to distinguish stationary objects only when they are quite close. One look at a live bear and its sensory organs is all we need to assess the level of its various senses. The nose is very flexible and when sniffing it turns in all directions. The ears, though small, are also very flexible and by turning them the bear can determine exactly what direction a sound is coming from. The eyes, however, are small and it is clearly evident the bear does not depend much on his sight.

People generally view bears as good-natured clumsy creatures. That is how they are depicted in fairy-tales, that is how they are presented to children in the form of toy teddy bears. Bears, however, are true beasts of prey and even those which have been hand reared from the time they were cubs are not fit companions for man. The difficulty with bears is that the poorly-developed facial muscles allow hardly any facial expression.

Kodiak bear
Ursus arctos middendorffi
The brown bear has a vast range which includes Europe, Asia and North America. The forest limit marks the northern boundary of its distribution and the Himalayas in Asia and the Mexican plateau in America its limit in the south. Naturally the widely different climates and environments through-

Kodiak bear

out its range mean it occurs there as several geographic races, each exhibiting specific adaptations to the given habitat. These can be divided into several groups, believed to be separate species by most laymen. First and foremost is the group found in western and middle North America which includes bears with greyish fur and grey claws. These are called grizzlies (grizzled means grey or streaked with grey) and are depicted as ferocious killers in novels of the wild west but in reality are no different from other brown bears. The only thing, perhaps, that may have been responsible for ruining the good reputation of other bears is that man let his domestic animals roam in their hunting grounds and these were naturally hunted by the grizzly.

The largest brown bears are found in Alaska and certain neighbouring islands such as Kodiak, Afognak, Montagne and Admirality. These bears are reddish-brown with black claws and their size is enormous. The bears from Kodiak reach a length of almost 3 metres (10 ft) and weigh approximately 800 kg (16 cwt), in rare instances even more. For example, the Berlin zoo had a kodiak bear that weighed 1,200 kg (24 cwt).

The size and weight of the adult bear makes the relative inadequateness of the cubs even more pronounced. At birth the cub measures about 20 to 30 cm (4 to 6 in), weighs approximately 0.5 kg (1 lb) and is relatively helpless. It is born nearly naked, blind and with closed ear canals. Not till the age of three weeks does it open its eyes and not till the age of seven weeks is it capable of hearing and finding its way about. Even that most important sense of all — the sense of smell, does not begin to develop until two months after birth. From this it is evident that the young are dependent on the mother's care for a long time. The cubs that emerge from the den at the age of three months or so look like toy bears and it is hard to imagine that they will one day be as large as their parents. From this point on, how-

ever, growth is very rapid. When they are a year old they are already about half as large as the adults. After that the rate of growth is slower and the bear attains its full size at about the age of 10 years.

American black bear or Baribal
Euarctos americanus

The American black bear is the one which generally appears in the pictures taken by tourists to Yellowstone and other national parks in North America. Formerly bears, and black bears in particular, were hunted as game by both the Indians and white men. All parts of the bear were put to good use; the skin as a cover, the meat as food, the fat to provide light or as ointment. The gradual felling of the forests forced the bear from its hunting grounds and even caused its extinction in many places. Laws limiting bear hunting issued at the beginning of the century came too late to save the less adaptable grizzly. The black bear, however, which became acclimatized to life in the vicinity of man, returned to its former hunting grounds and increased in number. Today its numbers are again an es-

timated several hundred thousand.

Though called black bears, their colour is varied and may even be cinnamon brown; often a single litter will include cubs of both colours. Rarest is the bear from certain parts of British Columbia which is coloured white or silvery blue-grey and is known as Kermode's bear or Glacier bear. American black bears live singly. Their den is located either in thickets or in old hollow trees. They feed chiefly on various forest fruits and small vertebrates.

As we have already said the black bear is most likely to be encountered in one of the national parks. Those which live there are tame, even brazen, and dangerous. Despite signs stating that it is forbidden many foolish persons feed the bears and then when such a bear becomes hungry it often becomes enraged and attacks. More than half the accidents in parks are caused by these black bears. Leaving the car to photograph them is equally dangerous. In some places the bears have become feared invaders of camp grounds because they steal things and even clamber into open cars.

Giant Panda, Coati and Genet

Giant panda

Giant panda
Ailuropoda melanoleuca

Ancient Chinese paintings depicted a strange, black and white bear. No one doubted that it was a creature born of the artist's imagination. But now and then reports leaked through about a white bear that lived somewhere in the heart of Asia. The curious and diligent Father Armand David, a missionary in China at that time who brought to the attention of Europeans many hitherto unknown plants and animals, set out in search of this legendary bear in the years 1868 and 1869. For a long time success eluded him until one day, when he had given up hope, he saw a bear skin in a farmer's house and immediately recognized it as the one he had been seeking for so long. After that it was not difficult to persuade the local hunters to bring him this strange animal. He sent the pelt and the skeleton to the museum in Paris and this is how the giant panda was discovered and described. In the meantime zoologists determined that this 'bear' does not belong to the bear family and put it in a separate fami-

ly by itself. How the first rare and handsome specimens came into captivity is a long story. Suffice it to say that difficulties were many and the cost enormous. $ 25,000 or DM 200,000 — that was the price paid for a single giant panda by zoos. Payment of such a sum, however, does not mark the end of all worry, but rather the beginning, for the giant panda feeds on bamboo shoots and leaves. It is therefore necessary either to grow bamboo or where this is impossible because of unfavourable conditions to provide a constant supply by air shipment. Nowadays the giant panda is being bred in a number of zoos. We have no information as to its numbers in the wild and because the territory it inhabits is very small and its existence is thus readily imperiled, the giant panda has been put on the list of endangered species. Because it is such an attractive and popular animal it has been also selected as the symbol of the World Wildlife Fund, which is part of the International Union for Conservation of Nature and Natural Resources (IUCN). The red panda, *Ailurus fulgens,* has a more wide-

spread distribution in the Himalayas. It closely resembles a rusty red raccoon with its bushy, striped tail and cat-like face.

White-nosed coati *Nasua narica*

Found in South and Central America is an unusual family of flesh-eating mammals which seems to be somewhere between the weasels and the bears. Some members are slightly reminiscent of bears — they are the raccoons. Others, for example cacomistles, look more like weasels and the peculiar kinkajou even has a prehensile tail found in only one other flesh-eating mammal. This family also includes the coatis. There are four species, the best known being the white-nosed coati. Coatis are distinguished by a long flexible snout which the animal can twist to any side a full 45°. The coati investigates every crack and crevice with its snout in search of food. It often forms large groups and bands numbering several hundred are not at all unusual. The members of such a band work in collaboration and it may be said that it is well organized and governed by rigid rules. Experiments have shown that coatis are capable of executing even quite complicated tasks and that their behaviour is surprisingly logical. The white-nosed coati is native to the forests of southern North America whence it spread to Central America and to Colombia. Coatis are active by day and so an exception in this family. During the breeding period males often engage in fierce combat over a female. At this time the band includes a single male and several females and their offspring. They are much more well-developed at birth than the young of bears or even the young of all other members of the family. They weigh almost 200 grams (8 1/2 oz) at birth, open their eyes on the llth day and begin eating solid food about the 20th day. By the time they are two months old they are fully independent of their parents.

Common genet *Genetta genetta*
Genets, civets, mongooses and related flesh-eating mammals form another separate family which includes 36 genera embracing roughly 70 species. Most are found in Asia and Africa. Only one, the common genet, has a range which extends as far as Europe, where it is found on the Iberian Peninsula and in southern France. All genets look more or less alike. They have short legs, a long body and long, ringed tail. They are very similar in colour and variously spotted. The common genet is readily distinguished from other similar species by the band of longish fur on the back which can be raised. Genets are delightful, agile creatures. Few people know that they were tamed by the ancient Egyptians long before the domestic cat appeared on the scene. The genet was used to hunt

mice but never became a domestic animal. All members of this family have large scent glands by the anal opening. Besides this, some species, including genets, have another large gland between the anal and genital opening. When irritated they eject the secretion produced by this gland. Otherwise the odorous secretion produced by these glands is used to mark the boundary of the territory occupied by the animal. Genets are active at night. In daytime they sleep in a concealed spot, generally in a cavity. They are not known to dig burrows, for which their soft paws are not suited. Their diet consists mostly of animal food, but genets also eat plant food.

The fur of genets is fine and thick. Nowadays it is not used much, but formerly it was used to trim ceremonial garb.

White-nosed coati

Common genet

Hares, Rabbits and Pikas

Many readers will probably be surprised that in this book hares and rabbits are not where they expect to find them — in the section on rodents. They have gnawing teeth it is true, but if we trace the history of hares and rodents we will discover that as far back as the early Tertiary period — about 60 million years ago — they were distinct species. Thus they cannot now be classed in the same group simply because they have certain characteristics which are similar. The rabbit moves its jaws, when chewing food, from side to side like a goat or a cow. A guinea pig or a mouse, on the other hand, moves the lower jaw forward and back! If you were to examine the skull of a hare or rabbit you would see that the upper jaw has 4 incisors, two on each side. In rodents there is only one incisor on each side. There are many more characteristics clearly indicating that the evolution of the two groups went entirely different ways and that therefore the hare

and related animals must be classed in a separate order by themselves.

European hare *Lepus europaeus*
The hare is an inhabitant of steppe regions and has adapted very well to life in cultivated steppes — in fields. It is also found, however, in deciduous woods. It is native to Europe (apart from Ireland and the Iberian Peninsula), east to the Urals and south through central Asia and Asia Minor to east and south Africa. It was introduced as a popular game animal to Australia, New Zealand and also parts of America. In some places it is called the brown hare.

The hare does not dig burrows but rests in a shallow depression in grass or under thickets. It uses its hiding place, which is called a form, for a long time, leaving it at dusk to forage for food and returning at dawn. On its departure and then again on its return it takes care not to leave any traces which might lead an enemy to its lair by making sudden turns at right angles and bounding with long leaps in various directions up to a distance of 4 metres (13 ft). It behaves in the same manner when threatened with danger. The doe bears the young in the form. They are furry at birth, able to see and very independent. As a rule there are two, sometimes even three, but rarely four. If conditions are favourable a single doe may have as many as five litters in a year. The leverets are fully independent after about a month. The hare is a vegetarian, feeding exclusively on green plants and nibbling the bark of trees.

Blue, Varying or Variable hare *Lepus timidus*
This species is called variable hare because it turns white in early winter — only the tips of the ears remain black. In summer it is brownish with a grey tinge, which is why

it is also called the blue hare. It is native to northern Europe, Asia and America. In the far northern parts of America it does not change colour in summer and remains white the whole year through, which is why it is known as the arctic hare in this part of the world. In some of Europe's mountains it occurs as a relic of the last glacial epoch, for instance in Scotland, Ireland and the Alps. In the eastward direction the limits of its range shift southward. There the blue hare is found from the Arctic Ocean to the high mountains of Central Asia. Unlike the European hare it does not make a form but hides under rock overhangs, between boulders, and the like. The doe bears several young, sometimes as many as eight.

Cottontail *Sylvilagus floridianus*
The cottontail is one of the commonest American rabbits. The term hare and rabbit is commonly inter-

European hare

Blue hare

Cottontail

changed. The general rule, however, even though there are exceptions, is that hares do not dig burrows and their young are furry and able to see at birth whereas rabbits live in burrows and their young are naked and blind at birth. The cottontail, then, is a rabbit. It is found in the woodlands, shrub country and prairies of the southern United States and northeastern Mexico. It does not avoid man and so has spread to gardens, city outskirts and even city parks. Its occurrence follows a ten-year pattern — five years when it is very plentiful and five years when its numbers are few. Because the cottontail is one of the chief foods of the fox it also affects the abundance of this animal. In winter when the leafless woods do not afford sufficient places of concealment it is common prey of the great horned owl. The cottontail is a territorial creature. The females occupy small, adjoining territories and there are several such groups of female territories within the territory of a single male.

European rabbit
Oryctolagus cuniculus
Formerly the European rabbit was

Pika

European rabbit

found throughout all Europe but was forced to abandon practically the whole of its territory during the Ice Age remaining only in the western Mediterranean region. The Phoenicians and Romans re-introduced rabbits to all of southern Europe. Their meat was considered a delicacy and they were therefore kept in preserves called leporaria. There were plenty of escapees from these leporaria for rabbits are very good at digging. As for their spread to more northerly regions, monks in monasteries were often responsible for this because unborn rabbit embryos, called laurices, were one of the foods they were allowed to eat during the fasting period. Naturally many rabbits escaped from the monasteries as well and the populations they formed spread still further. Nowadays the European rabbit ranges from Britain to the Ukraine. It was purpose-

ly introduced by man in many places throughout the world. In Australia and New Zealand, however, the results were disastrous. Having no natural enemies in that country they multiplied in such numbers that they became serious pests, forcing man to wage war against them with all the means at his disposal, including infectious diseases.

The European rabbit is the species from which the domestic rabbit is derived.

Pika *Ochotona rutila*
Unlike hares and rabbits, pikas, which belong to the same order, have all four legs the same length and short rounded ears. They are found in upland grasslands in Asia and North America. They form col-

onies with each of the members occupying its individual territory. The pika has an interesting habit of making and storing hay. From the pikas' burrows well-trodden paths lead to places where they gather grass and various herbaceous plants. They bite the plants off at ground level and carry them to their burrows where they spread them out to dry. If it looks like rain or bad weather they take the hay inside and when the sun comes out they carry it outside again to dry. The amount of hay is large and it is arranged in tidy heaps. There may be from 8 to 20 kg (18—45 lb) of hay all told, depending on the size of the colony. The Mongolian pika even holds the hay down with pebbles so the wind does not blow it away. Pikas do not hibernate and so in places where there is a great deal of snow they must have a sufficient supply of food on hand.

The Largest Order of Mammals — Rodents

Rodents comprise more than half of all living species of mammals. They include more than 300 genera numbering some 3,000 species, many of which occur as several geographic races. Altogether there are some 5,000 known forms of rodents and new ones are being described by scientists every year. Most are small animals and many people wrongly think a rodent is a creature like the brown rat both in size and appearance. Many rodents are like this but their number also includes types such as the squirrel, the hamster, the porcupine and others. Despite the differences in body structure rodents comprise a uniform group. Not only in the present day but already many millions of years ago, in the early Tertiary period, rodents differed markedly from all other orders of mammals. So markedly that not a single fossil exists representing a transitional form between the rodents and members of the other orders. Nowadays rodents are found practically everywhere. Even Australia has many native species. Rodents, together with the bats, are the only placental mammals native to that continent. Rodents include species which live underground as well as species which glide through the air, species which live in water as well as species which live in the desert. Other species inhabit regions where for more than nine months of the year the temperature does not rise above − 15° C.

Grey squirrel
Sciurus carolinensis
The grey squirrel is native to eastern North America but was introduced also to England. Since the last century it has spread from Woburn Park through the entire southern half of England and north as far as the Scottish border. In many places it ousted and replaced the less aggressive European red squirrel. Foresters are not fond of the grey squirrel because besides eating the seeds and fruits of forest trees it also strips the bark from branches and nibbles young shoots and may cause great damage in forest stands. In the course of the year it builds two types of nests. The winter nest of twigs is located beside a tree trunk and this is where the young are reared. In summer it builds leaf-nests, called dreys in Britain. These are untidy nests of the freshly-cut terminal shoots of broad-leaved trees placed freely in the branches. Each squirrel constructs several such nests.

Prairie dog *Cynomys ludovicianus*
Prairie dogs are related to squirrels and marmots. They live in large, often huge colonies — veritable towns. The burrows lead straight down into the ground. The entrance is ringed by a low mound on

Grey squirrel

Prairie dog

which the prairie dogs squat and guard the other, grazing members of the colony. When danger threatens the sentinels make barking cries which is how the prairie dog came by its name. Formerly, these colonies were much larger than they are today. The individual 'towns' spread and joined to form a large complex. An area of 65,000 square kilometres inhabited by a closed population contained an estimated hundred million prairie dogs! As man gradually changed the prairie into pastureland he killed its original inhabitants.

Canadian tree porcupine
Erethizon dorsatum
The tree porcupines from North and South America are very different from the porcupines of the Old World. They even belong to an entirely different sub-order. Their one

Canadian tree porcupine

the beaver's engineering skill, its ability to build dams from mud and branches and thus regulate the water level of streams and rivers. This is instinctive behaviour which provides beavers with an ideal environment for their way of life. With their strong incisors they can fell a tree up to 70 cm (2 1/2 ft) in diameter. In addition to the dams, beavers also build lodges with many entrances and chambers. The outside is plastered with mud.

The close, fine fur is very beautiful and long-lasting. It is therefore always much in demand on the market and at auctions. Beavers are still widely hunted.

Pygmy jerboa
Salpingotus crassicauda
Jerboas are among the most extraordinary of rodents. They have extremely long hind legs and travel in the same manner as kangaroos.

All are nocturnal creatures and live in dry steppe and desert regions with sparse vegetation. They do not reveal their presence in daytime, but with a spotlight one may see a great many in favourable localities. They feed on insects and the roots of desert plants, being provided with a veritable feast when it rains and the plants put out new green leaves.

The smallest of all is the pygmy jerboa, found in the semi-desert regions of central Asia. It digs burrows under clumps of saxaul or at the base of large rocks. Here the female gives birth to her young. The winter and periods of drought are spent in a torpid state when all the jerboa's functions are greatly slowed.

common characteristic, however, is their quills. The Canadian tree porcupine has quills only on the tail, the rest of the body is covered with long coarse fur. It is found in the forests of North America, being partial to coniferous woods, junipers and poplars. Its diet changes according to the season of the year. In spring it nibbles the flowers and catkins of willows, poplars and maples. Later, when the leaves appear, it feeds chiefly on the leaves of aspen and larch twigs. In summer its diet is supplemented by various herbs and in winter it feeds chiefly on spruce, pine and hemlock-spruce. It nibbles the bark and may thus cause great damage to trees by ring-barking.

Mating takes place in autumn or early winter. The male rubs the female with his nose and sprays her with his urine to keep rivals away. After 210 to 217 days the female usually bears a single large offspring. It has long, black hairs and short, soft quills.

Beaver *Castor fiber*
The beaver is the second largest rodent after the capybara and may weigh more than 30 kg (67 lb). Probably everyone has heard about

Pygmy jerboa

Beaver

151

Yellow-necked field mouse

Yellow-necked field mouse
Apodemus flavicollis
The yellow-necked field mouse is the commonest mouse in central Europe, reaching the edge of its range in England. It is very similar to the wood mouse or long-tailed field mouse which is abundant throughout Britain, but the yellow-neck is larger, weighing up to 40 kg (1 1/2 oz) compared with 25 g (1 oz) for the wood mouse.

The yellow-neck is an excellent climber and may appear high up on houses which are covered with ivy or wild grape. Besides plant food, field mice also eat insects, insect larvae and spiders. In the wild they often occupy abandoned birds' nests or nest-boxes. They line their nests with dry grass and moss and this is where the female bears the young. There are two to nine in a litter and there may be several litters in a year. You may come across 'expectant' mothers even in winter if the season is a mild one.

House mouse *Mus musculus*
Perhaps it is native to Asia but to-day the mouse is distributed throughout the whole world. Some populations still live as they originally did, in fields. Others have become fully adapted to life in man's dwellings. The house mouse can squeeze through narrow cracks and can climb through an opening only 9 mm in diameter. Even concrete flooring will not prevent its entering man's dwellings. The house mouse lives in colonies. It marks its trails with urine and the secretion from its scent glands. The house mouse's rate of reproduction is enormous. The female bears young at any time of the year. The average is five litters a year, each numbering five mice, so a new generation can occur every eight weeks. In congenial conditions and if no preventive measures are taken the mouse population may reach gigantic proportions and cause great damage.

The house mouse, however, also has its merits, mainly in the field of science. There are few medicines which have not first been tested on a mouse before being tested on other animals closer to man.

Brown rat or Common rat
Rattus norvegicus
People usually confuse the black rat, or ship rat, and the brown rat. The black rat favours the dry environment of granaries and attics, has a tail longer than its body, larger ears and a more pointed nose. The brown rat requires a damper environment and so is found in waste heaps, cellars, sewers and like places. The tail is shorter than the body, the ears are smaller and the nose more blunt. The black rat has been an associate of man from early history and was

House mouse

already known to the Romans. The brown rat came to Europe probably with the peoples of central Asia, the Huns and Mongols. It is more adaptable than the black rat and has ousted and replaced it in many places. Both species of rodents are hosts of the flea which carries bubonic plague.

European field vole
Microtus arvalis
To most people the field vole is simply a 'mouse' even though it belongs to an entirely different family. The field vole is readily identified at a glance. Unlike the

Brown rat

Eastern water rat

ing is the eastern water rat which is found in swamps and alongside rivers and sea inlets in Tasmania and along the eastern coast of Australia. It is superbly adapted to life in the water. The head is long and the nostrils are located at the very tip of the nose. The fur is short but very thick and the legs are modified into broad webbed paddles. The eastern water rat is active at night. It feeds mainly on mussels and molluscs, crabs and crayfish. However, it also eats fish, birds' eggs and bird nestlings. It digs long burrows parallel with the shore in which there are two chambers. One contains a nest of grass, bark and twigs, the other serves as a pantry or storehouse and is filled with bones and shells. The eastern water rat is a very useful animal, destroying the aquatic snails which are hosts of the parasitic sheep liver-fluke.

European field vole

mouse it has a short tail and short ears which are just barely visible. It inhabits open country. Its burrows are located just below the surface of the ground and are joined by well-trodden paths. In winter voles dig tunnels in the snow at ground level and line them with grass and moss. When the snow melts these pathways make a spreading network in meadows. Voles feed on plants and insects. They multiply at a remarkable rate for a single litter numbers up to seven young and there may be as many as six litters in a year. Theoretically, then, a single pair, its offspring and the offspring of its offspring could produce 25,000 voles in a single year. Under normal circumstances, of course, the number killed by their many enemies is so great that they never get out of hand. Weasels, foxes, cats, owls, buzzards, kestrels — all feed chiefly on voles. Many

are also killed by storks, gulls and other birds. Dry summers, however, which are particularly congenial for their multiplication, may bring on a population explosion with as many as 30,000 voles to one hectare (12,000 per acre). However, such populations disappear as quickly as they develop. Congenial conditions for the vole are naturally a favourable circumstance for its natural enemies as well. Sufficient food means that they, too, occur in large numbers.

Eastern water rat
Hydromys chrysogaster
Australia is a land of monotremes, marsupials — and mice, the only terrestrial placental mammals which have made their way to this continent. One of the most interest-

Greater mole rat
Spalax microphthalmus
The mole rat is a rodent which is excellently adapted to life underground. Its cylindrical body hasn't a single protuberance which would hinder its passage in the narrow tunnels. The ears are insignificant and do not jut above the fur. The small eyes, the size of poppy-seeds, are covered with skin. The flattened skull carries a large nose. The gnawing teeth, the incisors, are strong and project outside the mouth. The weak legs are not adapted for strenuous digging. The mole rat tunnels mostly with its head, removing soil with its large incisors. Its main, if not sole, food are roots and tubers of various plants. It consumes large amounts of food and may cause damage.

Greater mole rat

Hoofed and Horned

Several orders of mammals may be grouped together under a single heading as hoofed animals or ungulates. Their kinship varies but besides other common features they have one that is most striking — namely the hoof, the horny covering of the feet. Ungulates evolved from animals which lived in damp rain forests, gradually developing into animals of grassland. A five-toed foot which could be spread wide was well-suited for life in the soft ground of marshlands but the narrow, hard hoof was better suited to the hard ground under grass. That is why in ungulates the number of toes was often reduced, in horses down to a single toe on each foot. Ungulates walk on the tips of their toes. In some a soft fibrous pad developed on the sole of the foot, for example in the elephant, and the hooves are like toenails. In horses the horny hoof covers the entire terminal phalanx of the toe. When walking the horse touches the ground only with the hard protruding edge of the hoof, the softer sole section in the centre of the hoof is concave and does not touch the ground. Camels and llamas have an elastic pad on the underside of the third and fourth

toe and the hooves, which are of no assistance in walking, merely cover the terminal phalanx of the toes. Ungulates with a reduced number of toes are divided into two groups, the even-toed and odd-toed ungulates, according to the line followed by the long axis of the limb. In the even-toed ungulates the axis passes between the third and fourth toe and the animal walks on both toes. In the odd-toed ungulates the axis passes through the third toe which is larger and stronger and the animal walks only on this single toe (as in the horse). None of the ungulates feed solely on flesh. On the contrary, most are herbivores. Only pigs are omnivorous and certain deer and antelopes eat flesh on occasion. The alimentary tract of herbivores is adapted in various ways for the digestion of food containing large amounts of cellulose. This process is aided first of all by bacteria and protozoans present in great numbers in the digestive tract. At the same time these microscopic organisms are an important source of protein. Such protozoans are present, for example, in the appendix and large intestine of the horse and in the rumen of ruminants. The

stomach of ruminants is divided into four chambers, each with a special function in the processes of digestion. The food first collects in the rumen, then passes into the reticulum where it is partially digested and then regurgitated back to the mouth to be chewed slowly a second time. After this it passes into the omasum where water is absorbed and the cud is compressed. Actual digestion takes place in the abomasum. In camels and mouse deer the stomach has only three sections.

Many ruminants have hard projections — horns — on the head. These are outgrowths of the frontal bone and differ in the different families. They may be covered with skin, as in the giraffe, or with a horny layer, as in the antelope and buffalo, or they may be bony antlers as in the deer family. In the last case these are shed annually and replaced by new ones. The horns of the rhinoceros are of entirely different origin. They are outgrowths of the skin composed of long, horny epidermal cells and are not joined to the skull. The skin of ungulates has numerous openings leading to scent glands. Their secretion plays an important role in communica-

Indian rhinoceros

tion between the animals, be it animals of the same species or different species.

Indian rhinoceros
Rhinoceros unicornis

The rhinoceros has three toes on each foot, the second, third and fourth toe — the first and fifth are lost. The axis of the limb passes through the third toe and so the rhinoceros is an odd-toed ungulate. The Indian rhinoceros is a striking animal. The body is covered with thick skin usually thrown into folds which form characteristic plates. This is a typical characteristic of the Asian rhinoceros. The African species have smooth skin which does not form plates. There are now only five existing species of rhinoceros: the Indian with one horn and skin folded into five plates, the Javan, with one horn and three plates, the Sumatran, with two horns, three plates and relatively thickly-haired skin, and two African species. One is the wide-mouth or white rhino, the largest land mammal after the elephants, the other is the black rhino, which has the upper lip extended into a flexible projection. Both are two-horned.

The thick skin of the rhinoceros is not furnished with scent glands and so the animals mark their territories with urine and faeces.

All species of rhinoceros are in danger of extinction. They are hunted by poachers for their horns, which are considered an infallible medicine and stimulant. The chief threat, however, is the alteration and destruction of their natural habitats. The Javan and Sumatran species number only a few remaining specimens.

The rate of reproduction in this species is thus very slow. In most species the birth rate does not make up for the death rate and so the rhinoceros population is continually declining.

Plains zebra
Equus burchelli boehmi

Zebras are the striped horses of Africa. Do you know whether a zebra is black with white stripes or white

Plains zebra

with black stripes? The answer is easy. Take a look at the picture of one of the zebras which are no longer living, for example the quagga or Burchell's zebra. In both the ground colour is tawny with dark markings. The farther north one goes the whiter the ground colour and the more pronounced the stripes. Besides the extinct quagga there are three living species of zebras — the mountain zebra from southern and southwestern Africa, the plains zebra, native to eastern and central Africa, and Grévy's zebra from northeastern Africa. Burchell's zebra, the southernmost race of the plains zebra, is now extinct. Other races show slight differences in the stripes and in the structure of the skull. At one time, before the white man came to Africa, zebras roamed the plains in huge herds together with gnus and ostriches. However they were killed by man in vast numbers not only for their meat but often merely for sport. Nowadays herds of zebras, naturally not nearly as large as formerly, may be seen only in national parks and reservations. Each herd, headed by a stallion and composed of mares and their foals, forms

a unified group. Young stallions keep to themselves in separate 'male societies'. When their time comes, pregnant mares leave the herd and bear their offspring in the bush apart from the rest. The period of gestation is long — the plains zebra 11 months, the mountain zebra 12, and Grévy's zebra even 13. Not until some hours later or even the following day, when the foal is able to keep up with the herd, do the mother and offspring join the others. Stallions often engage in fierce combat for the position as leader of a herd and leave the battle-ground with numerous gashes.

Zebras are the main food of lions. However, other beasts of prey particularly leopards and hyaenas may also be dangerous, chiefly to newborn foals.

Man has often attempted to tame the zebra, mostly because he wished to use it in place of the horse in regions inhabited by the tsetse fly, which transmits the dreaded disease nagana to horses. Zebras are immune to this disease but attempts to tame them did not meet with much success, the zebra's kick being very powerful.

African elephant
Loxodonta africana

Every child recognizes the elephant by its trunk. This is a muscular, flexible organ, an extension of the nose, which may be used for several important purposes. The elephant can use it to smell at high levels, for instance above three-metre (10 ft) tall elephant grass; with the aid of the finger-like appendages at the tip it can feel and grasp small objects. Larger objects are grasped by coiling the trunk around them. The elephant has a short, relatively rigid neck so that it cannot reach the ground with its mouth. It therefore carries food to its mouth with the trunk, with which it also drinks. Very young elephants do not know how to use the trunk and drink from the mother's teats with their mouths.

Another interesting characteristic are the tusks, modified incisors of the upper jaw. They grow continually and are hollow at the base. These are used by the elephant to strip bark, crush the trunks of baobabs and to dig. Besides the tusks the elephant has one or two molar teeth on either side of each jaw. Whilst one is being worn down, it is being replaced by another. This may be repeated six times during the elephant's lifetime. Though the skin is very thick the elephant has special scent glands on the temples. From the age of two or so they produce a dark secretion which is a kind of personal identification mark for each elephant. When two elephants meet they put their trunks to one another's temples and sniff each other. Elephants form herds often comprising several hundred individuals. Females with young have a privileged position in the group. Besides the mother, the young elephant is cared for by 'aunts'. The gestation period is from 19 to 22 months. The female bears only a single offspring which measures about one metre (3 ft) and weighs approximately 90 kg (200 lb), and nurses it for two years. The young elephant reaches maturity at the age of 13 at the earliest, though generally much later. The African elephant may attain gigantic proportions. Specimens 3.5 metres (11 ft) high at the back are not exceptional.

Ivory has always been much in demand as a raw material and tusks were always precisely weighed and measured so that today we have ample detailed data. The longest known tusk was 3.45 metres (11 ft 3 in) long measured on the outside arc. The heaviest weighed 117.5 kg (263 lb).

African elephant

Wild boar

It is generally considered that elephants are very long-lived but according to the latest investigations the average age appears to be 40 to 50 years and the highest age, an exceptional instance, some 70 years. The adult elephant has no natural enemies other than man. Young elephants may very occasionally be attacked and killed by lions.

Wild boar
Sus scrofa

Pigs are even-toed ungulates with a simple stomach, that is, they are not ruminants. They have a more varied diet than most ungulates; besides plant food they eat anything edible — invertebrates and small vertebrates, birds' eggs and carrion. On record are instances of wild boars killing and devouring even a roe-deer. The habitat of the wild boar is the forested regions of Europe, north Africa, temperate and southern Asia, through Manchuria to Japan. In the south-east it is found on the greater and lesser Sunda Islands. Its presence in the lesser Sundas (Lombock, Flores, Timor etc) is probably due to its deliberate introduction as a game animal. With such a large range it naturally occurs in a number of geographic races of diverse size and coloration. The largest wild

boars live in Siberia — the adult male weighs up to 350 kg (790 lb).

The wild boar lives in groups, only old males are of solitary habit, joining the females in early winter, the rutting season for them. The female bears from six to twelve young, with attractive lengthwise stripes, after a period of 16 to 20 weeks.

The domestic pig is descended from the wild boar. Its ancestors must be sought chiefly amidst the races native to southern Asia. Man has developed many races of domestic pigs for various uses.

Collared peccary
Dicotyles tajacu

On the American continent wild pigs developed separately and independently from the pigs of the Old World. They form a separate family with many different anatomical characteristics. At first glance it is evident that in peccaries, as these American pigs are called, the incisors of the upper jaw point downward, not outward and upward as in Old World pigs. Another, very striking feature is the large scent gland on the hind part of the back. Peccaries rub this against branches and other objects and also spread the secretion on their own bodies, mainly on the head and cheeks. Two animals approach one another, each with head turned towards the back of the other peccary, and with up-and-down movements of the head rub against the scent gland of their partner. The whole herd is thus marked by a kind of group odour. If the herd is joined by an intruder who does not have the same smell he is immediately recognized and chased away. Peccaries are found in South and Central America but the collared peccary has spread north as far as Texas. Hunters there call it the razorback. The female bears fewer young than pigs, usually only two. They are not striped, merely lighter in colour than the adults.

Collared peccary

Hippopotamus
Hippopotamus amphibius

The hippopotamus is closely related to the pigs. It is found only in Africa, in large rivers and lakes outside the region of deep tropical forests. It prefers shallow water with gently sloping shoreline where it emerges to graze. The hippopotamus spends the day in water or sunning itself on the shore or on islets. It comes out at night to graze, continually using the same paths which are deeply trampled. From these extends the pear-shaped territory where the hippopotamus grazes. Males mark their territories in a very interesting way. They urinate and defecate at the same time and spray the urine and faeces over the area with sharp, swift movements of the tail. Two males engaged in combat behave in the same way. They spray their faeces and threaten each other with wide open mouths. Serious battles are uncommon as a rule, but during the

rutting season the males often inflict bloody wounds with their long, sharp incisors. The female gives birth to the offspring, weighing about 30 kg (70 lb), in the water and nurses it there too. The hippopotamus swims and dives expertly. It is capable of remaining submerged for 4 to 6 minutes and if necessary for more than a quarter of an hour.

Bactrian or Two-humped camel
Camelus ferus

Camels, together with the llamas of South Africa, belong to a special group of primitive ruminants. The llama is strikingly different from the camel at first glance but closer investigation reveals that this is chiefly because it lacks one of the most conspicuous features of Old World camels — a hump. There are two kinds of camels, distinguished by the number of humps. One is the one-humped camel or dromedary, the other the two-humped or Bactrian camel. The dromedary is a traditional riding beast of burden in north Africa, Arabia and the Middle East. The Bactrian, on the other hand, was native to the large area extending from Asia Minor across central Asia north from the Himalayas to Mongolia and northern China. Occasional wild specimens are still found in the Gobi Desert even today. Elsewhere the camel is kept only as a domestic

Bactrian camel

animal. The humps of the camel are reserve stores of fat. A camel which is well-nourished has erect humps, in undernourished or diseased beasts the humps droop like empty sacks.

Giraffe *Giraffa camelopardalis*

Few animals in the zoo attract the attention of visitors as does this ungulate, and no wonder, for it is the tallest of mammals, generally measuring about 4.5 metres (14 1/2 ft) in height; some males even reach 6 metres (19 1/2ft)! Probably the most typical feature of the giraffe is its long neck which enables it to feed in the tops of acacias and other African trees. It is interesting to note that the neck skeleton, as in practically all mammals, is composed of only 7 vertebrae. These, of course, are greatly elongated. The giraffe, then has the same number of vertebrae as a man or a mouse. Its long flexible tongue is also put to good use in the treetops; the giraffe uses it to grasp twigs and strip them of leaves. Because it is a ruminant the giraffe, when it is

Giraffe

resting, regurgitates mouthfuls of food from the first stomach back to the mouth to chew it again. We can clearly see these mouthfuls, or cuds, passing up and down the long neck. The keen observer will notice other interesting things. What is the reason for the giraffe's strange gait, for instance? This is because it walks differently from a horse or a cow. It simultaneously moves the two legs on the same side of the body. This gait is called a pace or rack.

Few people can say they have seen a sleeping giraffe. Giraffes rest frequently, either standing or lying down, but they fall into a deep sleep only briefly — for only 1 or 2 minutes. The giraffe sleeps lying down with its head on its thigh.

If we observe a herd of giraffes we will discover they are peacable beasts. Their communities do not have a rigid structure and individuals often move from one herd to another. If two males engage in combat during the rutting season then this is only a symbolic contest in which the rivals butt one another on the shoulder. Young giraffes are without doubt some of the most delightful of young animals. The gestation period is 13 to 14 months. Usually the female bears only a single offspring but there have been several instances of the birth of twins. The mother drops her young while standing on all fours, which means it falls from a considerable height. It measures approximately 170 cm (5 1/2 ft) and weighs 40 to 50 kg (90 to 110 lb) at birth. The giraffe is a very quiet creature, only rarely does it sound a brief bellow or snort.

Red deer and Wapiti
Cervus elaphus

Antlers are a typical characteristic of the deer family, even though in certain primitive species such as musk deer and Chinese water deer they may be absent. In most deer only the male has this ornament on the head, but in reindeer both sexes sport antlers. The commonest deer species in Europe is the red deer. Its range extends to north Africa, the temperate regions of Asia and North America. The American and Siberian races are called wapiti. When the first Europeans reached America, they saw this large deer and named it after the largest of the European deer, the elk, and this is the name by which it is commonly known in North America.

The antlers of the red deer are greatly branched. Branching from the main stem, called the beam, are numerous points which may form a crown at the tip. The purpose of the antlers has not been fully explained as yet but it seems that they are a symbol of rank rather than a weapon. The males, or bulls, use the antlers in combats, which are mostly ritualized jousts. Red deer cast their antlers between February and April, young deer later than old bulls. In spring and summer, when their new antlers are growing, the bulls remain apart from the herd. In autumn the antlers are fully grown and hard and the bull cleans them. The rutting season begins and the bulls roar. From September till October deer visit peat bogs and muddy pools to wallow more than at any other time. Mating takes place during the rutting season in the autumn. In spring the cows nearing the end of their gestation separate from the herd and drop their young between the end of May and mid-June. The young calf is able to stand and follow its mother within a few minutes after birth. Its coat is covered with spots which do not disappear until it matures. The calf remains with its mother until the following autumn.

Elk or Moose *Alces alces*

The elk is known as the moose in America, the name elk having been previously applied to the wapiti by the early colonists. It is the largest animal of the deer family. It occurs in several geographic races in northern Europe, northern Asia and North America. The largest moose are found in Alaska and neighbouring parts of the Yukon and British Columbia. The Alaskan moose is up to 2.20 metres (7 ft) high at the withers and weighs more than 900 kg (200 lb). European elks are much smaller, about 170 cm (5 1/2 ft) high with a weight of approximately 600 kg (1350 lb). The elk has huge palmate antlers; those of a record Alaskan specimen measured 194 cm (6 ft 4 in) across! The record for the European elk is a 'mere' 122.5 cm (4 ft).

Elks are of solitary habit, only occasionally do they form small groups. They are fondest of dense stands of trees near water. In summer they feed on aquatic vegetation, chiefly the stems, leaves and flowers of water-lilies. Otherwise the elk feeds on the leaves of all deciduous trees, primarily the willow, poplar, aspen and birch.

The rutting season is in autumn, from mid-September to mid-October. Rival males (bulls) fight fiercely over a female. The victorious bull remains with the cow while she is in heat. Then he leaves her for another cow. The young are generally born in late May and June. A young cow usually bears a single offspring, older cows often have twins and sometimes even triplets.

The young elk is a defenceless creature and for a while, usually two or three days, remains in hiding. Soon, however, he ventures into water and death by drowning is not uncommon among young elks. Quite a few are also killed by beasts of prey, mostly bears. The mortality of young elks is high; scientists have discovered it to be between 25 % and 43 % in the first year.

It is interesting to note that in Sweden and the USSR efforts were made to tame the elk and use it as a draft beast to pull sleighs, as a beast of burden, and even as a saddle animal in the forests. The strength, endurance and speed of elks is extraordinary and the results of the experiments appear good and promising for the future.

Elk

Greater kudu

Bighorn sheep *Ovis canadensis*
The bighorn sheep was at one time found in large numbers from Alaska to Mexico. However it was hunted indiscriminately for its meat as well as for sport and in many places was either exterminated or forced to retreat into the inaccessible mountains. A great catastrophe for the bighorn were diseases brought to America with domesticated sheep. The bighorn was not immune to these diseases and its numbers were greatly reduced, all the more so because man had exterminated also the beasts of prey which had previously attended to the removal of weakened, diseased individuals.

The bighorn sheep is a large animal. The adult male is up to 105 cm (3 1/2ft) tall at the withers. The females have short, erect horns but those of the males are strong and curve downwards and outwards from the head. The horns of large males are about 1 metre (3 ft) long measured on the outside curve. The record is held by a trophy

Bighorn sheep

Greater kudu
Tragelaphus strepsiceros
There are many species of antelopes. Besides the few found in India, all live in Africa. The pronghorn of the American prairies, which is commonly called an antelope, is not a real antelope but a member of a closely related family. The smallest antelope is the royal antelope from the jungle forests of west Africa which is barely 30 cm (1 ft) high at the withers. The largest antelope is the Cape eland which is almost 180 cm (6 ft) high and weighs up to one tonne. However, the antelope with the largest horns is the greater kudu. The handsome, double-spiraled horns, a longed-for prize of hunters, measure 100 to 120 cm (3−4 ft). The world record, however, is a length of 178 cm (5 ft 9 1/2in)! The greater kudu is found in bush country throughout Africa south of the Sahara. In some regions it is very scarce, in others abundant.

measuring 125 cm (4 ft 1 in). During the rutting season the rams engage in combat. They approach each other on their hind legs, then drop down on all fours and butt each other with such force that the crash of the colliding heads may be heard from as far away as 1.5 kilometres (1 mile).

American bison *Bison bison*

The history of the American bison is one of the tragedies which fully proves how man's recklessness can destroy nature. At one time huge herds of bison roamed the North American prairies. It was estimated that they numbered 60 million beasts at least and that this may be a low figure. Bison were hunted by the American Indians but this did not threaten the existence of the herds. The real catastrophe for this animal was the arrival of the white man who began to kill bison in large numbers with modern weapons. The westward advance of white settlements and the building of a railway across the conti-nent brought with it professional hunters who supplied various companies, first and foremost the railway company, with the meat of these animals. One such, for example, was Billy Tilghman, who shot and killed 3,300 bison in seven months, another was the famous Buffalo Bill Cody, who chalked up 4,280 bison. The peak of the slaughter came in 1867 when the Union Pacific Railroad reached the town of Cheyenne in Wyoming, thus splitting the herds in two and preventing their migration. It furthermore made access to the herds easy for anyone who wished to try his hand at shooting. In the year 1893 only about 1,100 bison remained of the once immense herds and two years later their number was an estimated 800. A proposal for the preservation of the bison was placed before Congress as early as 1871 but was not approved. The year 1905 saw the foundation of the American Bison Society as a result of the efforts of William T. Hornaday, first director of the New York Zoological Park. He gathered together all the bison in captivity and offered to the government of the United States a small herd of 15 to be placed in the Wichita Forest Reserve in Oklahoma. The offer was accepted and the reserve was laid out for the nucleus of the herd. In October 1907 the 15 bison were let loose in a fenced-in area measuring 6,200 acres. The herd at Wichita thrived and slowly increased in number. Finally it was possible to move many bison to reservations founded in other states. Nowadays there are some 8,000 bison in the USA and there is also a large herd in Canada which was spared the worst ravages of hunters. The bison has now been saved, but its history should forever remain an ugly blot on the white man's conscience.

The American bison, like the European bison or wisent, is heavily-built, from 150 to 180 cm (5 to 6 ft) high at the humped shoulders and weighing between 360 and 890 kg (800—2,000 lb).

American bison

Return to the Water

Life had its beginnings in the sea. That is where the first living organisms originated and that is where most evolution took place. There came a time when some groups of animals left the water and gradually began to occupy the land and the air. We have seen in many examples the paths taken by this evolutionary development. Many organisms became extinct in the course of evolution, but many are at the peak of their development now. The tendency of most organisms is to adapt to the greatest possible variety of environments in which they can live and reproduce. Fish are typical aquatic animals but still we find species that also leave the water for land. Likewise some birds, which are at home in the air, have left it for other environments, returning, for instance, to the water from which their distant ancestors came. The same is true of some mammals. The new way of life so changed the shape of their body that at first glance we don't recognize their true kinship.

Penguins

Birds in tailcoats — this is how most people see the penguin. These are birds which have lost the ability to fly but have compensated for this by being excellent swimmers and divers. They can even compete with the dolphins. They are much more at home in water than on land. Their wings have been transformed into powerful paddles covered with small, scaly, close-fitting feathers.

Penguins are native to the southern hemisphere. Most species in-

Penguins

habit the cold waters of the Antarctic but some occur along the coast of Australia, South America and Africa far to the north. The largest is the emperor penguin, which reaches a height of 120 cm (4 ft). It lays only a single egg which is kept warm in a fold of the belly skin which overlaps the bird's feet. The male and female take turns incubating. This penguin does not build a nest. The Gentoo penguin, which inhabits islands in the Antarctic, builds a nest on top of a mound of earth and lines it with grass and other available plant material.

Californian sea lion
Zalophus californianus

Seals and sea lions and related animals belong to the order Pinnipedia, which means 'having winged feet' in Latin, exactly describing their fin-like feet. It may be said without exaggeration that they are flesh-eating mammals which have adapted to life in water. The

Californian sea lion

Manatee

body is torpedo-shaped and the fin-like feet or flippers so flexible that they can almost twist through a complete circle.

Manatee *Trichechus manatus*

At first glance you might think the manatee was some kind of seal, so similar is the body. However, it feeds on aquatic vegetation, including the water hyacinth which often clogs waterways. The manatee has nothing in common with seals and flesh-eating mammals. It is an un-gulate which has returned to the water. Its appearance as well as the shape of the body were moulded by its aquatic environment. It is

a member of a separate order of mammals which exhibits a marked kinship to the elephants.

Blue whale

Balaenoptera musculus

Almost all mammals which left the land and returned to an aquatic en-vironment spend at least a short while on land. The whales are an exception. Their adaptation to life in water has gone so far that if they are accidentally cast up on shore they die. The blue whale is the largest living mammal. The male,

which is generally smaller than the female, reaches 32 metres (104 ft) in length. The largest female ever caught measured 33.27 metres (108 ft). The weight of such a colos-sus is about 200 tonnes. This gigan-tic animal, found in all the oceans between the Arctic and Antarctic, feeds on very small animals. It strains the water-borne food through the whalebone in its mouth and can swallow small crustaceans,

Blue whale

molluscs, squid and fish up to the size of herrings.

The newly-born whale measures 6 to 9 metres at birth. The female bears one, very occasionally two offspring after a gestation period of one year.

Botany

The world of plants is immensely rich and varied and there is a vast range of size, shape and complexity. The simplest plants consist of single cells that can be seen only under a microscope. Contrast these with the massive giant sequoias of California, which grow to a height of over 80 m (262 ft) with girths of more than 20 m (66 ft), and the tall coast redwoods of California, which grow to over 100 m (328 ft) in height.

Like animals, plants are classified by biologists into several groups. At the lowest level there are the algae, which include plants ranging from single cells to the more advanced seaweeds. Such plants are distinguished not only by their simple body forms but also by simple methods of reproduction. Going up the scale, there are the bryophytes (mosses and liverworts), ferns, clubmosses and horsetails. Each of these groups demonstrates evolutionary advances that took place hundreds of millions of years ago. Their plant bodies are progressively more complicated, as are their methods of reproduction.

An important factor in plant evolution was the gradual change from being dependent on water. For example, most algae must have a wet environment in order to survive and reproduce. A fern, on the other hand, has a plant body that does not easily dry up and a moist environment is needed for only a short period during its life cycle — when a male sex cell has to swim to a female

sex cell in order to fertilize it. The climax of this evolutionary change can be seen in the gymnosperms and the flowering plants. They have evolved reproductive methods that ensure that fertilization takes place without the need for water and many are able to survive drought conditions.

Gymnosperm is a term which means 'naked seed' and gymnosperms include the familiar conifers and a number of other plants that produce seeds that are exposed to the air. However, there are few gymnosperms in comparison to the number of flowering plants, or angiosperms ('enclosed seeds'). This group alone contains more species than all the other groups put together, and they have colonized most parts of the world.

How plants work

A typical flowering plant consists of a root, stem and leaves, each of which has its own functions.

The root provides the plant with an anchorage and takes water from the soil. Typically, a plant has a main, or primary root that extends downwards from the stem and divides into smaller secondary roots. The finest divisions of all are only one cell thick and these are called root hairs. They penetrate the gaps between the soil particles and absorb water, which contains dissolved minerals. The water passes into the conducting vessels (xylem) of the root and then into the stem.

The stem supports the leaves and provides a link between them and the roots. Water passes

Botany

up the xylem of the stem to the leaves and dissolved organic substances pass up and down the other conducting vessels — the phloem.

The leaves are the most important part of the plant. It is here that two vital processes — transpiration and photosynthesis — take place.

Transpiration is the evaporation of water from the leaves. This occurs via small pores, called stomata (singular: stoma) on their undersides. Transpiration is necessary because it draws water up from the roots, thus keeping the whole plant well-supplied with water and minerals. At the same time transpiration helps to keep the plant cool — on hot days plants lose much more water than on cool days.

Photosynthesis is the process in which plants make their own food. Only plants can do this and, in fact, this very special process is the basis of all life on Earth — plants create food; animals eat plants or other animals.

Photosynthesis, like most other processes that occur in living organisms, is a long chain of chemical reactions. However, it can be summed up simply as the building up (synthesis) of carbon dioxide (CO_2) and water (H_2O) into simple sugars, such as glucose ($C_6H_{12}O_6$). Oxygen (O_2) is released during the process and leaves the plant via the stomata.

Energy is needed to drive this chemical process

The products of photosynthesis are carried down from the leaves to the roots. Water passes up the stem to the leaves.

and this is supplied in the form of light energy from the sun (hence the 'photo' part of the word 'photosynthesis'). Inside the cells of the plant's leaves are tiny bodies called chloroplasts which contain a green pigment called chlorophyll. This pigment converts the light energy into chemical energy which can then be used to drive the reaction.

Much of the sugar made during photosynthesis is used to make starch, a storage compound. This is held in reserve and can be used to make cellulose — the material in the walls of plant cells — or it can be converted back into sugar for use in respiration.

Respiration is basically the reverse of photosynthesis. Oxygen from the air is made to react with the sugar to form water and carbon dioxide. The chemical energy released during the process is used to drive other chemical reactions that take place in the cells of the plant.

The processes of photosynthesis and respiration result in a continuous exchange of gases between the atmosphere and a plant — via its stomata. As respiration takes place both plants and animals take in oxygen and release carbon dioxide. But during the day, when it is light and photosynthesis is taking place, plants take in carbon dioxide and release oxygen. They release considerably more oxygen than they take in and so the net result is that the plant life of the Earth replenishes the oxygen supply of the atmosphere.

Sugars, starch and cellulose are all carbohydrates — chemicals composed of the elements carbon, hydrogen and oxygen. But other chemicals are also essential to plants. In particular, nitrogen is needed in order to build up proteins. Unfortunately, none of the higher plants can use the vast quantities (78 per cent) of nitrogen in the atmosphere and therefore most of them have to rely on nitrates from the soil. Leguminous plants (such as peas and beans), however, have nodules on their roots. These contain bacteria of the genus *Rhizobium* that can 'fix' atmospheric nitrogen and help convert it into amino acids — the building blocks of proteins. Other important elements that plants take from the soil are potassium, calcium, phosphorus, magnesium, sulphur and iron.

Using the chemicals available to them, plants manufacture a number of products in their cells. Many of these are waste-products, but some of them actually benefit plants — poisons, such as alkaloids, and bitter substances, such as tannins and glycosides, discourage grazing animals. Some plants specialize in the production of certain chemicals and such plants can be useful.

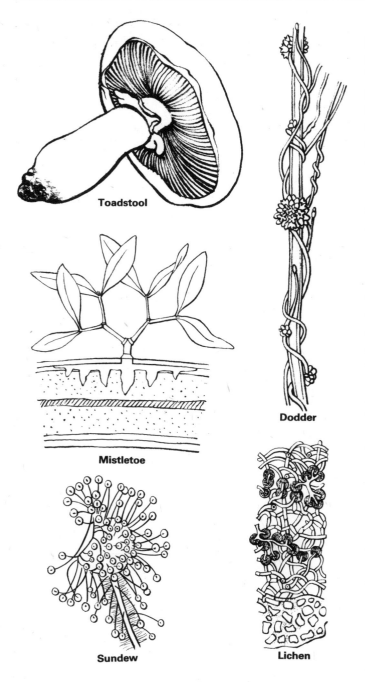

Toadstool

Dodder

Mistletoe

Sundew

Lichen

Toadstools are the fruiting bodies of saprophytic fungi. Dodder and mistletoe are parasites. Sundews are carnivorous plants. A lichen is a symbiotic association between an alga and a fungus.

Other ways of obtaining food

Not all plants are green. Notable examples of plants that lack chlorophyll are the fungi. In fact, some biologists argue that they are not plants at all and classify them apart from the animals and plants in a third kingdom, the Protista, together with the bacteria, the protozoans (single-celled animals) and single-celled algae. Whatever the case, fungi cannot make their own food and have to obtain ready-made organic substances.

Many fungi and bacteria play an important role in the cycle of life. They are saprophytes, organisms that live on dead plant and animal tissue. To do this, they produce enzymes (organic compounds that assist chemical reactions without themselves being used up) and break down the organic compounds into compounds they can absorb. As a result of their activities they produce simple inorganic compounds that can be used by living plants.

Some saprophytes form special relationships with other plants that are of mutual benefit to both partners. Such an association is called symbiosis. The relationship between a leguminous plant and the bacteria *(Rhizobium)* in its root nodules is a symbiosis — the bacteria make nitrogen available to the plant and receive carbohydrates and other food substances in return. Fungi also form symbiotic relationships. For example, a lichen may look like a single plant, but it is in fact a combination of an alga and a fungus. The alga makes food for both of them by photosynthesis and in return receives minerals and protection from bright light and dry conditions.

A number of fungi form symbiotic relationships with trees and other flowering plants. The mycelium of the fungus, which consists of a mass of tiny threads, or hyphae, becomes inextricably entangled with the roots of the plant, forming a mycorrhiza. The benefits of this kind of association, which is often found on poor soils, are not fully understood. However, the fungus appears to receive simple organic substances and assists in the growth of the tree.

Many plants cheat in the way they obtain food by taking it from other living plants. Such plants are called parasites. They include a number of fungi and several flowering plants. Dodder, for example, attacks a variety of other plants. It grows normally as a seedling until it finds a suitable host. Then it buries club-shaped structures called haustoria in its victim, its roots wither and die and the dodder plant then relies entirely on its host for food. Mistletoe is another flowering plant parasite. However, as it makes its own food by photosynthesis and takes only water and dissolved minerals from its host tree, it is described as a partial parasite.

Another group of plants has come up with an ingenious solution to the problem of living in nitrogen-poor habitats — they have become carnivorous. Plants such as Venus' fly traps, sundews, butterworts, bladderworts and pitcher plants trap insects, digest their tissues and absorb the resulting chemicals.

Botany

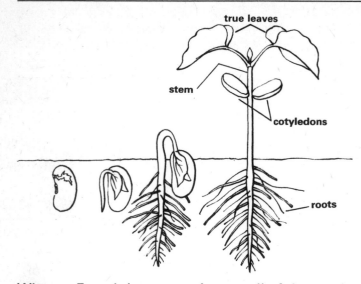

When a French bean germinates, all of the seed, except for the radicle, emerges from the soil. Although the cotyledons turn green, they do not become leaf-like and play no part in photosynthesis. Instead, the seedling quickly grows true leaves and the cotyledons shrivel.

From seed to plant

Most flowering plants reproduce by means of seeds, which are produced by sexual reproduction as opposed to asexual, or vegetative, reproduction. A seed consists of an embryo inside a seed coat. The embryo comprises a radicle, which will grow into the first root, a plumule, which will grow into the first shoot, and one or two cotyledons, or seed leaves. The seed may also have a separate food store, known as the endosperm, in which case it is described as endospermic. On the other hand, the seed's food reserves may have been absorbed by the cotyledons, in which case the seed is described as non-endospermic. In either case, the growing seedling uses the food reserves until it is able to make all its own food requirements.

Germination requires suitable conditions of temperature and moisture. When such conditions are right, the seed absorbs water, swells and begins to burst out of its seed case. Then the radicle begins to grow down into the soil and eventually branches and anchors the growing seedling. The plumule starts to grow upwards. Seeds germinate in one of two ways. In hypogeal germination the cotyledons remain inside the seed case (e. g. broad bean). The term hypogeal means 'underground', but the cotyledons are only likely to remain underground if the seed is planted or accidently covered with soil. In epigeal germination, on the other hand, the cotyledons emerge from the seed case, turn green and begin photosynthesis (e. g. castor oil plant).

The radicle forms the first root. The tip of the root, the point where growth takes place, is covered with a thimble-like structure called the root cap. This protects the root tip as it forces its way between the particles of soil. The part above the root cap is thickly covered with fine root hairs, which absorb water from the soil. These function for only a few days. The plumule forms a shoot, which quickly develops into a stem bearing the first two foliage leaves. Germination is now complete and growth of the young seedling begins.

Roots

The roots of a plant anchor it in the ground. Some plants have strong primary roots and such roots often penetrate deep into the ground in search of water. Plants such as the carrot have few or no secondary roots at all and their thick fleshy tap roots contain large food reserves. Roots that store food may also be round and fat and the roots of dahlias and orchids are swollen into tubers. Some plants, on the other hand, do not have obvious primary roots. For example, the primary root of the spruce is greatly reduced and its root system is shallow; the tree is held in position by means of strong lateral roots. Grasses and many other monocotyledons have bundles of fine roots.

Various kinds of roots.

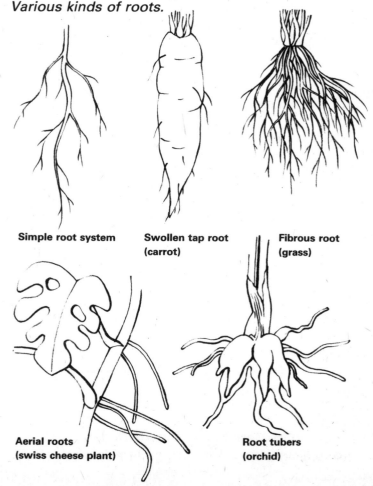

Simple root system　　**Swollen tap root (carrot)**　　**Fibrous root (grass)**

Aerial roots (swiss cheese plant)　　**Root tubers (orchid)**

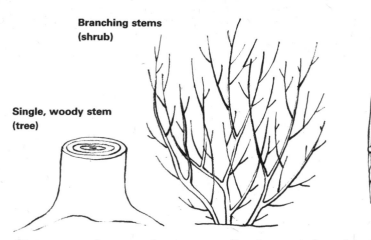

Single, woody stem (tree)

Branching stems (shrub)

Green stem, hollow except at nodes (grass)

Runner, or stolon (strawberry)

Fleshy, swollen stem (cactus)

Stems may be woody or green (herbaceous) and there are a number of modifications of both types.

The main function of roots is the absorption of water and nutrients from the soil. The roots of many plants, however are variously modified and serve other functions as well. They may, for example, serve as a buttress or they may breathe. The breathing roots of the swamp cypress protrude like pegs from the mud. The aerial roots of some tropical epiphytes (plants that live in the branches of trees) are filled with spongy tissue that absorbs water from the atmosphere.

Some climbers have special roots that can grow into walls or the bark of trees. Parasites such as mistletoe and dodder have roots called haustoria that they bury in the tissue of their hosts. Lower plants have only root-like filaments called rhizoids by which they anchor themselves.

Stems

The size and structure of a plant stem depends on the plant's particular needs. It grows at the tips of its shoots and at the same time becomes thicker and stronger. The strongest stems of all are the trunks of trees.

Young stems and the stems of herbaceous (non-woody) plants are green and supple. However, to keep them upright they are often strengthened in places by columns of thick-walled fibres. Collenchyma fibres are living cells thickened only in the corners; sclerenchyma fibres are dead cells with very thick walls. Plants that have perennial stems above the ground generally form wood. Such plants include trees, which have a single stem arising from the ground, and shrubs, which have several stems arising from or near the ground. In sub-shrubs only the base of the stem is woody and the upper, herbaceous parts die every year.

Wood is actually the xylem (water conducting tissue) of the plant. Each year another layer of xylem is added around all parts of the tree and the older, more central xylem vessels become blocked up. If a tree is cut, these annual rings of xylem can be seen — counting the annual rings in a tree trunk therefore gives its age.

There are many modifications of stems. Cacti have thick, fleshy stems for storing water. Some plants have weak stems that creep along the ground, twine round a support or cling on by means of tendrils. The spines of hawthorn and blackthorn are modified twigs. Some plants produce long, thin stems called runners, which serve as a means of vegetative reproduction. Modified stems may even be found underground. The 'root' of celeriac is actually the swollen base of the stem combined with the swollen upper part of the root. Rhizomes are long, horizontal underground stems that are both organs of vegetative reproduction and storage organs. The long, thin rhizomes of the potato plant swell up at intervals to produce potato tubers. Bulbs are short stems with thick, fleshy leaves and corms are swollen stems whose leaves are reduced to scales. Both of these are also organs of vegetative reproduction.

Leaves

Considering that the main function of a leaf is the same in nearly all plants there is an amazing variety of leaf shapes. Generally, however, dicotyledons (plants that have two cotyledons in their seeds) have broad leaves with a network of veins; monocotyledons (plants with one cotyledon in their seeds) have long, thin leaves with parallel veins. Also a dicotyledon leaf is usually attached to the stem by a leaf stalk, or petiole. The leaves of monocotyledons generally do not have petioles. The leaves of grasses are divided into two parts, the blade and the sheath, which surrounds the stem and swells at the base to form a node. Where the blade joins the sheath there is often a ligule present and sometimes the blade divides to form two pointed auricles.

Various terms are used to describe the shapes

Botany

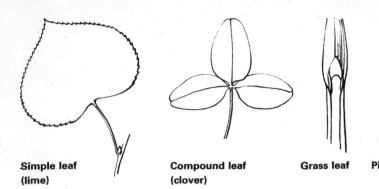

Simple leaf
(lime)

Compound leaf
(clover)

Grass leaf

Pine leaf

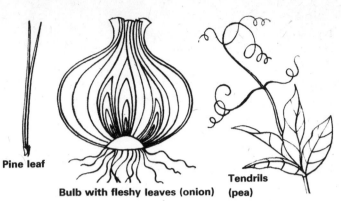

Bulb with fleshy leaves (onion)

Tendrils
(pea)

Leaves are basically organs of photosynthesis, but their shape varies considerably. There are also many leaves modified for purposes other than photosynthesis, such as storage, climbing and protection.

of leaves. For example, the long, thin leaves of grasses are described as linear. However, the greatest variety of leaf shape is found among the dicotyledons. Terms used to describe these include lanceolate (thin, spear-shaped), ovate (oval), orbicular (round) and palmate (deeply divided into large lobes). The term pinnate is often used. This indicates that the leaf is divided into a number of separate leaflets. Some leaves have one or more projections from the base called stipules.

The arrangement of leaves on the stem also varies. They may be opposite (two arising from the same point, or node, on the stem, one on each side), alternate (one arising from each node, first on one side then on the other side of the stem) or whorled (arranged in rings up the stem). Many plants have a rosette (leaves arising very close to each other on the stem) near the ground.

There are also many modifications of leaves. The leaves of an onion bulb store food. The two fleshy leaves of a pebble plant store water. The thorns of barberry, the tendrils of the pea and the 'root filaments' of the water fern are all modified leaves. Small modified leaves in which photosynthesis is only a secondary function, if it occurs at all, are called cataphylls. These include the outer

protective leaves of a bud. A small cataphyll at the base of a flower stalk is called a bract and one that is borne on a flower stalk is called a bracteole.

Flowers

Flowers are also modified leaves. The ancestors of flowering plants had special spore-bearing leaves which gradually evolved into the various parts of the spectacular flowers that exist today.

At the end of a flower stalk (pedicel) is a struc-

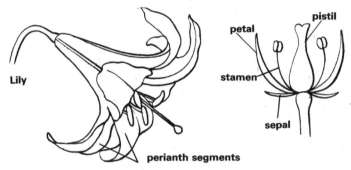

Lily

pistil

petal

stamen

sepal

perianth segments

The sepals and petals of a lily flower are indistinguishable and form a perianth. Other flowers have green sepals and coloured petals.

ture called the receptacle. Attached to this are the sepals, petals and the reproductive parts of the flower. The sepals are the outer, leaf-like structures which protect the flower when it is in bud. Collectively, they are known as the calyx. They open to reveal an inner ring of petals, which are often brightly coloured and are known collectively as the corolla. Sometimes the sepals, too, are

Various corolla shapes.

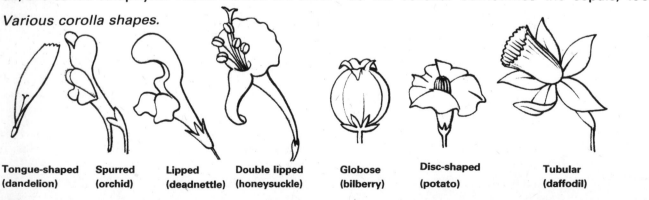

Tongue-shaped
(dandelion)

Spurred
(orchid)

Lipped
(deadnettle)

Double lipped
(honeysuckle)

Globose
(bilberry)

Disc-shaped
(potato)

Tubular
(daffodil)

coloured and it is impossible to distinguish between them and the petals, in which case the whole structure is known as the perianth.

Flowers may be grouped according to their shape. Those that can be cut vertically in half in any direction are described as regular, or actinomorphic. Those that have only one plane of symmetry are described as irregular, or zygomorphic. The petals of the corolla, or even the whole perianth, may be separate. In many flowers, however, they are united and form a tube.

The reproductive parts of the flower are the reason for its very existence. The male parts are the stamens. Each stamen consists of a long filament on the end of which is an anther. This is a two-lobed structure and in each lobe there are

two sac-like chambers that contain pollen. Each pollen grain contains a male nucleus. The female parts of the flower are the carpels, collectively known as the pistil. At the top of each carpel there is a stigma – a pollen-receiving surface – which is often on the end of a long style. Inside the carpel is an ovary, in which there are one or more ovules, each contains a female sex cell, or ovum.

Not all flowers, however, have both male and female parts. Single-sex flowers are quite common. Male or staminate flowers have only stamens; female or pistillate flowers have only carpels. Plants that have separate male and female flowers on the same plant are called monoecious. Those that have separate male and female plants are called dioecious.

Flowers may be single, but more often they are arranged in groups, or inflorescences. The names of inflorescences are given according to the way the branches are arranged. There are basically two main types – racemes and cymes.

A raceme has a main growing axis that bears stalked flowers. However, there are a number of variations. For example, a spike is a raceme in which the flower stalks are almost non-existent; a spadix is a form of spike with a fleshy main axis; and a catkin is a hanging spike with single-sex reduced flowers. A panicle is a compound raceme; a corymb is a raceme with flowers borne at the same level; an umbel is a raceme in which the main axis is reduced to virtually nothing; and a capitulum is a raceme in which the axis has become flattened and has expanded sideways. The inflorescence of a grass is among the most complicated of all. Basically it consists of a number of spikelets, each of which has several parts. At the base there are small leaf-like structures called glumes. These are sometimes equipped with bristles, or awns. Above the glumes are

Various types of inflorescence.

RACEME-TYPE INFLORESCENCES

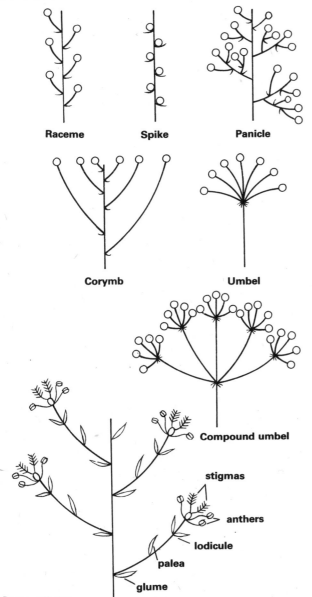

Raceme **Spike** **Panicle**

Corymb **Umbel**

Compound umbel

stigmas

anthers

lodicule

palea

glume

Grass spikelet

CYME-TYPE INFLORESCENCES

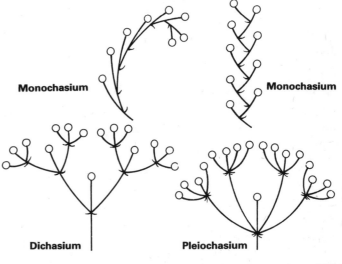

Monochasium **Monochasium**

Dichasium **Pleiochasium**

Botany

bracts called lemmae and in the axils of these are the flower stalks. Each flower stalk has a bracteole, or palea. The perianth segments, or lodicules, are small and insignificant. Above these are the pendulous stamens and the pistil with its feathery stigmas.

In a cyme there is no growing main axis. Instead growth takes place by branching and the name given to the inflorescence depends on the number of branches that arise from a single point — one (monochasium), two (dichasium) or several (pleiochasium).

Pollination

Pollen is transferred from an anther to a stigma by one of several agents. In some tropical countries birds, such as hummingbirds, and mammals, such as bats and bushbabies, are pollinating agents. A few plants are pollinated by water. But the majority of plants are pollinated by wind or insects.

Wind-pollinated plants have hanging anthers that produce vast quantities of pollen. This is to ensure that at least some pollen grains will land on the stigmas, which are often large and feathery so as to present the largest possible surface area for the pollen to land on. However, the flowers of wind-pollinated plants, such as grasses and catkin-bearing species, are generally small and insignificant. Insect-pollinated plants, on the other hand, have brightly coloured flowers to attract their pollinators. Insects are particularly attracted by shades of blue and mauve and some insects respond to ultra-violet colours that are invisible to humans. The reward a visiting insect receives is usually nectar, which is produced by glands called nectaries at the bases of the petals.

When a pollen grain lands on a stigma, it grows

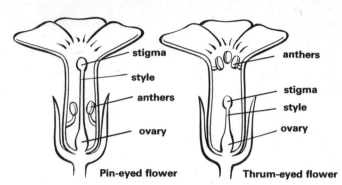

There are two kinds of primrose flower. When an insect visits a pin-eyed flower, it gathers pollen on the front end of its body. When it visits a thrum-eyed flower, this pollen is rubbed on to the short stigma. At the same time it collects pollen on the hind end of its body.

a long tube down through the style to the ovary. The male nucleus then passes down the tube and fuses with the female nucleus in an ovule. This process is known as fertilization. But the strongest and healthiest plants result from the fusion of sex cells from different plants. Thus cross-pollination is more desirable than self-pollination.

Accidental self-pollination undoubtedly does occur, but many plants take steps to prevent it or, at least, to encourage cross-pollination. Many wind-pollinated plants have single-sex flowers — either on the same plant (monoecious) or on separate plants (dioecious). Insect-pollinated plants have a number of ways of ensuring cross-pollination and some of them are very elaborate. For example, the parts of a pansy are organized into a complex structure. An insect can enter the flower in only one way. As it enters, it moves

Indehiscent, dehiscent and false fruits.

INDEHISCENT FRUITS

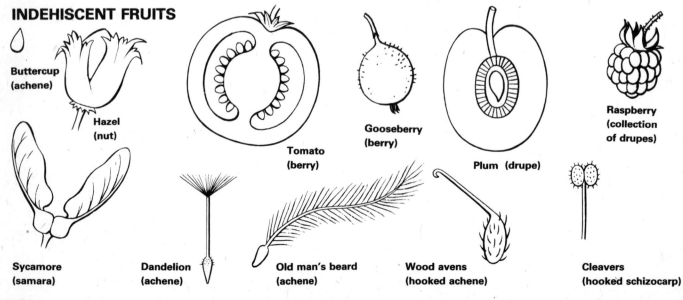

Buttercup (achene)

Hazel (nut)

Tomato (berry)

Gooseberry (berry)

Plum (drupe)

Raspberry (collection of drupes)

Sycamore (samara)

Dandelion (achene)

Old man's beard (achene)

Wood avens (hooked achene)

Cleavers (hooked schizocarp)

a flap, exposing the stigma — and any pollen on the insect is deposited on the stigma. As the insect leaves, the stigma is covered again. At the same time the insect is showered with pollen, which it then carries to another flower. Other insect-pollinated plants produce more than one kind of flower.

Some plants have flowers in which the male and female sex organs ripen at different times. Pollination occurs when pollen from the ripe anthers of one plant is deposited on the stigma of a flower with a ripe ovary.

Fruits and seeds

After fertilization an ovule develops into a seed and the wall of the ovary ripens into a fruit. Sometimes other parts of the flower are involved in fruit formation, in which case the structure is known as a false fruit.

There are a number of types of true fruit. They differ in the origin and structure of the pericarp (the outer part of the fruit derived from the ovary) and in the number of seeds they contain. There are two main groups: indehiscent fruits, which are dispersed with the seeds inside them; and dehiscent fruits, which open in an organized way to release their seeds.

Indehiscent fruits may be dry or fleshy. An achene is a one-seeded, dry fruit formed from a single carpel; the fruits of buttercup and sycamore are achenes (the pair of winged fruits of the sycamore is called a samara). The dandelion fruit is also known as an achene, although this is not strictly true; its origins are slightly different from the others. A nut is also a one-seeded, dry fruit, but it is derived from two or more carpels and usually has a woody pericarp; the fruits of hazel and oak are nuts. A berry is a fleshy fruit, usually

many-seeded, in which the pericarp consists of three layers — a skin or epicarp, a fleshy or pithy mesocarp and a membranous endocarp. Examples include gooseberry, bilberry, tomato, cucumber, banana and orange. A drupe is a one-seeded, fleshy fruit in which the seed is surrounded by a hard endocarp (the stone). Examples include plum and peach, but many so-called berries and nuts are also drupes. For example, the blackberry is a collection of tiny drupes and the coconut is a drupe with a fibrous epicarp and mesocarp (usually removed before the coconut is exported and sold).

All dehiscent fruits are dry when ripe. The most common type is the capsule. Most capsules split down their sides, but poppy capsules have pores through which the seeds are shaken out, and the top of a plantain capsule comes off like a lid. The fruits of the cabbage family have capsules in which a membrane forms down the middle. When ripe, the capsule splits into two halves, exposing the seeds on the membrane. Long, thin capsules of this type (e. g. wallflower) are called siliquas; short, pouch-like ones (e. g. shepherd's purse) are called siliculas. Pods, or legumes, are fruits that split along two seams; well-known examples include the pea and laburnum. Follicles, on the other hand, split down one side only; the fruits of monkshood are follicles. Schizocarps are fruits that break into pieces; each piece consists of a seed with a portion of the pericarp.

Examples of false fruits include the apple, the strawberry and the rose hip. Only the core of an apple is the true fruit; the remainder of the pulp is formed from the swollen receptacle. In a strawberry the red, fleshy receptacle bears tiny achenes on the outside. In a rose hip the achenes are contained within the receptacle. Sometimes com-

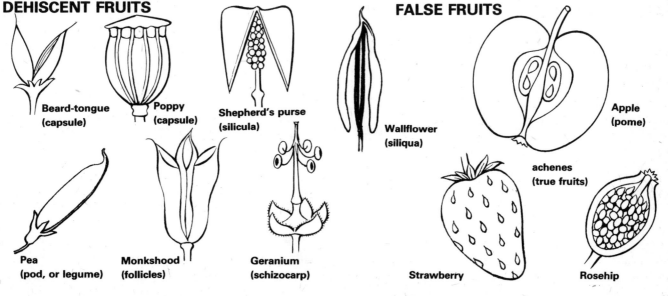

DEHISCENT FRUITS

Beard-tongue (capsule)

Poppy (capsule)

Shepherd's purse (silicula)

Pea (pod, or legume)

Monkshood (follicles)

Geranium (schizocarp)

FALSE FRUITS

Wallflower (siliqua)

Apple (pome)

achenes (true fruits)

Strawberry

Rosehip

pound fruits are formed from all the flowers of an inflorescence. The resulting common fruit mass may include part of the stem, the bracts, sepals and petals. Examples of such compound fruits include the pineapple and fig.

Seed dispersal

To colonize new areas and give seedlings sufficient space in which to grow, plants have to disperse their seeds as widely as possible. Wind is an excellent dispersing agent and many plants have seeds that are small and light enough to be blown by the wind. Some fruits have parachutes or wings. The hairy pappus of the dandelion fruit and the long plume of the fruit of old man's beard are both forms of parachute. Winged fruits include those of sycamore, maple and ash.

Many seeds are dispersed by animals. Juicy fruits are eaten and the seeds pass through the animals unharmed. Hazel nuts are stored by squirrels; those that are forgotten germinate in the store. A number of fruits have hooks that catch on to animals' fur.

Some plants disperse their seeds by using explosive mechanisms. When a pea pod dries out, its two halves twist apart and hurl the seeds out. The pods of the touch-me-not burst open as a result of water pressure building up inside.

The distribution of plants

The two main factors that govern plant distribution are climate and geography. The climate of a particular region, which is itself affected by local geography, includes such physical factors as temperature and rainfall. These determine the nature of the environment and hence the type of vegetation that can be found there. Biologists divide the natural world into a number of biomes — major ecological communities of plants and animals. Land biomes include tropical, deciduous and coniferous forest, tropical and temperate grassland, montane regions and polar regions.

Temperatures on the Earth's surface tend to increase from the Poles towards the Equator and thus it is not surprising that the hottest environments are found in the equatorial region. Tropical forests are found in areas where the temperature is over 27°C all year and there is a high rainfall. In tropical rain forests the growth of plants is continuous and the trees are usually evergreen. Many exotic species of plants, such as epiphytic orchids, are found in such forests (epiphytes are plants that live on the branches of trees).

Where the rainfall decreases, rain forests give way to tropical grasslands, or savannah. Here, there are many species of herbaceous plants, together with clumps of bushes and shrubs. However, areas in which the rainfall is less than 250 mm (10 in) a year become desert or semi-desert. This type of environment is one of the world's harshest; the temperature of desert sand can reach 70°C. Only plants that have special adaptations can survive such conditions.

The temperate regions of the world have more moderate rainfalls, but they also have distinct seasons. Depending on the local geography, summer temperatures vary from 10°–30°C and winters may be mild or very cold. Temperate forests are therefore either deciduous or coniferous. Deciduous trees lose their leaves in autumn and overwinter in a state of dormancy (sleep).

Coniferous trees are found throughout the northern hemisphere, but most coniferous forests form a wide band, sometimes called taiga, below the Arctic Circle. Here the rainfall is low, the winters are long and cold and the summers are short and cool. However, conifers can tolerate poor soil conditions and lack of water.

The coldest areas of the world are the polar regions. The Antarctic is largely covered with ice, on which no plants can grow. However, a small number of plants manage to survive on the few ice-free coastal strips. The ice-free region around the North Pole is known as the tundra. Here, a number of plants are found, including mosses, lichens and several flowering plants. They survive temperatures as low as −60°C.

Mountains have zones of vegetation, which, because temperature decreases with altitude, parallel the main vegetation zones found in the rest of the world. At the bottom the vegetation is the same as in the surrounding region. Higher up there are bands of coniferous forest and alpine grassland. Above these and below the snow line there is a region of montane vegetation, which is similar to tundra. Alpine grassland plants have distinct adaptations. They often form rosettes that grow close to the ground to avoid the harsh winds and to benefit as much as possible from the warmth that the ground absorbs from the sun. Also they are often covered with hairs, which help to retain warmth.

Today each species has its own area of distribution into which it has spread from its place of origin. Some species are confined to particular areas, where they are said to be endemic. However, most plants have spread to a number of areas. Only a few land plants have a world-wide distribution, and most of these are weeds that have been spread by man.

PERIOD	MILLIONS OF YEARS AGO
QUATERNARY	1.8
TERTIARY	
	65
CRETACEOUS	
	135
JURASSIC	
	195
TRIASSIC	
	230
PERMIAN	
	280
CARBONIFEROUS	
	345
DEVONIAN	
	410
SILURIAN	
	440
ORDOVICIAN	
	530
CAMBRIAN	
	570
PRE-CAMBRIAN TIMES	

BACTERIA, ALGAE LICHENS AND FUNGI
MOSSES
PSILOTES
CLUBMOSSES
HORSETAILS
FERNS
GINKGOS
CONIFERS
BENNETTITALES
CYCADS
FLOWERING PLANTS
CORDAITALES
SEED FERNS (PTERIDOSPERMS)
PSILOPHYTES

•••••••• = POSSIBLE ORIGINS
GEOLOGICAL TIME SCALE IS DRAWN TO SCALE.
WIDTH OF COLUMNS INDICATES APPROXIMATE RELATIVE ABUNDANCE.

Spring Flowers

The arrival of spring in Europe is heralded by several flowers. Some steal a march on the others by appearing when the temperature is still below freezing point. But, as the sun warms the ground they are soon followed by others that demonstrate that winter is really over.

A common characteristic of all spring flowers is that they have reserve supplies of food stored in tubers, rhizomes or bulbs. It is this that makes it possible for them to grow, blossom and bear fruit as quickly as they do. Early-flowering species generally like the damp soil of riverine woods and deciduous forests. Before the trees put out new leaves, spring plants have had time to flower and finish growth. Their leaves and stems then dry up and their life processes are again confined to their underground parts until the following spring.

Lesser celandine *Ficaria verna*
The growth and development of this plant is very rapid, usually from April, but sometimes as early as March, until May. During that time lesser celandine manages to form clumps of leaves and bear a profusion of flowers. It produces only a small number of seeds and sometimes none at all. However, this drawback is offset by its ability to multiply readily by vegetative means, by small bulbs or bulbils. These are of two kinds. One kind is produced on the root system underground, where some of the roots swell into club-shaped storage organs. The other kind are white bulbils formed in the axils of the lower leaves. They look like cereal grains. When the leaves and stem die, they fall to the ground and are washed away by rainwater. They overwinter and the following spring grow into new plants.

Young plants are tender and delicate and the leaves of lesser celandine are sometimes eaten as salad. Older plants have a bitter flavour and are poisonous.

Lesser celandine

Spring cinquefoil

178

Common snowdrop
Galanthus nivalis
The flowers of the snowdrop emerge when the ground is still covered with patches of snow. It is a popular plant, poetically dubbed 'the first herald of spring'. The three, long, snow-white petals may be regarded as a symbol of departing winter and the three, broad, white petals edged pale green as a symbol of approaching spring.

Common snowdrop

Spring snowflake

'flower' is actually a flowerhead, or inflorescence, of the same type as the daisy and other members of the family Compositae. The huge leaves do not appear until later.

Butterbur figured importantly as a medicinal plant in medieval times during plague and cholera epidemics. Nowadays, use is made chiefly of the medicinal properties of the rhizome in treating coughs.

Worthy of note is the ancient myth about the invulnerability bestowed on heroes by the use of magic herb mixtures, said to protect warriors against enemy missiles. The root of butterbur was an important component of such herb mixtures, even as late as the Middle Ages.

The bell-like flowers conceal a pistil and six stamens with yellow anthers awaiting the arrival of bees and other insects to distribute their pollen. As for the distribution of seeds, it is ants that help to do this, being rewarded for their efforts with a tasty meal provided by the fleshy appendage on each seed.

Spring snowflake
Leucojum vernum
The spring snowflake produces its yellowish-white blossoms shortly after the snowdrop alongside brooks and in woodlands. It is closely related to the snowdrop, but is more robust. Like the snowdrop, it has an underground bulb from which rise three or four narrow leaves with a single flower in the centre. However, its petals are all alike and are yellow-green at the tip. The snowflake soon dies down after flowering.

The related summer snowflake flowers later. It is more robust than the spring snowflake and has several broadly bell-shaped flowers.

Spring cinquefoil *Potentilla verna*
In March, on sunny banks and in grassy places, the golden flowers of spring cinquefoil appear. Its broadly-spreading clumps of leaves form a cushiony carpet dotted with numerous buds and flowers. The flowers are short-stalked and set close over the foliage. The ground leaves have five to seven leaflets and are covered with simple hairs on the underside. This distinguishes the spring cinquefoil from the similar sand cinquefoil, which has grey-green leaves densely covered with star-shaped hairs underneath.

Common butterbur
Petasites hybridus
In the mountains spring comes belatedly but with a rush. As soon as the last patches of snow are melted by the warmth of the sun, bright mountain flowers push up through the soil. Butterbur is one of the first to appear. Its bunches of tiny flowers form sheets of pink on the banks of brooks and streams. Each

Common butterbur

A Walk in the Meadow

In late spring and early summer meadows are flooded with the bright colours and varied hues of myriads of flowers — white daisies, bluebells, pink ragged robins, and yellow dandelions. This is also the flowering time of grasses, which are the dominant feature of meadows. Meadow flowers have one major thing in common — love of the sun.

There are many different kinds of meadow. The flowers that can be found in a meadow in any particular area reflect the soil conditions (e. g. chalk, sand or clay) and the amount of water available (e. g. marshy or dry).

Meadow buttercup

Field eryngo

Meadow buttercup
Ranunculus acris
The meadow buttercup adds a special glow to the flowering meadow with its golden-yellow blossoms. Grazing livestock, however, avoid it, for it contains a poison.

Ragged robin
Lychnis flos-cuculi
Late spring, when meadows are in full bloom, is when the ragged robin flowers. It takes its name from the petals of its ragged, rose-red blossoms, which form a small tube at the base and are then cleft into four long narrow lobes. Viewed from the side they resemble a cock's comb.

Common milkwort
Polygala vulgaris
Several species of milkwort grow in the meadows of Europe. Most widespread of them all is the common milkwort. It is so diversified that its subspecies are sometimes regarded as separate species. The flowers are generally blue-violet, occasionally pink or white.

Plants of the milkwort family are distributed throughout the world. Besides herbaceous kinds their number includes sub-shrubs and climbers.

Field eryngo *Eryngium campestre*
The field eryngo of central Europe is interesting in that, although it looks more like a thistle, it actually belongs to the parsley family (Umbelliferae). Field eryngo is a perennial, even though in autumn it looks as if it is entirely dry and dead. By that time the top part of the 'small

German pink

Ragged robin

shrub' has cast off from its moorings and is blown by the wind like a spiny ball over dry land, fallow land, pastureland and grassy places, scattering its fruits (double achenes) along the way.

German pink
Dianthus carthusianorum
The German pink is sure to figure in the verses of poets recalling the beauty of their native countryside — its sunny hills, meadows and

Common milkwort

pastures. It has crowded heads of red flowers enclosed by red-brown bracts. As it is a very variable species, however, the flowers may be coloured pink as well as red and on occasion also white. This species of *Dianthus,* which translated from the Greek means 'divine flower', was named *carthusianorum* after the 18th-century naturalists, the Karthauser brothers.

Common cotton grass

Common cotton grass
Eriophorum latifolium
The fruits (achenes) of the cotton grass, with their silvery-white hairs, float like tufts of wool above wet bogs and damp meadows. That is why ancient Greek naturalists named the plant *Eriophorum,* which means 'wool carrier'.

181

Water Plants

Water plants are affiliated with water in various ways. Some have little more than their roots underwater; others are completely or almost completely submerged; and others float on the surface.

Duckweeds are examples of floating plants. The leaves and flowers of water lilies and the broad-leaved pondweed also float on the surface, but not freely, for they are attached by stalks to rhizomes anchored in the mud. Their stems and leaf stalks contain air spaces to aid buoyancy and are as long as is necessary to reach the surface of the water. Another adaptation shown by these aquatic plants is that their stomata are on the upper side of the leaves.

Plants of muddy banks that grow only in shallow water, for example reedmace, reed, sedge and sweet flag, have grass-like leaves. Their tangled roots play an important part in trapping and holding mud.

White water lily *Nymphaea alba*
The flowers of the water lily begin closing in the late afternoon and by nightfall they are submerged. Early in the morning the closed blooms surface once again and open if the sun is shining. Facing east when they emerge, they follow the sun's path through the heavens until evening, when they face westward and bid the sun farewell.

Water lilies have lovely white flowers and large circular leaves. Besides being attractive, however, they are also botanically peculiar. Their flowers show a gradual transition from the sepals, which are green below and white above on the outside of the flower, to the inner petals. They are arranged in a spiral and are narrower and smaller towards the centre of the flower and their whiteness is highlighted by the yellow of the stamens that cover the ovary up to the stigma.

The close proximity of the anthers to the stigma may result in self-pollination, but as a rule these striking flowers are pollinated by the flies and beetles they attract. The fruit is an inflated capsule that is well adapted for dispersal by floating away on the water.

Broad-leaved pondweed
Potamogeton natans
The still water of ponds, quiet pools and backwaters is where the flowers of the broad-leaved pondweed may be found. Arranged in slender spikes, they jut above the surface of the water. After being pollinated by the wind the spikes submerge. The fruits are achenes that ripen and then sink to the bottom or else are dispersed by animals. Some achenes remain in the spike, which floats after the stalk rots.

White
water lily

Great reedmace

Great reedmace *Typha latifolia*

Shown in the illustration are the well-known 'pokers' of still and slow-flowing water. These are the inflorescences of reedmace. The female flowers form a velvety brown spike, or spadix, which may be up to 30 cm (12 inches) long and is terminated by a pale 'tail' of male flowers. After being pollinated by the wind the female flowers develop into white downy achenes. The down keeps the ripe seeds afloat for two or three days, after which they sink.

Reedmace is a decorative and striking plant. But it also has several uses. The rhizome contains starch, proteins and sugar. Because it is quite large — 2.5 cm (1 in) thick and up to 60 cm (24 in) long — it can be dried and crushed to yield a flour from which bread, pancakes, biscuits and even gingerbread can be made. The bread is very good eaten together with young, boiled reedmace shoots, which are very delicate and taste like asparagus (they were a favourite food of the Chinese as early as 200 B.C.). This can then be finished off with a cup of coffee made from roasted pieces of rhizome. Any raw leftover rhizomes may be added to feed for pigs.

Other parts of the reedmace are also useful. The firm, flexible leaves are excellent material for making baskets, matting, hats, bags, wine-bottle jackets, slippers and other useful objects. The stems may be used as fuel, and the down from the seeds as packing and stuffing material — in former times it was also added to rabbit underfur in making felt hats.

Lesser bladderwort
Utricularia vulgaris

The bladderwort takes its name from the translucent bladders on the leaves, the leaves themselves being the thread-like parts that resemble roots. From their centre rises a stem bearing several yellow flowers above the surface of the water. The entire plant floats freely in warm, still water.

The bladderwort is a carnivorous plant and each bladder is an efficient trap. When a small animal touches a trigger hair, the door of the trap flies open and the animal is

Lesser bladderwort

sucked in. The trap door closes and the animal is digested by enzymes. Glands pump water out of the bladder, thus resetting the trap.

Common duckweed *Lemna minor*

The surface of a pond, forest pool, sluggish stream or even a large shaded puddle may be covered with a green film. This is composed of minute flowering plants — 2.5 to 4 mm (0.6 to 1.5 in) in diameter — and is called duckweed because it is eaten by ducks. The plants are mostly common duckweed, but they sometimes include other related species. Each plant looks like a circular leaf with a single root

Broad-leaved pondweed

below. Duckweed, however, does not have any leaves and what looks like a leaf is actually a very simplified stem adapted to floating on water. This smallest of all angiosperms rarely bears flowers; it multiplies and spreads readily and rapidly by a kind of budding.

Common duckweed

Grasses, Sedges and Rushes

Grasses are a large family of plants with a characteristic appearance common to all. The inflorescence is either a panicle or a spike and their wind-pollinated flowers are inconspicuous. Their uniformity is mainly in the leaves, which are typically long and narrow. Some plants belonging to other families show a marked resemblance, for instance sedge, woodrush and thrift. These, however, do not have the jointed, hollow stems that are the most striking feature of grasses.

Wood sedge *Carex sylvatica*
Wood sedge, like other sedges, is similar to many grasses. Unlike grasses, however, sedge stems are triangular, solid and without joints.

Wood sedge has loose, more or less nodding spikes on long, thread-like stalks and a single terminal spike which is erect. This is the male spike. The lateral female spikes produce small, beaked fruits.

Wood sedge is widespread in woods, alongside streams, in thickets, and in woodland meadows.

Smooth meadow-grass

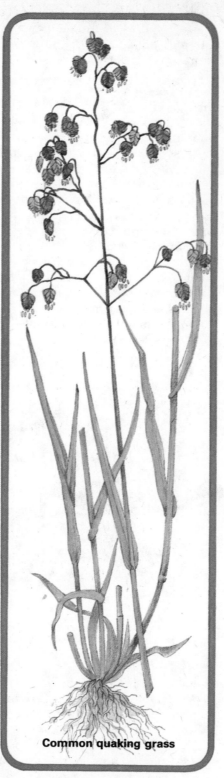

Common quaking grass

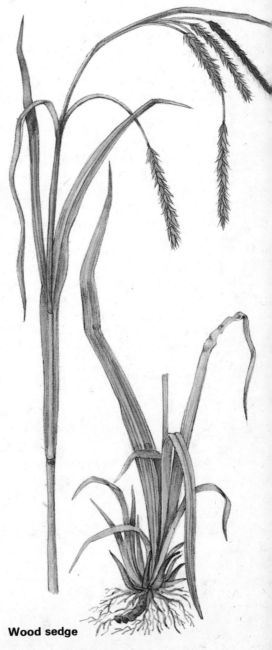

Wood sedge

Smooth meadow-grass
Poa pratensis
Smooth meadow-grass is one of the commonest of the meadow and pasture grasses. It is one of the best forage grasses for livestock. It has no special soil requirements and grows rapidly again when cut. It forms firm, thick clumps in meadows.

Common quaking grass
Briza media
Quaking grass is an attractive grass with its slender-stemmed, open panicles of heart-shaped, awnless spikelets dancing in the breeze. It is not used much for forage, but is often dried for winter decoration in the home. Quaking grass of southern Europe is ideal for this.

Greater woodrush *Luzula sylvatica*
The greater woodrush grows in the woods and forests of central Europe and in mountain meadows and grasslands. The large tufts of grass-green leaves and clusters of tiny brown flowers are an attractive feature of mountain spruce and mixed woods up to the subalpine belt. When it grows at lower levels, it is usually in damp and shady places.

The brown seeds, enclosed in a globular capsule, have a yellow fleshy appendage. Ants eat this and are their main distributors.

Greater woodrush

Drooping brome

Timothy grass

Timothy grass *Phleum pratense*
The inflorescence of timothy grass is not a true compound spike as in wheat, rye and barley. Instead it is a very dense panicle that merely resembles a spike.

Timothy grass is quite easy to distinguish from other grasses. Just bend the spike over the finger — it does not form lobes along the outer edge and may break off.

Drooping brome *Bromus tectorum*
Little vegetation is found on today's flat and highly-placed roofs. However, a few plants struggle to grow in the gutters of old, low houses, where seeds have been deposited by the wind or by birds and then taken root in the dust blown up from the streets. Among these is drooping brome, whose Latin name *tectorum* means 'of roof-tops'. Gutters, however, are not the place where it usually grows. Being an annual plant it produces a great number of grains — as many as several hundred — which, like those of other weed plants, germinate and grow in places where they have little competition.

185

Plants and Civilization

The development of useful plants is closely linked with the development of civilization. Most are well known and many are now essential. Their modern appearance, however, is generally a far cry from what they looked like originally. They have undergone marked changes as man has striven to increase the yield of their useful products.

time immemorial. Wild species of wheat were discovered in Asia, in the high-mountain steppes of Pamir. From there wheat spread to other places and was continually developed by breeding and selection until it acquired its present form. Nowadays wheats are grown in rich soils throughout the whole world.

Rye is a younger cereal than

other two are also both cereals – wheat and rice). It has been cultivated for over 7000 years and originated in tropical America. There the Indians are believed to have begun developing it by cross-breeding an early form of maize. Since then selective breeding has produced all the modern varieties.

One very popular variety is popcorn, whose hard-pointed grains open into a white, puffy mass when heated. Another is sweet corn, which is eaten as a table vegetable. One that is of botanical interest is pod corn, considered to be the earliest form of maize, where not only the female spike but also each separate kernel is enclosed in a husk. Other interesting types of maize are dwarf varieties, maize that is hairy all over, particularly on the leaves and husks, and the extraordinary form with a spike branched into smaller lateral spikes. Besides these there are also forms with striped leaves grown for garden decoration.

The grains are usually yellow or white. However they may also be orange, violet, blue, blue-black, rose to red, and variously striped. Very striking is a variety called harlequin that has spikes with multicoloured grains.

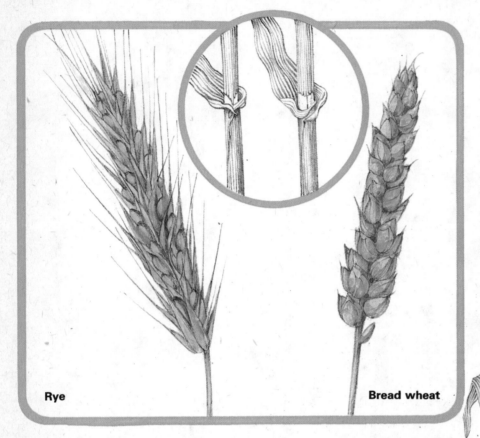

Rye

Bread wheat

Bread wheat *Triticum aestivum*
Rye *Secale cereale*
Cereals are food plants that are used all over the world. They are grasses, but over the years breeding and selection have created forms that are very different from the uncultivated meadow grasses. However, the structure of their spikes, flowers and grains, their fibrous roots and hollow, jointed stems confirms their membership of the grass family.

Cereal grains (the plants' fruits) contain large amounts of starch, proteins, sugars and fats. Besides flour, cereals are also the raw material for the production of starch, gluten, malt and spirits.

Wheat has been cultivated since

wheat. It, too, originated in Asia, in the Caucasus region, and made its way westward into Europe as an undesirable weed of wheat and barley fields. In Europe's cooler, more northerly regions, however, man soon discovered its good points. It does better in more rugged climates, where it also gives larger yields. Flour from rye is dark and not everyone likes it, but in Europe rye bread is a popular item. Of all the cereals rye yields the longest straw, which is used for bedding, thatching, papermaking, mats and straw hats.

Maize, Indian corn *Zea mays*
Maize is one of the world's three most important food plants (the

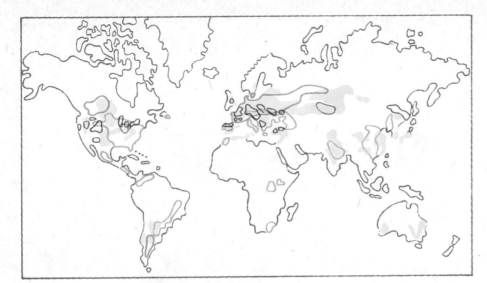

Areas of cultivation of beetroot (red), maize (green), rye (blue) and bread wheat (yellow)

Beetroot

Sugar beet

Maize

Beetroot *Beta vulgaris*

Beetroot, sugar beet and the several other varieties of beets are all believed to have been developed from the same plant — wild beet (*Beta vulgaris,* subspecies *maritima*). This plant is a salt-loving, perennial plant native to the seashores of Europe (including Britain), North Africa and parts of Asia. It has a slender tap root that is sweet to taste — it contains about five per cent sugar. The sugar is produced, as in other plants, by photosynthesis, but beets store a high proportion of the sugar they make. Intrigued by this characteristic, people set about developing it. Breeding and selection over a period of almost 200 years saw a continual increase in weight and, above all, an increase in the sugar content of the root. Today's sugar beet (*Beta vulgaris,* subspecies *cicla*) has an average sugar content of about 18 per cent.

The fat, conical root of sugar beet and the round, red root of beetroot show marked differences from the slender root of their ancestor. Today, sugar beet is the most important source of sugar in countries where the climate is too cold to grow sugar cane. Sugar is extracted by pulping the roots and heating them in running water. Crystals of sugar are then obtained by evaporating the water.

Beetroot is used as a salad, generally cooked, allowed to cool and then pickled. However, it can also be eaten hot and is the principal ingredient of Borsch, the famous Russian soup. Other varieties of beet have also been cultivated for a number of years. Mangel, a large variety with a white, fleshy root, is used as food for cattle. Spinach beet, of which only the leaves are used, is another great favourite.

Sunflower *Helianthus annuus*

The sunflower is the golden sun reflected on Earth. It resembles the sun in both the shape and colour of the ray florets (little flowers) that border the central disc. Throughout the day its blooms follow the sun's path across the sky. The sunflower is a robust plant reaching a height of up to three metres (ten feet) and its daisy-like blossoms, composed of about 2,000 florets, can measure up to half a metre (20 in) in diameter.

Nowadays the sunflower is a very useful cultivated plant, but this was not always so. It has a complex and interesting history. Its native land is Mexico, where images of the flower were hammered out from pure gold and worshipped by the Indians. In 1510 it was brought from South America to Europe by the Spaniards for garden decoration. In the 18th century it found its way to Russia where it was bred and cultivated by growers who produced increasingly larger flowerheads and larger seeds.

The flowerhead has an ingenious structure. The golden-yellow petals of the ray florets round the margin attract insects to pollinate the flower. Having alighted, the insects look about for nectar. This, however, is contained only in the youngest of the tubular disc florets, which open in a spiral pattern from the edge towards the centre of the flowerhead. Following pollination, the disc florets close, wilt and fall, and are replaced by a dense mass of achenes that ripen successively from the edge to the centre.

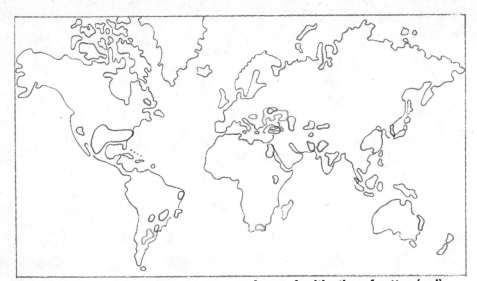

Flax *Linum usitatissimum*

The delicate blue flowers of flax, followed by capsules with seeds, are borne on long, slender stems. The plant has not one, but two uses. Some varieties are grown for their fibres, which are used in the textile industry. Others are cultivated for their oily seeds. Growers have also succeeded in combining both characteristics in one oil-fibre variety.

Flax is one of the oldest plants cultivated by man and has been used to make clothing for about four or five thousand years. Linen, the fabric made from flax, followed on from animal furs, and was thus the first fabric. Even mummies were wrapped in linen cloths and remnants of these ancient fabrics have survived to this day. That linen fabrics have proved their worth and passed the test of time is borne out by the fact that the linen shirts once popular in ancient Greece and Palestine are in great demand nowadays, in the age of man-made fibres. Flax has one great rival, however, and that is cotton. In some countries it has caused flax to be relegated to the realm of such items as bed linen and tablecloths, for linen fabrics are stronger and more durable than cotton.

Oil varieties of flax differ from the tall, little-branched fibre varieties by having shorter, branched stems and, above all, larger seeds which contain as much as 50 per cent oil. Our ancestors used this oil in preparing food, but nowadays it is used only in industry — to make

Areas of cultivation of cotton (red) and sunflower (green)

varnishes, sealing compounds, and impregnating compounds; it is also combined with ground cork to make linoleum.

Flax is also a medicinal plant. Externally, crushed seeds mixed with limewater are used to treat burns; internally, they have a mild laxative effect.

Cotton *Gossypium*

Take a flax and a cotton fibre and compare the two. Leaving measurements aside, try tearing one and then the other. Whereas cotton fibre tears fairly easily, flax resists even quite strong tension. As well as this difference in strength there are many differences between the two as regards quality. The reason is that the fibres have different origins. Flax fibres are from the stem of the flax plant. There they form strengthening bundles that run the whole length of the stem. Cotton fibres, on the other hand, are the

Flax

long hairs — only one cell thick — that surround the seeds in a cotton boll.

There are more than 50 species of cotton, including large herbaceous plants and shrubs, as well as small trees. Only a few, however, are used by man, namely *G. barbadense, G. hirsutum* and *G. herbaceum.*

Cotton originated in Asia and Central America. The large flowers are yellowish or white and resemble mallow, clearly indicating cotton's relationship to that plant. They are followed by capsules (bolls) which burst to release a tuft of yellow to white (depending on the quality) cotton and seeds. The fibres are used to spin yarn, weave cloth and make cotton for medicinal use. The seeds yield an oil used in the food industry (e. g. in canning sardines) and in other industries. The cakes that remain after the oil has been pressed from the seeds make food for livestock.

Unwelcome Intruders

In fields and gardens man wages a battle with 'weeds' — plants that have not been put there by man himself. Such plants can be destroyed mechanically with a hoe or else by spraying with chemical agents called herbicides. In time they generally retreat. Sometimes the eradication of a 'weed' is so thorough that it may result in the species being placed in danger of extinction.

However, 'weeds' are not all undesirable plants. On the contrary many perform extremely useful work. With their large quantity of seeds and their adaptability they readily colonize new areas. They can thus reclaim land that has been scarred by man's activities. Corn cockle and cornflower release carbonic acid and phosphatides through their roots, thereby enriching the soil. Many 'weeds' have medicinal properties, for example stinging nettle, white deadnettle, dandelion, couch grass, field bindweed, and others.

Fat hen *Chenopodium album*
Shining orache *Atriplex nitens*
These two members of the goosefoot family look alike. There is little difference between the leaves that cover the tall 1 to 2 metres (40 to 80 in) stems. Both plants are covered with a whitish bloom (tiny glandular spherical hairs), except for the upper side of the leaves of shining orache, which are glossy. The tiny flowers of both are arranged in small round heads in dense racemes. The flowers of fat hen have a distinct perianth, which later dries up and protects the achene. Shining orache has no perianth; the single flowers, and later the achenes, are enclosed in persistent bracteoles resembling the two shells of a bivalve.

Both species readily colonize new, fertile ground, such as compost, waste heaps and road verges. The number of seeds they produce is huge — a single fat hen plant may bear as many as 100,000 seeds.

Remarkable is the fact that these plants, which are definitely regarded as weeds, are closely related to useful species — sugar beet and annual spinach.

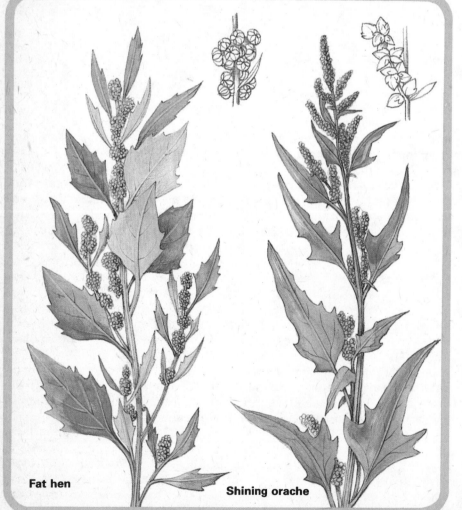

Curled dock

Fat hen

Shining orache

Curled dock *Rumex crispus*
If you are unsuccessful in ridding the garden or meadow of curled dock then it is probably because you are not removing all the roots. If even a small piece of root is left in the ground then the plant will grow again. Worse still is dock that is overlooked and allowed to ripen. Vast numbers of triangular winged achenes are dispersed by the wind.

The leaves have a sour taste because they contain oxalic acid, which man as well as animals tolerate in small quantities. The tender spring leaves, particularly those of

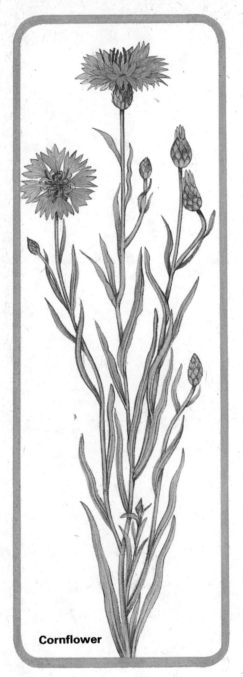

Cornflower

skin — just a minor irritation. In Java and India, however, there are some species of nettle whose sting is as dangerous as a snake bite.

The flowers of the nettle are somewhat unusual. They are of two kinds, both green and very tiny. One kind are only male with stamens and the other only female with carpels. And what is more, each kind is produced on separate plants. Stinging nettle is dioecious.

Nettle is not just an unpleasant weed, it is also useful. It contains a large amount of the green pigment chlorophyll. This is obtained from the dried plant and used to colour food products, soap, per-

White deadnettle flowers from spring until autumn and during those few months yields ample quantities of a medicinal drug obtained from its white blossoms. Only the petals are collected and they should not be allowed to turn dark during the drying process.

Cornflower, Bluebottle
Centaurea cyanus
The flowerhead of the cornflower is composed of two kinds of floret. The larger florets on the outside attract insects. Generally, these florets are blue, but they can be violet, rose or white. The central disc is composed of blue-violet

White deadnettle

Stinging nettle

common sorrel (*Rumex acetosa*) are eaten as a salad or else used to flavour soups.

Stinging nettle *Urtica dioica*
Because of the unpleasant effects of brushing against it, the stinging nettle is regarded by most people with dislike. The whole plant is covered with delicate stinging hairs. These stinging hairs contain silicic acid. The hollow hairs stand ready to pierce the skin like hypodermic needles. The point breaks off and the acrid juice inside spills out, usually causing only tiny blisters on the

fumes and fabrics. Nettle grows in early spring and is rich in vitamins. This makes it a valuable spring vegetable. Salad or soups made from spring nettles are popular foods.

White deadnettle *Lamium album*
Although the leaves of the deadnettle resemble those of the stinging nettle they do not sting. The whole plant is softly downy and the lip-like flowers are quite different from those of the stinging nettle. They are white and arranged in whorls on the angular stem, always in the axil of the upper paired leaves.

tubular florets, which contain nectar and possess an interesting characteristic intended to aid cross-pollination. The filaments of the anthers are fused into a tube and react to mechanical stimuli. When an insect alights on the cornflower and the pollen is rubbed off on its body, the tube contracts. This exposes the stigma. However, fertilization does not occur immediately because the stigma of this floret ripens later, after the anthers. The pollen-covered insect then transfers the pollen to a floret on another flower with a ripe stigma.

191

Green Healers

The medicinal properties of plants were originally linked with the activities of magicians and witches. There was no scientific basis for their use and superstition and faith played a large part in the results they achieved. Magic potions and 'universal medicines' abounded.

Not all the healing effects of medicinal plants, however, were due solely to magic and superstition. Gradually, people learned which of the potions were really effective. The knowledge of the medicinal plants that proved their worth and stood the test of time was passed on from generation to generation and came to be used in genuine medical practice.

With the development of chemistry, man-made medicaments became more popular and medicinal plants were pushed to the background for a time. It was discovered, however, that the effects of many man-made medicines could not compare with the effects of some natural substances.

Common mullein
Verbascum thapsus

The golden spikes of mullein adorn dry stony hillsides, pasturelands and railway embankments from July to September. The entire plant is adapted to life in places with insufficient moisture. The woolly leaves are arranged on the stalk so that rainwater runs down them to the roots, and the long tap root reaches deep into the soil.

One of the loveliest of all mulleins, the common mullein, is a biennial. The first year it forms only a ground rosette of large shaggy leaves. The second year a tall upright stem rises from the centre of the rosette to a height of more than 2 metres (80 in). It is covered with a profusion of large yellow flowers up to 5 cm (2 in) across. In the centre of the circular corolla (composed of five fused petals) are five stamens, three of which are shorter and covered with white wool.

The flowers open in succession

from the bottom upwards. The flower petals and stamens are collected in dry weather and then dried. They can be used to brew tea that is a good cough medicine.

Sweet woodruff
Asperula odorata

The low, dark green masses of whorled leaves of sweet woodruff are brightened by tiny white flowers. It thrives chiefly in deciduous, particularly beech, woods. As the plant fades the delicate aroma of coumarin becomes stronger. The leaves contain a drug from which can be extracted a substance that has a sedative effect on the nervous system. However, large doses cause

Sweet woodruff

192

Common mullein

poisoning. Sweet woodruff is also popular for making May wine.

Chicory *Cichorium intybus*

According to folk legend, chicory is an enchanted maiden that waits patiently by the wayside for her beloved. Its bright blue flowers adorn waysides, hedgerows, ditches and dry meadows. The flowers are the typical ones of the daisy family (Compositae) except that all the florets are ray florets.

The flowers open in the early morning and close after midday. Several other plants show this kind of movement, which may be due to changes in light intensity, temperature or both. The ground leaves of chicory resemble those of dandelion, but are usually already dried up by the second year when chicory flowers.

The cylindrical root contains a large amount of bitter white milk with up to 20 per cent healing inulin and the bitter substance intybin. This aids digestion and improves the appetite. It also increases the secretion and flow of bile and urine.

In the 18th century breeders developed forms of chicory with thick roots, which are dried, roasted and ground and used as a substitute for, or addition to, coffee. Another form is grown with the total exclusion of light and is eaten as a salad or vegetable.

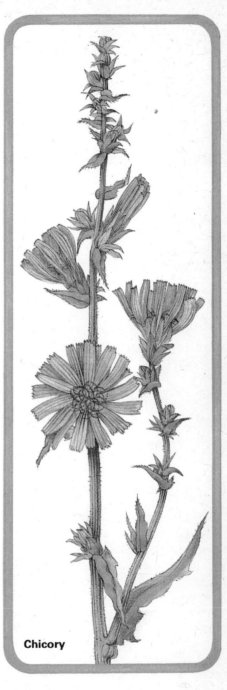

Chicory

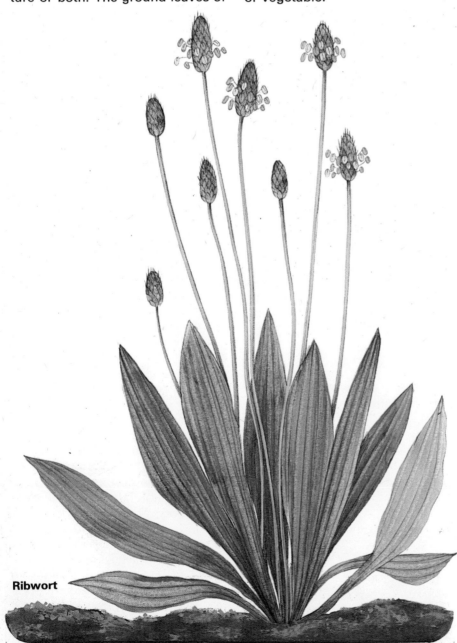

Ribwort

Ribwort *Plantago lanceolata*

The flowers of ribwort are very tiny and brownish with yellow anthers on long filaments protruding from the four petals. The long, angular stems bearing the short, oval spikes of flowers are deeply furrowed and rise from a rosette of leaves that taper to the leaf stalk. The leaves have medicinal properties. Externally they can be applied fresh to wounds and swellings caused by wasp stings. Juice from the crushed leaves can be mixed with sugar and used as a cough syrup, mainly for children.

Common lady's mantle

Common comfrey
Symphytum officinale
This medicinal plant has a remarkable ability. It helps to mend broken bones, and heal bruises and stubborn wounds. It played an important role in medieval medicine and was an important component of healing ointments. Comfrey poultices were applied to swollen gums and varicose veins and provided relief for arthritic and gouty joints. Nowadays, although it contains mucilages which are effective in the treatment of coughs and sore throats, it is used fairly rarely.

The 'black root' is the chief medicinal part of the plant. It is a thick rhizome coloured black on the outside and white inside. From this rises a rough angular stem bearing leaves, covered with bristly hairs, that are arranged like wings tapering towards the stem. The stem branches at the top and is terminated by dense, downcurved clusters of bell-like flowers.

Common stonecrop

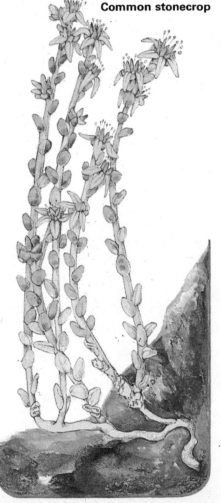

Common comfrey

Common lady's mantle
Alchemilla vulgaris
It is hard to believe that this plant with its small, unattractive yellow-green flowers is closely related to the rose. And yet, lady's mantle is a member of the rose family. The tiny flowers have no petals and the receptacles bear fruits (achenes) without fertilization.

Lady's mantle has another unusual characteristic. Its striking, round leaves stand out amidst the meadow vegetation. When young the leaves are knife-pleated. Later the folds spread out and the leaves form a shallow funnel, in the centre of which glitters a drop of water, sometimes the whole day long. This drop is not ordinary dew but excess water the plant has pressed out on to the surface through pores in the leaf margin. Medieval alchemists (hence the scientific name *Alchemilla*) used this 'heavenly dew' in their endeavour to produce philosophers' stone and the elixir of perpetual youth.

Common stonecrop *Sedum acre*
The fleshy leaves and stems of succulent plants store supplies of water. In dry weather these plants are capable of limiting transpiration, but they are slow growers. One example is the common stonecrop. Forming cushions 5–15 cm (2–6 in) high, these plants make low, thick green carpets which are covered with yellow, star-like flowers.

Common stonecrop is poisonous. It has an unpleasant, acrid, sharp taste. The leaves contain alkaloids. Small doses of the drug in the form of an infusion are used to lower blood pressure. Larger doses of the drug cause vomiting; for this reason it is administered in cases of poisoning. The fleshy leaves contain large amounts of mucilage and therefore have a smoothing, cooling effect. Since time immemorial they have been applied to wounds, ulcers, corns and fungal diseases.

Common St. John's wort
Hypericum perforatum

The Latin word *perforatum* means perforated. If the leaves of St. John's wort are held up to the light, they can be seen to be covered with tiny transparent dots that look like perforations. However, these are not perforations but tiny vesicles made of silica. More striking, however, are the black dots on the golden-yellow flowers. These are stalked glands, which contain tannins and flavonoid glycosides and a red colouring matter called hypericin that dyes the fingers a dark red. When the plant is eaten by cows and goats, their milk is likewise coloured red, which was formerly believed to mean that they were bewitched.

St. John's wort has been used to treat diseases of the gall bladder, bile ducts, digestive system, kidneys, lungs, glands and nerves; also rheumatism, bleeding, irritability and disturbed sleep. Popular is the oil from St. John's wort, which is applied to burns, wounds that refuse to heal and haemorrhoids.

The panicles of yellow flowers with their many stamens may be seen throughout the summer on dry river banks, meadows and

Common St. John's wort

Yarrow

hedgerows. The main difference between this plant and the similar imperforate St. John's wort is that the latter has a square stem and lacks the dots on the leaves.

Yarrow, Milfoil
Achillea millefolium

Two parts of the yarrow can be collected for the drug market; either the entire top parts, together with the basal leaves, or the richly-branching flowerheads with a short stalk. This is not easy, as yarrow has a very tough stem. Generally, it successfully resists all efforts to tear off part of the plant; one is more likely to pull it up whole with a piece of the branching root.

Yarrow has tiny, dingy white flowers, which may occasionally have a pinkish tinge. The delicate leaves and whole, woolly stem have a distinctive aroma.

Yarrow is one of the most widely used herbs in the home. However, if taken over a long period or in large doses it may have unpleasant effects, such as causing rashes and dizziness. The drug, used in the form of tea, stimulates the flow of gastric secretions and has a beneficial effect on the blood circulation. It also soothes coughs. Externally it is used in the form of a bath in the treatment of rashes, festering wounds and cracked skin, and as a gargle in gum inflammations.

195

Plant Chemicals

Besides precious medicinal substances plants also contain substances needed by healthy people, such as vitamins and flavouring principles. Vitamins and flavouring principles are present not only in the well-known plants that we eat such as lemons, horseradish and carrots but also in many other plants. Vitamin C, for example, is contained also in the leaves of primrose, which to most of us is primarily the first flower of spring.

Sweet flag *Acorus calamus*

Sweet flag is a subtropical marsh plant native to southern and eastern Asia. It arrived in Europe in the 16th century. When planted, it grew into a sturdy plant and flowered – but proved a disappointment. The fruits (red berries) decorate the plant only in its native lands, but do not develop in the northern hemisphere. The fleshy spike of tiny greenish flowers is called a spadix and grows on a leaf-like stalk.

Because it readily multiplies by means of its creeping rhizomes, sweet flag soon spread throughout the entire northern hemisphere. A broken-off piece of root carried by water to a suitable place was enough to found a whole new colony.

The rhizome contains a pungently bitter essential oil, the bitter principles acorine and acoretine, phytoncides, a large amount of starch and mucilages. It promotes the body metabolism and is used in the treatment of stomach and intestinal disorders. Also popular are the liqueurs made by steeping the rhizome and other herbs in alcohol. Dried young rhizomes are added as an aromatic substance to beer, custards, cakes and compotes, and are also candied or made up into preserves.

The fragrant volatile oil is used in perfumery and cosmetic preparations, such as soaps.

Cowslip *Primula veris*

Primulas, such as the cowslip and primrose, are usually thought of in relation to spring. Indeed, they are among the first flowers of spring, their vivid yellow brightening woodland groves, damp meadows, pastures and brooksides throughout Europe. However they are also medicinal and kitchen herbs.

Tea made from the flowers may be used to treat dizziness, cramps and migraine. The flowers and roots contain saponins, which in small doses are used to treat lung and kidney diseases. Also worthy of note is the root, which smells like anise and is gathered for the preparation of medicines. The fragile, wrinkled leaves contain vitamin C.

Primulas produce two types of flowers. Some plants have flowers with long styles and short stamens

Sweet flag

Cowslip

Carrot *Daucus carota*
Wild carrot can be found growing in meadows, hedgerows and fallow land as a weed. Its long, slender, white root is quite different from the fleshy orange root of cultivated strains.

Carrot is popular not only for its use in cookery as a root vegetable but also for its beneficial effects. The raw root is a rich source of vitamins A and C and also contains pectin substances which promote digestion, as well as anti-bacterial substances, minerals, sugars and carotene. The fruits (achenes) contain an aromatic oil.

Chives

Ramsons

and others have short styles and long stamens — nature's clever way of preventing self-pollination and encouraging cross-pollination by insects.

Ramsons *Allium ursinum*
The leaves of *Allium ursinum* grow from a long, greenish-white bulb. Emerging between the leaves is a flat cluster (umbel) of white, star-like flowers borne on a triangular-shaped stem. As the seeds ripen, the stem lengthens and bends toward the ground.

Ramsons often grows in large masses in riverine and shaded woodlands, mainly beech woods, up to subalpine elevations. It is

readily located by its odour — the pungent aroma of garlic. Members of the genus *Allium* contain substances that are used in the treatment of intestinal disorders.

Chives *Allium schoenoprasum*
Chives is closely related to garlic. The slender, hollow, dark green leaves grow from a small, white bulb which looks like garlic and also has a similar pungent aroma. The underground bulbs can be used in cooking. However, it is generally the grass-like clumps of leaves that are used. These are chopped up and, while still fresh, used to flavour various foods. They have a mild onion flavour and scent and are a rich source of vitamin C.

Carrot

Plant Poisons and Medicines

Corn cockle
Agrostemma githago

Corn cockle is a very attractive weed. Its leaves have silky grey hairs and the single flowers are a soft reddish-violet colour. Noteworthy are the sepals. In the corn cockle they are fused at the base to form a cup and the narrow

Corn cockle

tips extend beyond the petals. After the flower has been fertilized, only the petals fall and the calyx remains to shelter the capsule until it is ripe. Then the tips of the sepals dry and the teeth of the capsule open and curve backward to release the seeds. There may be as many as several hundred seeds on a single plant and all are poisonous, for they contain glycosidic saponins. The seeds are often present in grain that has not been properly cleaned. Flour milled from such grain is then bitter and useless.

Herb Paris *Paris quadrifolia*

Tempting and treacherous are the berries herb Paris flaunts in woodland settings. The large, glossy blue-black fruits are nestled in a whorl of leaves, usually four, but sometimes only three or as many as five to seven.

All parts of the plant are extremely poisonous, but the berries are the most dangerous. Herb Paris contains several effective poisons, chiefly saponins. In folk medicine it was at one time believed to provide protection against contagious diseases, but in view of its toxicity this was a very dubious treatment!

Purple or Common foxglove
Digitalis purpurea

The purple foxglove is well-known in today's pharmaceutical industry, but it has been widely used in folk medicine for hundreds of years. This poisonous plant yields the chief drug used in the treatment of serious heart diseases. The effective substances are the glycosides digitoxin and digitalin, found mainly in the leaves. These are gathered during the afternoon when the concentration of glycosides is greatest.

Herb Paris

Purple foxglove is often cultivated as a crop plant for medicinal purposes, as is *Digitalis lanata*, whose leaves are about four times more potent. It is often grown also for garden decoration. In the second year a tall stem with one-sided raceme of drooping, thimble-shaped flowers grows from the ground rosette of leaves. The purplish-pink flowers, marked with darker, pale-edged spots on the lower lip, brighten the open woodlands and clearings where they grow, but it is best to avoid them for they are poisonous. The capsules that follow are full of tiny seeds and a single plant may produce as many as 350,000!

Deadly nightshade, Belladonna
Atropa belladonna

The use of a poisonous plant as a beauty aid may seem surprising. But Roman beauties used belladonna to enlarge the pupil of the eye to make their eyes more striking. They chewed the leaves of the plant or drank juice from the berries, even at the cost of chronic poisoning, only to be pretty. They also applied the juice to their cheeks to make them red. The generic name of this plant is derived from Atropos, the

Purple foxglove

eldest of the three Fates in Greek mythology represented as cutting the thread of life. The second, specific name refers to the beauty of these vain lovelies, for *bella donna* is the Latin for beautiful lady.

The whole plant is extremely poisonous, not just the berries. It contains very toxic alkaloids (hyosciamine and atropine). These substances affect the nervous system. Atropine is used by physicians for its excellent sedative properties and in eye examinations and operations. It contracts the eye muscles

Deadly nightshade

and dilates the pupil, making the surgeon's task easier.

Take care, then, when you come across this large herbaceous plant with brownish-violet, bell-shaped flowers and glossy black berries!

Meadow saffron
Colchicum autumnale
The crocus-like flower of the meadow saffron gleams brightly in the grass in autumn. The plant first appears in spring as a clump of broad, bright green leaves. However, these wither and die by midsummer and the wilting grass of autumn meadows is dotted with dainty, pinkish-purple flowers, which are much longer than they appear to be. The lower part of the flower is a long, white tube, which extends down into the ground to the corm, where the ovary is located. Next spring the meadow saffron again produces leaves and a short stem with a capsule. Medieval naturalists called this phenomenon *filius ante patrem*,

which means 'son before the father'. They believed that the plant produced its fruits before it produced its flowers.

All parts of the plant are extremely poisonous, but most of all the seeds which contain a large amount of the alkaloid colchicine. Colchicine has an interesting effect on cell division — it retards the rampant division of cells. The meadow saffron is therefore a plant with properties that stop the growth of tumours. It is also an effective medicine used to relieve severe pain in gout.

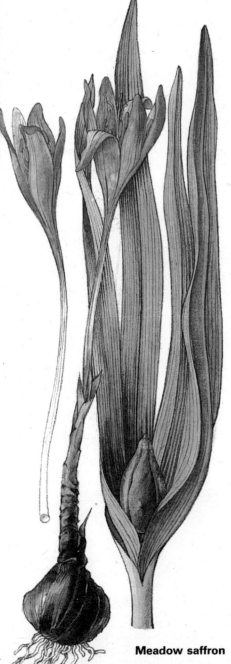

Meadow saffron

Field poppy

Field or Corn poppy
Papaver rhoeas
In early summer, the bright red flowers of the field poppy may stand out in a field of grain. Although this is a sight that gladdens the hearts of some people, farmers consider the plant a nuisance — a weed that depreciates their crops. Systematic eradication by mechanical and, above all, chemical means has forced this poppy to the edges of fields, hedgerows, roadsides and fallow land.

The drooping bud of the field poppy 'stands up' as the flower opens. The bristly sepals open to expose the four overlapping bright red petals. The inside of the flower contains a great many stamens, whose black anthers form a striking contrast with the red petals, and an ovary which develops into the fruit after the flower has faded. The fruit is a hairless capsule that opens at the top to release tiny, oily seeds. One capsule contains about 30,000 seeds!

The field poppy is poisonous, but it is also a medicinal plant. The drug is obtained from the dried petals, which contain mecocyanine. It is used in cough syrups, as a sedative and to colour liquids.

199

Seeking Space to Live and Grow

In order to spread their species over a wide area, plants have invented many ingenious methods of seed dispersal. Some are able to shoot their seeds out; others use parachutes or wings to catch the wind; others use hooked fruits that catch onto the fur of passing animals; others use juicy fruits to tempt animals.

A number of plants reproduce by vegetative means as well. Tubers (e.g. potato), rhizomes (e.g. iris), bulbs (e.g. daffodil), corms (e.g. crocus), bulbils (e.g. lesser celandine, see page 178) and runners (e.g. strawberry) are all organs of vegetative reproduction.

Wild strawberry *Fragaria vesca*
Strawberries, both wild and cultivated, are typical of the juicy fruits used to tempt animals. But a strawberry is different from other fleshy fruits in that it is not in fact a single fruit. Instead it consists of many hard fruits (achenes) which nestle in the enlarged fleshy receptacle.

Wild strawberries can be found in June in woodland clearings, forest margins and hedgerows. The ground is likely to be carpeted with them because they have an extremely effective means of vegetative reproduction. Each plant puts out several slender runners that take root and produce new plants.

Touch-me-not

Touch-me-not
Impatiens noli-tangere
Many surprises await someone walking in a forest. Among them may be the delicate touch-me-not plant, which shoots its seeds out. Just try touching its ripe, slender pods filled to bursting. The seams, unable to withstand the great pressure inside, burst and the separate valves coil inward in a spiral. Like springs, these strike the seeds

Common wood sorrel

and hurl them out, scattering them far and wide.

Common wood sorrel
Oxalis acetosella
Touch-me-not is not the only 'shooting' plant! The much smaller (at most it is only 15 cm (6 in) high) wood sorrel also ejects seeds. Its capsules measure only about 1 cm (0.4 in); they split on the upper side and the seeds burst out by themselves (due to a change in tension in the seed coat) and may land as far as 1 metre (40 in) from the plant.

The wood sorrel is remarkable also for a movement of a different kind which may be described as a 'sleep' movement. In inclement weather and in the evening its flowers and leaves droop and fold up. Wood sorrel also has flowers which remain unopened and are self-pollinated in the bud.

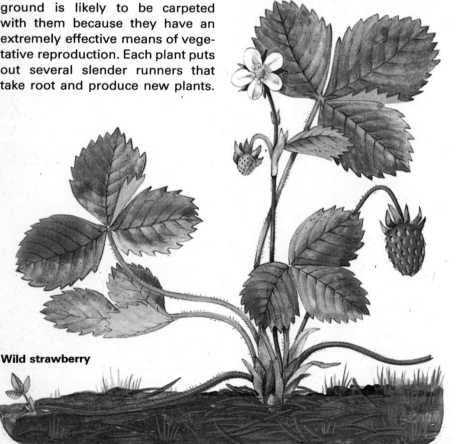

Wild strawberry

Dandelion

Common stork's bill, Heron's bill
Erodium cicutarium

This member of the cranesbill family (Geraniaceae) makes sure that its offspring will be provided with the proper conditions for growth. It does not leave the burial of its seeds in the ground to chance — its fruits bore into the soil. For this purpose the fruits, which follow the rosy blossoms, are equipped with simple but very sensitive mechanisms. They are composed of several sections, each with a very long beak which curls in a spiral in dry weather and unfurls in damp weather. It is by this means, by the alternation of curling and unfurling, that the sections bore into the ground as the weather changes, as damp alternates with drought. Movements caused by sensitivity to atmospheric moisture are called hygroscopic movements. Therefore a ripe stork's bill can be used to foretell the weather.

Common stork's bill

Dandelion *Taraxacum officinale*
Dandelions are a treat to the eye, particularly that of the vegetarian, who prizes the tender young leaves in salads. The flowers also have their uses. Prepared according to an old recipe, they make an excellent honey and are also used by the home winemaker. What interests the botanist, however, is the dandelion after it has finished flowering. Then each of the flowers (there are some 200 to a flowerhead) produces an achene fitted with a kind of parachute, or pappus, made of tiny hairs. The achenes are then dispersed far and wide by the wind.

The root, which oozes a milky juice, and the top parts are collected for medicinal purposes. The plant contains the bitter substances taxacin and taraxosterin, tannins, inulin and substances that destroy various bacteria species.

Water chestnut

Water chestnut *Trapa natans*
This aquatic annual ensures its propagation by means of its four-horned nuts. The shell is composed of the hardened remainder of the flower and protects the tasty kernel inside, which besides being rich in proteins, contains fat and starch. The plant's existence is threatened by the excessive consumption of water chestnuts and above all by its eradication in intensive pond husbandry. Fortunately, it is a protected species in some countries.

When the flowering period draws to an end, the large, heavy fruits begin to form in the water. The plant is prevented from being pulled under by their weight by the air sacs on the leaf stalks. When ripe, the nuts fall to the bottom. In spring they germinate and 'anchor' the new plants in the mud until they are fully developed. Fully-grown plants then cast off and rise to the surface.

Plant Diet

Most plants make their own carbohydrate foods in the process called photosynthesis. Even so, they still have to obtain certain essential minerals from the soil. Sometimes, unusual aspects of their natural environment, however, lead to deviations in the diets of some plants.

For plants growing in nitrogen-poor mud and peat soils there are insufficient minerals present to provide adequate nourishment and so some supplement their diet with the flesh of small animals, which is rich in nitrogen. There are about 500 such carnivorous plant species.

Another green plant, mistletoe, is a parasite. It absorbs water from the wood of trees, and with it dissolved inorganic salts.

Orchids have the least demanding food requirements of all plants. Their store of water and food is supplied only by stem or root bulbs. Also typical of these plants is the symbiotic association (known as a mycorrhiza) of their roots with the mycelium of certain fungi. The fungus invades an orchid seed and the association is important in the young stages.

Common sundew
Drosera rotundifolia
The common sundew is a carnivorous plant that grows on wet moors and peat bogs. The long hairs on its leaves produce sticky drops. An insect that lands on these becomes stuck and tries to free itself. However its movements only serve to cause more sticky hairs to bend towards it until the trapped animal is almost completely surrounded. Then the plant exudes a strong digestive fluid that dissolves the soft parts of the insect's body and absorbs the liquid. Indigestible remnants are blown away by the wind when the leaf opens again. The sundew thus supplements its diet with nitrogenous substances.

Venus' flytrap *Dionaea muscipula*
Venus' flytrap grows in the bogs of Florida and the Carolinas. Its leaves consist of two lobes with fringed margins, lined on the inside with digestive glands. Also on the inside of each lobe are three stiff bristles that serve as triggers. When an insect alights and touches two of these bristles, the two lobes snap shut and trap the victim.

Common sundew

Venus' flytrap

Mistletoe *Viscum album*

Mistletoe is a small dense bush that grows on the branches of trees. Its branches and leaves are green, but it does not rely solely on its own resources for food. It penetrates the branches of the tree with its roots, through which it absorbs water and dissolved mineral substances from its host. It is a partial parasite.

Mistletoe has tiny, insignificant flowers which are dioecious. The fruits are gleaming white berries the size of a pea which ripen in December and are eaten by birds. The seeds are distributed by the birds to the tops of other trees such as pine, apple and poplar.

In many countries mistletoe was (and still is in some places) believed to have magical powers. Whether or not this is so, the twigs

Mistletoe

and leaves of mistletoe do have medicinal properties. The effective constituents (viscotoxin, cholin, acetylcholin, and others) have a beneficial effect on the heart.

Cymbidium
Cymbidium canaliculatum

The large family of orchids is represented in this book by the Australian species *Cymbidium canaliculatum*. It is not among the most beautiful orchids, but its fragrant flowers are very long-lasting and they are able to wait long for tardy pollinators.

Cymbidium canaliculatum is one of 120 species of cymbidiums which are native to the tropical and subtropical regions of Africa, Asia and Australia. The genus includes epiphytic species that grow on trees, as well as ground species with short stems and numerous linear, leathery leaves. What distinguishes orchids from other plants are the green bulb-like structures called pseudo-bulbs from which the leaves and flower stems rise. These serve the plants as a reserve store of food in times of need. Nutrients are stored inside in the form of a mucous liquid that does not evaporate as rapidly as water.

Cymbidium is a small ground-living orchid with a profusion of small dainty flowers arranged in racemes. All the petals are dark, nearly black, edged with yellow. The lip is three-lobed with carmine markings.

Cymbidium

Life without Water

Flowering plants have colonized almost all parts of the world, even the inhospitable places such as deserts. There is very little moisture in such places, but many plants have become adapted to the lack of water. Their bodies are fleshy, their leaves are considerably modified or reduced and many desert plants have developed spines to protect their bodies. Such fleshy plants are called succulents (the word *succus* means 'juicy'). Stored in their fleshy tissues is a large amount of water, which is bound to mucilaginous substances. They are very careful in the way they use this water, for the moisture they absorb during the brief rainy season must last them for many months of hot, dry weather.

Mexico is a typical country with such a dry climate. It is the home of the cacti, which spread to other parts of the world after the discovery of America. The succulents that developed in Africa, though much like the American cacti, belong to other plant families. These include various members of the spurge family, the orpine family (e. g. houseleek, stonecrop) and the milkweed family (e. g. *Stapelia);* also groundsels of the daisy family, aloe of the lily family and agave of the amaryllis family.

Prickly pear *Opuntia*

Opuntias are plants with flat, fleshy lobes, which may be smooth or prickly and which grow one on top of the other, branching at intervals. They are native to South America, particularly Mexico, where they can grow to a height of more than 2 metres (80 in). However, they have become established in other warm regions as well. They arrived first in Spain and then in the remainder of southern Europe, Africa and Australia. In Australia they spread to such an extent that they became a bane to farmers, until brought under control by an insect parasite. In the Mediterranean region and Central America they are sometimes planted in a row to form an impenetrable wall.

Some opuntias are found in temperate zones or high up in the mountains. Such species as *O. humifusa* tolerate even the cool climate of central Europe and their handsome red and yellow flowers brighten the rock garden of many an avid gardener.

The flowers of opuntias are generally large and ornamental. They are followed by oval, pulpy berries that are eaten as fruit and used mainly in making sweets.

In northern Brazil spineless species are sometimes used as food for livestock, particularly when green fodder is in short supply. Elsewhere the 'wood' of tall, dry cacti is used as construction material, because it is light, strong and tough. Mexicans consider cactus 'wood' the best fuel of all.

Prickly pear

Pebble plant

Pebble plant, Living stone *Lithops*

The pebble-like body of this well-camouflaged plant is composed of a pair of fleshy leaves. They are pressed together tightly and the plant looks so like two pebbles that it is difficult to find in the stony ground in which it grows.

Signs of life are more evident during the flowering period, when the gap between the leaves opens just wide enough to let the long bud emerge. This then opens close above the leaves into a beautiful white or yellow flower.

Houseleek
Sempervivum soboliferum

This houseleek grows on sunny cliffs, rocky screes, hills and walls throughout central and eastern Europe. Its tightly-closed, globose rosettes enable it to survive long periods of drought. It often forms great clumps because it can multiply and spread by means of side rosettes. During the flowering period it develops a thickly-leaved stem, 10—20 cm (4—8 in) high, topped by a spreading cluster of stalked, six-petalled flowers coloured greenish-yellow. These are rich in nectar.

When a side rosette rolls away from the parent plant and comes to a stop on its side or upside-down with roots upward, it then turns itself the right way up.

Spurge

Houseleek

Spurge *Euphorbia*

The spiny plant in the illustration looks like a cactus. But it is easy to be misled by appearances. The succulent spurge *Euphorbia horrida* in fact belongs to a different family and its resemblance to a cactus is the result of convergent evolution; both types of plant have quite separately evolved the same adaptions for desert life. The spurges adapted to the dry climate of South Africa. Photosynthesis takes place in the fleshy stem.

Spurges, like cacti, have a wide variety of shapes. Some keep their fine leaves for at least part of the year; others are leafless. Some, such as *E. horrida,* have prickly spines, which are modified stipules or residual flowers. They also differ greatly in size. Some measure only a few centimetres whereas others are trees several metres high.

Climbers

The need for light makes all plants reach upwards. For this they usually need strong stems. A number of plants, however, cheat and climb the nearest support. Some coil round their supports, others catch hold with tendrils, suckers, special rootlets or spines. Such climbing plants include a number of herbaceous types and even woody plants, such as tropical lianas.

Trumpet creeper *Campsis radicans*
The trumpet creeper, a high-climbing plant of the southern United

Trumpet creeper

States, does not make new stems every year like hop or wild pumpkin. Its stem becomes woody and from it grow new shoots bearing clusters of large, brilliantly-coloured, trumpet-shaped flowers. This creeper climbs high up walls and fences, to which it attaches itself with its aerial rootlets. These rootlets are absent in the even lovelier but more tender *Campsis grandiflora,* a related climber native to China and Japan with larger flowers.

Creeping jenny
Lysimachia nummularia
The inconspicuous creeping jenny can be found in the grass in damp meadows alongside woodland streams and at the bottom of ditches. The attractive, yellow, star-like flowers grow on short stalks close to the leaves, which are set opposite each other in pairs. Because it has neither tendrils nor a twining stem it has to remain resigned to creeping along the ground. Its round leaves give it the name *nummularia,* which means 'coin-like'.

Common hop *Humulus lupulus*
Hop is a plant that is essential in the brewing of beer. It gives beer its typical aromatic, slightly bitter taste and at the same time prolongs the life of the beer. Other uses of hop are as food and as a drug.

The stem of a hop plant always twines around a support in a clockwise direction. The strong, hooked hairs that cover its skin are an additional climbing aid. Only female plants are grown in hop fields. Male

Common hop

plants are not desirable because they do not bear catkins containing lupulin, the bitter resinous substance important in the brewing of beer. Furthermore, they are destroyed to prevent pollination. If fruits develop, the female catkins disintegrate and their precious lupulin is lost.

Creeping jenny

Field bindweed
Convolvulus arvensis

The metre-long, slender stem of field bindweed coils anti-clockwise round its support, which can be anything from a blade of grass to a young sapling. The stem is covered with spearhead-shaped leaves and produces single white or rosy, trumpet-shaped flowers that open in early morning and close again after midday. This plant is very difficult to eradicate because it spreads readily by vegetative means. Even a tiny piece of the long, white underground rhizome will give rise to a new plant.

Field bindweed

Australian desert pea *Clianthus*

The beautiful plants of the genus *Clianthus,* with flowers shaped like a butterfly, are native to the warm Australian region, but are also grown in the temperate regions of the north. Like many members of the pea family they climb by means of tendrils. Their large blooms in varied shades of red are gems of the greenhouse, particularly *Clianthus dampieri* of West Australia. *Clianthus puniceum* is native to New Zealand.

Australian desert pea

Wild pumpkin *Cucurbita pepo*

This annual of the gourd family is a recumbent plant, even though it could climb, being furnished with branching tendrils that are sensitive to the touch and twine round a support.

The pumpkin as a vegetable is noteworthy for its size. The rough bristly stem is two or three and sometimes as much as ten metres long. It is covered with shallowly lobed leaves and golden-yellow flowers measuring 10—14 cm (4—5.5 in) in diameter. The ovaries of these flowers develop into fruits — orange-yellow berries of various shapes and sizes, some of which weigh 70—100 kilograms!

The flowers are a striking example of monoecism, where separate male and female flowers are borne on the same plant — the former have only stamens and no carpels, the latter only carpels and no stamens.

Wild pumpkin

Garden Captives

Many plants can be grown in the garden in beds or rockeries. However, very few garden plants are ever found growing in the wild. This is because most garden plants are specially bred and are often hybrids. Thus they need careful tending and, although they may have been derived from wild plants, they are not all particularly hardy.

Coneflower *Rudbeckia*

North American coneflowers are widely grown in European gardens and can be propagated with ease. The most widespread is *R. laciniata,* particularly the double form up to 2 m (80 in) high. In moist soils it readily reverts to the wild and makes whole masses. The tiny, dark-coloured florets of the cone-like central disc together with the long ray florets, generally coloured bright yellow, make lovely daisy-like flowers that are produced in profusion from late summer until autumn. They are great garden favourites.

Peach-leaved bellflower

German iris

Coneflower

208

German iris, Bearded iris
Iris germanica

The sword-like leaves of irises grow from a thick, branching rhizome. The elegant violet-blue flowers are each composed of six segments. The three outer segments droop and have a tuft of yellow hairs at the base; the three inner segments stand upright and form a vault above the three petal-like styles that conceal the stamens. The fruit is a capsule.

Peach-leaved bellflower
Campanula persicifolia

Peach-leaved bellflower is a wild species of the open woodlands, deciduous forests and thickets of Europe and Asia. Its large, wide-open flowers have made it a popular plant for garden decoration. Though it bears only a few flowers at a time, they are large and coloured blue or sometimes white. Some cultivated varieties produce semi-double flowers.

stems of some turn woody at the base. The showy bell-like flowers open in succession from June until the first frost, making them one of the longest-flowering of perennials.

Swan river daisy
Brachycome iberidifolia
This dainty Australian flower, a shrub-like plant almost 30 cm (12 in) high, is smothered with some 200 star-like blossoms during the flowering period.

Beard-tongue

Shooting star

Shooting star, American cowslip
Dodecatheon
Shooting star has long leafless stems terminating in a cluster (umbel) of three to twenty small flowers in shades of pink to red.

Beard-tongue *Penstemon*
These pretty North American perennials are true ornaments of the garden. There are a number of low-growing as well as tall species. The

Californian poppy
Eschscholtzia californica
The warmth-loving Californian poppy is a native of the Pacific region of North America where there are both annual and perennial types. The type species had shining flowers coloured yellow with an orange blotch. Nowadays this plant is grown in a number of cultivated forms, both double and semi-double, in a variety of colours.

Californian poppy

209

Plants and Animals

Some animals are beneficial to plants. For example, insects (and some birds and mammals) often act as pollinators. Flowers are equipped to attract insects, and to ensure that the insects perform the right actions, by their colours, the presence of nectar and their shapes. Animals also help to disperse fruits and seeds. Small mammals and birds eat fleshy fruits and eliminate the seeds with their droppings. Hooked fruits catch on to mammals' fur and sticky fruits are often dispersed on birds' feet.

However, many animals are harmful to plants. For example,

herbivores eat green plants, as do newly-hatched caterpillars and some birds.

The relationship between a plant and an animal may be fairly loose, in which case neither relies on the other for its survival. In other cases, a plant and an animal may be more dependent on each other, as in many symbiotic relationships. In extreme cases, such as that of *Yucca filamentosa* and the

Yucca moth *Pronuba yuccasella,* the relationship may be essential to the survival of both partners.

White clover *Trifolium repens*
If you look long enough you may come across a four-leaf clover amidst the many three-leaved ones. As everyone knows, the four-leaf clover is a symbol of good luck. And you will certainly be lucky to find one, as four-leaf clovers are very few and far between.

White clover is a creeping species with clusters of whitish to pinkish, honey-scented flowers. The scientific name for such a cluster is a capitulum. The flowers have a deep calyx and loose petals; there are about 50 to a cluster. They are

Fireweed

Evening primrose

Sweet violet

generally pollinated by bumblebees, which suck nectar from the flowers with their long probosces.

Clover and other leguminous plants are very important to agriculture because they influence the fertility of the soil. The reason for this is that the strong root is often covered with nodules containing nitrogen-fixing bacteria *Rhizobium* — bacteria capable of converting atmospheric nitrogen into organic nitrogen compounds. They can produce 20 kg (44 lb) of nitrogen in organic substances per hectare per year, which about equals the amount contained in 150 kg (330 lb) of cow manure.

White clover

Wild cabbage

Wild cabbage *Brassica oleracea*
An example of an undesirable association between a plant and animal is the relationship between both wild and cultivated varieties of cabbage and the cabbage white butterfly. In summer this butterfly may be seen in large numbers in fields and gardens where it lays its eggs on the underside of the leaves. When the caterpillars hatch they feed with gusto on the soft parts of the leaves thus damaging the plant.

Wild cabbage is native to the Mediterranean region and possibly southern England. It has been cultivated since ancient times, thereby resulting in a great number of varieties, including cauliflower and sprouting broccoli, which look quite different from the original wild plant.

Most like wild cabbage in appearance is kale, which forms only a loose rosette of spreading leaves. The leaves of cabbage and Savoy are formed into a round, compact head on a short, thick stalk. Cauliflower, on the other hand, has only

211

a few leaves enveloping a head formed by a dense fleshy mass of yellow-white flower heads on a short, branching stalk. Brussels sprouts, developed in Belgium in the 18th century, bear miniature cabbage-like heads on an erect stem up to 1 m (40 in) high, topped by a tuft of larger leaves that are also edible. Kohlrabi has a stem swollen at the base and a tuft of loose, long-stalked leaves at the top.

All varieties are good vegetables containing vitamins C, B, and A, sometimes E and K, mineral substances and trace elements. They are prepared and eaten in many different ways.

Evening primrose *Oenothera*
Some plants have flowers that do not open until evening, waiting to be pollinated by hawk moths. These nocturnal pollinators behave much the same as tropical hummingbirds. When they visit a flower, they hover above it by rapidly moving their wings and with their long probosces suck the nectar inside, at the same time pollinating the flower. The delicately scented, golden-yellow flowers of the evening primrose remain open until the following day and then they close. The fruit is a cylindrical capsule containing a great number of tiny smooth seeds which are readily dispersed.

Evening primrose is native to America, where it rapidly spread by roadsides and waterways. Nowadays it may be found by waysides and on banks, alluvial deposits and waste ground throughout most of the northern hemisphere. Gardeners have also helped it to spread. In some places the evening primrose is grown also as a vegetable; its sweetish root may be eaten as a salad called 'rapontika'.

Yucca *Yucca filamentosa*
The existence of this robust plant depends on the tiny moth *Pronuba yuccasella,* which is the only insect that can pollinate its flowers. In fact, the insect lays its egg inside the ovary of the flower and then deliberately pollinates it to ensure that a fruit will develop. The larva of the moth feeds on part of the fruit (leaving the seed intact) after it hatches. In this way the survival of both species is ensured, but if one were to become extinct, then the other would too.

Yucca filamentosa is an American species that tolerates the extreme climate of deserts. Its stiff, tough leaves are up to 50 cm (20 in) long and they form a rosette from the centre of which grows an erect stem up to 2 m (80 in) long. At the tip of this there is a panicle of nodding, bell-shaped greenish-white flowers.

Fireweed, Rosebay willow herb
Chamaenerion angustifolium
Butterflies and moths have their favourite plants, to which they remain faithful. The females entrust them with their offspring, laying their eggs on the leaves, thus ensuring a supply of food for the caterpillars when they hatch. This matters to man in the case of cabbage, but fireweed has no economic use. Fireweeds and balsams are 'home' to the deep-green, 'big-eyed' caterpillars of the elephant moth.

Fireweed, along with groundsel and raspberry, is a typical plant of woodland clearings. During the flowering period such clearings are flooded with its lovely dark pink blossoms, which are followed by longish, angular capsules filled with tiny seeds, each of which has a long, white pappus – a kind of hairy parachute. The seeds are dispersed far and wide by the wind

Heartsease

212

and in mountain regions the fireweed thus readily spreads even up to elevations of 2000 metres (1.2 miles).

Sweet violet *Viola odorata*
Violets produce two kinds of flower. The first to appear are the familiar fragrant violet blossoms that can be found in woods in spring. These flowers are well-adapted for cross-pollination by insects and have elaborate mechanisms for preventing self-pollination. But they produce few seeds. The important role of preserving the species has been taken over by the insignificant, green, closed flowers that appear after the striking large ones have faded. They are fertilized by their own pollen and the closed flowers produce capsules filled with seeds without opening at all.

The seeds of sweet violet have a fleshy, oily appendage which is a favourite food of ants, which benefit the plants by dispersing their seeds in the neighbourhood. Sweet violet also spreads by means of its creeping runners, which take root and give rise to new plants.

Violets are often grown in nurseries as well as in the garden in a wide variety of forms (large-flowered, double-flowered, repeatedly-flowering) as well as colours (violet, white, pink, yellow).

Heartsease, Wild pansy
Viola tricolor
The hardy and easy-to-grow pansy is a common flower of parks and gardens. It was not always as we know it today, for it is the result of the cross-breeding of a number of species. One of its ancestors is the wild pansy or heartsease, a common weed plant of fields, gardens and waste places that flowers practically the whole year long. Its blossoms are half as large and not nearly as beautifully coloured as those of garden pansies. The petals are in combinations of yellow, white, violet and blue.

The green top parts are used in folk medicine to treat skin irritations.

Meadow clary *Salvia pratensis*
Meadow clary, with its dark blue-violet flowers and pleasant fragrance, is a common plant of warm regions. It grows on sunny banks and meadows. A peculiarity of the flowers is that they have a movable style and movable stamens. This arrangement is used to ensure cross-pollination. The flowers are pollinated by bumblebees, whose long probosces enable them to reach the nectar at the bottom of the flower tube. As soon as a bumblebee alights and begins penetrating to the heart of the blossom, it moves a lever-like mechanism that causes the stamens to bend and dust its body with pollen. The pollen is then transferred to the stigma of another, older flower whose anthers have already shed their pollen. In such a flower the pistil, with its stigmas, curves in an arc so that the bumblebee cannot avoid brushing against it.

Meadow clary

Deceptive Appearances

Plants are not always what they seem. For example the succulent spurges look like cacti, but they are unrelated (see page 205). Nor are flowers always what they appear to be. Rafflesia has flowers that resemble mounds of decaying meat and the 'flowers' of the cypress spurge are far less simple than they seem.

Cypress spurge

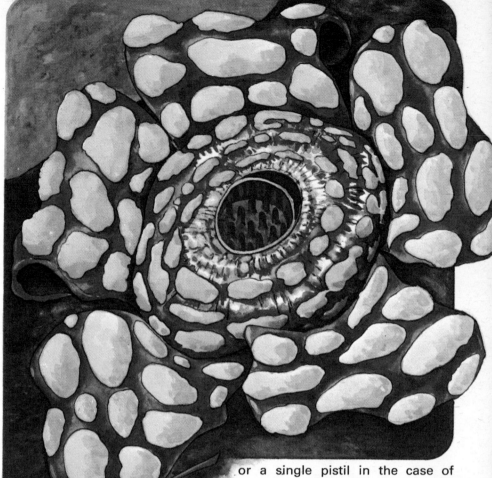

Rafflesia

Rafflesia *Rafflesia arnoldii*

This species of rafflesia has the largest flowers in the whole of the plant kingdom. Its strikingly large fleshy blossoms measure up to 1 m (40 in) in diameter and weigh nearly 5 kg (11 lb)! The plant has no stem; it consists only of the giant flower, which grows and feeds as a parasite on the roots of woody kangaroo vines by means of a system of root-like outgrowths called haustoria. The flowers are of separate sexes and have a fleshy disc containing either carpels or stamens surrounded by five or more brick-red petals with prominent warts. It is impossible not to notice

a rafflesia; its pronounced stench resembles the smell of decaying meat. This stench, however, has the same function as the pleasant fragrance of many flowers — it attracts swarms of flies which pollinate the flowers.

Cypress spurge
Euphorbia cyparissias

Cypress spurge is a persistent European meadow species found in grassy and sunny places. Its flowers are among the simplest and most insignificant in the plant realm. They are of separate sexes and a single flower is either a single stamen in the case of male flowers

or a single pistil in the case of female flowers. However, the seemingly simple 'flower' displayed to our view is actually a multiple inflorescence in which the minute flowers are clustered. In its centre is a single female flower — a round pistil with three ovaries and three stigmas. The long-stalked pistil juts far above the surrounding groups of male or staminate flowers. The complex inflorescence, called a cyathium in spurges, is protected by a cup-like involucre made up of five bracts. The false flower has yellow nectaries and the whole is set off by two enlarged, opposite bracts. These are yellow-green at first, turning red after flowering has finished. In some spurges, such as poinsettia, the large spreading bright-red bracts are often mistaken for petals.

The fruit of the female flower is a warty capsule containing three seeds. The seeds, like those of violets, have an oily fleshy appendage on which ants feed. The poisonous, milky juice of spurges protects them from most herbivores.

Grass tree *Xanthorrhoea*

The grass tree looks like a huge clump of grass set on top of a column. Often this column is not even visible because it is covered with old dry leaves. In cultivation the old leaves are trimmed to reveal the tall (3—4 metres) stem and the tree-like character of this plant. The leaves resemble the leaves of lilies and irises. This is not surprising for the Xanthorrhoeaceae family and the Liliaceae (lily family) both belong to the order Liliales. Only its flowers are not as lovely. They are small and arranged in racemous clusters. There are several species of grass tree growing in the savannahs of Australia, Tasmania and New Caledonia. They produce the acaroid resin called yellow gum which is of commercial value as a paper coating. The leaves are used as fodder.

Grass tree

Darling pea *Swainsona*

Swainsona is a large Australian genus with 45 species. With the exception of several New Zealand species, all grow in Australia. It is the southern hemisphere's equivalent of the northern hemisphere's milk vetch, from which it differs mainly in that its flower has a broader and more developed keel. The shape of the keel and fruit (pod), on the other hand, show a close relationship to the bladder sennas of Europe and Asia.

Swainsona species are herbs with an upright or prostrate stem, which sometimes turns woody at the base. The leaves are composed of many leaflets. The rosy, red, purple or yellow flowers are arranged in stalked racemes. Larger flowers up to 2 cm (0.8 in) across are only few to a cluster, whereas small flowers may number as many as 30 in a single long raceme. The structure of the flower is characteristic of the pea family, with five sepals and five petals. The bottom two petals are fused and form a long beak, which in *Swainsona* is the same length or longer than the narrow wings on either side. These petals are covered on top by a broad, recurved fifth petal called the keel. The stamens are fused, except for one which is free, and form a tube round the pistil. The ovary develops into a fruit characteristic of this family — a pointed, inflated pod with many small seeds. The ripe pod dries and splits open, hurling the seeds out.

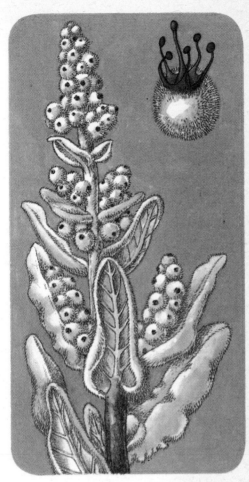

Blanket plant

Blanket plant, Lamb's tail *Lachnostachys verbascifolia*

The rare *Lachnostachys verbascifolia* is one of the members of this genus found only in the sand flats of West Australia. It is protected from the sun's heat by a thick white coat of felt.

Darling pea

215

Cone-bearers

A cone is a characteristic feature of conifers, that is trees such as pines, spruces, larches, firs and cedars. It is a dry, woody reproductive structure composed of a central axis upon which is borne a spiral arrangement of overlapping scales. The scales are pressed tightly together for a long time. Only after the seeds they carry are ripe do they open. The seeds are then carried away by the wind.

Norway spruce *Picea abies*
In spring the golden pollen of spruces is shed in vast quantities by the male cones and dispersed by the wind. The pollen grains are equipped for flight with two air sacs that act like wings. Large amounts of pollen go to waste, but some grains find their way to the ovules on the scales of the female cones. These stand like small erect 'candles' at the tips of the branches in the top of the tree. After pollination the cone becomes larger and heavier until the ripening cone tips over and hangs downward. During this process the cone changes colour from reddish-purple, when young, to green, yellow and finally the brown colour of maturity.

On dry days, occasionally in winter but usually not until spring, the cones open up and that is when the winged seeds fill the air as they are carried by the wind. This occurs only once in four to six years; at higher elevations even once in ten years. Trees do not begin to bear seeds until they are 30 years old.

The spruce does not shed its needle-like leaves in the autumn. However, it does not retain the same leaves throughout its lifetime. They are shed gradually and each leaf has a life of five to seven years.

The spruce is cultivated for its wood, which is widely used in the building industry and also for the production of paper and making musical instruments. The bark yields tannins, rosin and turpentine, and the needles a fragrant essential oil used in cosmetics.

Common or European larch
Larix decidua
The needles of the larch are soft, supple and coloured a delicate green. They do not survive freezing weather and so, like the leaves of broad-leaved trees, they turn yellow and fall to the ground in the autumn.

When 15—20 years old, the larch produces cones at the same time as it puts out new leaves in spring. The yellow male cones hang downwards and the carmine-red female cones are upright. By autumn the female cones have ripened and are woody, but they do not open to release their winged seeds until spring. Seeds are produced at intervals of four to five years.

The common larch is native to

Norway spruce

Europe from two different places: from North America and from East Asia. They do not have prickly needles like the conifers native to Europe; instead they have small scale-like leaves pressed tightly to the twigs. On the upper side of each leaflet is a resinous gland, which emits a pleasant scent when the twig is rubbed between the fingers. The fragrant volatile oil *oleum thujae* is distilled from young twigs.

The fruit is a woody cone composed of several pairs of spine-tipped seed scales. The seeds are either winged (American arborvitae, *T. occidentalis*) or wingless (Chinese arborvitae, *T. orientalis*).

The American arborvitae or white cedar is a native of eastern North America, where it forms large stands. The trees, which reach a height of 20 m (66 ft), are slender and conical and the branchlets are more or less horizontal. Europeans, however, are more accustomed to the lower, generally shrubby, cultivated forms.

The Chinese arborvitae differs from the American arborvitae by having the branchlets more or less vertical and larger cones. It is a native of northeastern China, whence it spread to Japan and Europe.

Common larch

the Alps where it can grow at high mountain elevations. Native to the Carpathians is the related species *L. polonica*. Nowadays the common larch is widely grown throughout central and northern Europe for its high-quality wood.

Only rarely does this conifer form woods and when it does the stands are thin. Being a light-loving tree, it usually grows at forests' edges.

The wood of the larch is reddish-brown, light, tough, flexible and resistant to damp. It is used to make planks for ships, bridges and buildings. It contains a large amount of a honey-coloured resin that was used in folk medicine to treat coughs and as a healing ointment.

Arborvitae *Thuja*

Several species of this genus have found their way to the gardens of

Arborvitae

Tasty Fruits

Many trees and shrubs are prized by both man and animals for their fruits and some have fruits that are not only tasty but also decorative. The bright colours of the fruits are intended to attract animals, who then help in dispersing the seeds. However, man has also discovered that many of these fruits are edible and has gathered and eaten them for thousands of years. In the course of time, many ways of preparing them have been discovered.

Dog rose

Dog rose *Rosa canina*

The rose is described as the queen of flowers because of its beauty, fragrance, colour and elegance. This applies only to cultivated roses, not to the wild relative that is one of their ancestors. Rather than beautiful, the dog rose is modestly pretty with its profusion of delicately-scented pink blossoms that in spring transform the unpleasantly prickly shrub into a veritable bouquet. These shrubs adorn field paths both during the flowering period and in the autumn, when their pink blossoms are replaced by coral-red hips.

Hips are the false fruits of the rose. The true fruits are hard, hairy achenes located inside the receptacle, which is the part that ripens into the fleshy hip.

Hips are collected and dried for the food and pharmaceutical industry and in smaller measure also for home use. The taste of rose hip tea is not only refreshing but also a pleasant change from regular tea. Rose hips are also an excellent source of vitamin C (up to 40 %) and contain sugars as well as citric and malic acid. During World War II, when oranges were scarce, thousands of tons of rose hips were collected to make rose hip syrup. This and rose hip wine are still made in many country areas.

The arching twigs of the dog rose make dense thickets that serve as a shelter for birds and their offspring. The hips also provide them with a rich source of food.

Besides flowers and hips, the shrubs are also dotted here and there with rosy-red tasselled balls with a hard marble inside. These are made in the same way as the smooth brown marbles (oak apples) found on oak trees. They are galls caused by tiny gall wasps (e. g. *Rhodites rhosae),* that pierce a leaf or bud and lay an egg inside. The egg changes into a larva which you will find inside the gall. False galls are formed also by the larvae of flies, mites, aphids, some beetles and other insects.

Blackthorn, Sloe
Prunus spinosa

The white blossoms of blackthorn appear in May or April, some time before the leaves. Later in the year small, plum-like fruits, known as sloes, develop. When ripe they are blue-black and coated with a waxy bloom, and the pulp is green and extremely bitter to the taste. This bitterness becomes slightly less after the first autumn frosts and, although not good for eating, sloes can be used in a number of ways. They may be used to make jam or wine and they can also be dried. Soaked in gin with sugar they make sloe gin — a liqueur-like drink. They contain sugars, vitamin C, tannins, colouring matter and several other substances. The dried fruits or fresh pulp can be used for treating diarrhoea and diseases of the urinary passages. The small, ovate leaves are used as a substitute for Chinese tea.

Most effective for medicinal purposes are the small white flowers of the blackthorn. They have a slightly bitter fragrance, for the blossoms contain the glycoside kamferin and a small amount of prussic acid. That is why the flowers must be dried rapidly so that they do not turn brown. They are used internally in the form of an infusion to treat nausea, regulate the body metabolism and increase the flow of urine.

Blackthorn

Hazel *Corylus avellana*

The dog rose has lovely large blossoms, the blackthorn small ones in great profusion, but the hazel does not seem to have any at all. And yet it does flower – very early in spring, in March and April, thus providing bees with their first food. The bunches of male flowers, the drooping catkins, are formed in the autumn. Each catkin consists of hundreds of tiny flowers. The female flowers, shaped like small buds, appear only at the end of winter; jutting from their tips are thread-like red styles with brush-like stigmas. Both types of flowers are very simple and appear before the leaves. And because the hazel is a typical wind-pollinated plant the catkins sway readily in the breeze, causing their pollen to fall on the female flowers.

Following pollination, the catkins are shed and the ovaries of the female flowers develop into fruits – woody oval nuts. Botanists describe the hazel nut as a one-seeded fruit with the kernel placed freely inside a hard woody pericarp. When ripe, the brown nuts fall from their green leafy cups. They are a favourite food of squirrels, mice, jays and woodpeckers, which thus aid in their dispersal (nuts stored away and forgotten may then germinate).

The leaves of the hazel have medicinal properties (for treating rashes and inflammation of the intestines), but even more important are the tasty oily kernels which contain 50-60 % oil, 15 % proteins and 2-5 % sugar. The oil is yellowish, very nourishing and tastes a little like almond oil.

Hazel shrubs are widespread in Europe where they grow in groves, hedgerows, on banks and in open woods. They reach a height of 6 m (20 ft) within a few years. Because of their great vitality, even the constant nibbling of wild animals or grazing cattle poses no problem.

Archaeologists have found remnants of hazel nuts in the dwellings of Neolithic man. The hazel has been cultivated since ancient times, not only for its nuts (chiefly in Turkey) but also for its soft, flexible wood, used to make sticks, hoops and pipes. Young shoots are used to weave baskets and the bark is used in tanning and to make a yellow dye.

The hazel is also important in gardening. Breeding and selection have yielded many ornamental forms with leaves of various shapes and colours.

Plants that could be used by man always attracted his attention and it is no wonder that they became the subject of strange legends and

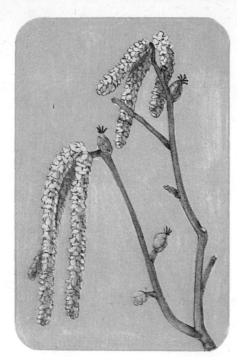

Drooping catkin

superstitions. Of Greek origin is the belief that the kernels of hazel nuts promote a man's growth. The nuts were also recommended for relieving headaches. This sounds reasonable, but the nuts were also ascribed magical powers that could make a person invulnerable, halt a bullet in flight, extinguish fire and avert a storm.

The hazel was believed to have miraculous powers and so was considered sacred in days of old, as was the oak, and felling it was forbidden. People believed that a hazel twig could show the way to buried treasure.

Mountain ash, Rowan tree
Sorbus aucuparia

The mountain ash is a close relative of the apple and pear tree. Dense clusters of tiny white flowers deck the tree from May to June, but the mountain ash is at its loveliest in late summer, in September, when it is covered with bright red berries. Their pulp is bitter-sour and so the fresh berries are inedible. They are also slightly poisonous, due to the fact that they contain sorbic acid. However, they also contain numerous organic acids, sugars, tannin, pectin, vitamins B and C and carotene, which is what gives the berries their coral-red hue. Being

Hazel

rich in vitamin C, the mountain ash is sometimes called the lemon of the north. Some types of cultivated mountain ash contain up to 180 mg vitamin C in each berry, most of which is retained even when the berries are dried.

Higher elevations provide the best conditions for its growth and that is where the mountain ash is most widespread, occurring practically up to the forest limit. It grows slowly and thus cannot compete with other trees to form a closed stand; in the forest it does poorly. Inasmuch as it is a light-loving species, it does best planted alongside roads and on banks at higher elevations. The berries remain on the tree for a long time after they are ripe and are thus welcome food for birds, mainly of the thrush family. The birds, in turn, distribute the seeds throughout the countryside.

Besides the wild mountain ash with bitter-sour berries, there is also a cultivated variety (*S. aucuparia* var. *edulis*) which has larger, sweet berries that are used to make compotes and preserves. It contains twice as much vitamin C as the lemon, a large amount of sugar and provitamin A. The berries are also used to make a liqueur and formerly also vinegar.

Mountain ash

Common elder *Sambucus nigra*

Elder has been used in folk medicine since days of old. All parts of the plant have healing properties. The twigs were used to relieve toothache, young leaves mixed with wheat flour were used to heal burns and bites from mad dogs and powder from the dried leaves stopped nosebleeds. The parts mainly collected for medicinal use nowadays, however, are the flowers and fruits. The small yellowish flowers, borne in broad flat heads, appear in June and July. At this time the shrub has a very intoxicating scent, but it has a distinctive aroma even after the flowers have faded. The flat flower heads can be dipped in batter and fried like cauliflower. They contain glyco-sides, volatile oils, tannins, resin, mucilages and certain acids. Tea from the dried flowers promotes sweating and lowers the temperature during bouts of influenza and tonsillitis; it also relieves coughs. The flowers can also be used to prepare a refreshing effervescent lemonade. Elderflower wine, with its distinctive aroma and taste, is often called 'the queen of home-made wines'.

The fruits, black berries, are round and soft. The juicy pulp contains a dark red juice formerly used to colour textiles. The fruits also contain organic acids, sugars and vitamins A and C. All parts of the plant contain phytoncide sub-stances that kill or check the growth of bacteria and numerous fungi. Juice from the berries has a beneficial effect in the treatment of neurological diseases, migraine and neuralgias, particularly trigeminal neuralgia. The fresh fruits have a laxative effect, the dried berries the opposite effect. Preserves from ripe fruits have a diuretic and soothing effect on the nerves. The fruits are also used to make wine, jellies and vinegar. Any remaining fruits are eaten by birds, who thus distribute the seeds.

Elder is also spread by man. Since days of old he has planted it near his dwellings, formerly often behind the cattle shed because it was said to prevent plague in cattle. In the wild it also grows in abundance in open woods, in coastal thickets and in nitrogen-rich soils.

Red or Scarlet-berried elder
Sambucus racemosa

The red or scarlet-berried elder differs from the common elder in the colour of the fruits. Another striking difference between them is in the colour of the pith. The branches of red elder have a cinnamon-brown instead of white pith.

The flowers of red elder are small and yellowish-green and appear at the same time as the leaves. They do not last long. The red berries, unlike those of common elder, have no particular use. Juice from the berries is cloudy and the seeds contain a small amount of the poisonous glycoside amygdalin. This, however, in no way affects the tea rich in vitamin C that can be prepared without fear from the skin and flesh of the fruits. All that is necessary is to scald the fresh fruit.

Red elder prefers high, mountainous areas. It often grows in thin woods, mountain pastures and on slopes.

Red elder

223

Blackberry, Bramble
Rubus fruticosus

The blackberry is a shrub with arching shoots up to 2 m (80 in) long covered with straight or hooked spines. Where they touch the ground, the shoots readily take root thereby forming impenetrable thickets. Spines cover not only the branches but also the stalks of the leaves and leaflets and the primary leaf veins. The first year the shoots grow only in length; not until the second year do they branch and bear flowers and fruits. The white to pinkish flowers characteristic of the rose family are followed by black fruits. These are not simple fruits but clusters of small drupes. In a northern location or shaded spot some flowers open later and then the shrub is covered with white blossoms and black fruits.

The blackberry is a collective species composed of a group of small species showing only slight differences. Some of the species readily cross-breed and interbreed so that identifying them is sometimes a difficult problem.

The soft pulp of blackberries is full of a dark juice that colours both hands and mouth when eating them. The violet pigment is used to colour foods. The fruits are used to make beverages, syrups, jams, jellies and pies. The leaves contain mainly tannins. The drug has a soothing effect as a component of gargles and is used in bath preparations for skin diseases and in the form of an infusion in the treatment of diarrhoea and gastric disorders.

Blackberry

224

Wintergreen *Gaultheria*

The large genus of wintergreens, comprising some 100 species, is widely distributed in the mountains of North America, in South-east Asia and in the southern hemisphere in Australia and Tasmania.

The shady woods of North America are the home of the species *Gaultheria shallon*. It is a spreading shrub up to 1.5 m (5 ft) high with broad, leathery leaves and insignificant but pleasantly scented, pale pink flowers. The drooping blossoms are followed by dark fruits borne in large clusters.

Another species that is well known in Europe and cultivated there is the American wintergreen (*G. procumbens*). Its round, red

Common bilberry, Blueberry
Vaccinium myrtillus

The ripening fruits of the woodland bilberry turn black and sweet in the sunlight of late summer. The small leaves turn red and slowly fall.

Berries grown in gardens are easier to pick than woodland berries. These are cultivated forms obtained by breeding and selection.

The common bilberry is native to Europe and northern Asia. It is plentiful throughout the central and northern part of Europe, forming large masses in damp woodlands and on moors. It grows from lowland to alpine elevations, to the dwarf-pine belt. It has also become widespread in North America and northern Asia.

Wintergreen

berries, called checker-berries, are edible. It is a low shrub with creeping stems that form spreading carpets in its native home — eastern North America — and in summer it bears single pink blossoms.

The leaves make an astringent drug. They contain tannins, the glycosides arbutin and ericolin, sugars and enzymes. The fragrant, volatile oil (called oil of wintergreen), obtained by distillation from the leaves, is used for pharmaceutical preparations and for flavouring confections.

The shrub is prostrate and grows to a height of half a metre or less. From the woody base grow angular green branchlets with thin, ovate leaves, in the axils of which appear greenish-pink flask-shaped flowers. The fruits are black berries covered with a bluish bloom. They can be eaten fresh and also can be used to make compotes, preserves, syrup and liqueur. The juice is used to colour foods. Dried fruits and leaves are used in the form of tea to treat infections of the intestines and bladder and also gastric disorders.

Common bilberry

Useful Timber

Unlike the trees and shrubs that bear tasty fruits, many broad-leaved trees are of little interest to gardeners except, perhaps, as decoration. However, for foresters, cabinet makers, wood carvers, architects and members of many other professions, wood is an important raw material. It is needed in practically all branches of human activity and has been put to good use by man since time immemorial. In the beginning it served to heat the dwellings of primitive man and it is still used as a fuel. Later, wood came to be used in the construction of houses. Many other things are made of wood, including furniture, instruments and toys. Even in this age of plastics and metals, man cannot do without wood.

Aspen *Populus tremula*

The aspen, though it belongs to another family (the willow family), slightly resembles the birch. This is true particularly of the young trees before the bark of the birch turns white. The two, however, are readily distinguished by their leaves. Those of the aspen are round with long, thin, flattened leaf stalks, and are set in motion by even a slight breeze. The aspen is just as undemanding as the birch in its soil requirements. Often the two occur together. In the forest the aspen rarely forms tall trunks; not until the age of 80 years or so does it reach a height of 30 m (98 ft).

The aspen is useful in places requiring fast-growing trees, for example in a devastated landscape. The soft wood is a source of high-quality cellulose and is used to make matches and plywood.

The aspen bears a profusion of flowers in late March before the leaves appear. It is a dioecious species and the wind-pollinated flowers are produced at the end of winter. They are plump, green, feathery catkins. The brownish-red male catkins fall after they have shed their pollen. After pollination the female catkins, which have purple stigmas, produce fruits — capsules containing small seeds covered with a white cottony down. These ripen in May.

Silver birch *Betula alba*

White birch trunks with their dark pores are a beautiful sight in the forest in summer and even more so in winter when other trees have

Aspen

Silver birch

shed their leaves and their beauty is offset by the dark green of the surrounding spruces. Also beautiful is the light, airy structure of the crown of the silver birch before the slender branches are covered with pale green leaves. The birch flowers in April and May, at the same time as it puts out its small sticky leaves. The simple monoecious flowers are borne in separate male and female catkins. The male catkins are usually brown and hang downward from the tips of the branchlets. The female catkins are shorter, coloured green and stand upright. They are pollinated by the wind, which later also disperses the seeds. The cone-like female catkins disintegrate and the separate three-lobed scales carry the winged achenes (three to each scale) through the air. The achenes are tiny and very light. An adult birch annually produces several hundred thousand seeds, of which only a few find a suitable spot for germination. Lovely birch trees, however, may be found growing on cliffs, ruins and unfertile soil. The birch is a very undemanding tree with no special soil requirements; all it needs is plenty of sun. It can even grow near the Arctic Circle, where it either forms separate stands or grows together with aspen and alder.

Birch groves cover a large part of Europe and North America. The birch is a valuable tree both in forestry and industry. A fast grower, it is an important colonizer of barren areas, clearings and fallow land damaged by industry. Its wood, though not very durable, is tough and flexible and used for interior woodwork, furniture and wheels. The twigs are used to make brooms. In spring the birch yields a sweet sap, which is boiled down into syrup or fermented into a beverage; it is also used by the cosmetic industry in hair tonics. The leaves yield a drug used in pharmaceutics — the decoction promotes the flow of urine.

The slender white trunks and loose crowns with their pale green leaves that turn gold in autumn are an ornamental element in every garden, be it the original birch species or a cultivated form.

Pussy willow, Great sallow
Salix caprea

The pussy willow is a shrub or small tree with broadly ovate leaves. The male and female catkins are borne on separate trees. In winter they are covered with scales which break off in the warmth of the sun's rays. Before they open no one can tell whether the shrub is male or female. However, once open, the male catkins can be seen to have long silky hairs (from which the tree gets its common name of pussy willow) and long stamens with yellow anthers. The grey-green female catkins are less conspicuous. The flowers do not have a perianth. In both male and female flowers there is an organ that secretes nectar which attracts insect pollinators and provides them with their first feast of the year. The fruit is a capsule with small, downy seeds.

The pussy willow is distributed throughout all of Europe and far into Asia. It is a light-loving species with no special soil requirements. It grows in riverine woods, alongside streams and, together with the birch, readily covers forest clearings.

The wood is of no particular value. It is soft, flexible and not very durable. It is the twigs rather than the trunks that are used for various purposes.

Pussy willow

Nowadays pussy willows are among the first spring flowers to decorate our homes, but in former times both flowering and non-flowering willow twigs were put to various uses. They were important in telling fortunes.

False acacia, Black locust
Robinia pseudoacacia

A few sunny days in spring are enough for the buds of most trees to begin swelling and for the new leaves to emerge. However, in late April and early May, when everything is turning green, flowering and smelling sweet, some trees look sad and stark in their winter garb amidst all this delicate greenery. Not only are they bare, but here and there on their spiny branches sway the remains of the previous year's fruits — flat brownish pods. These are false acacias — latecomers that hesitate a long time in spring. What they are waiting for no one knows, but their young leaves appear much later than those of other trees, in about the middle of May.

Then, however, the false acacia quickly catches up with the rest, for it is a tree with a great lust for life and at the same time with no special requirements. It grows practically everywhere. It is a native of eastern North America, chiefly of the hill country of Pennsylvania and Georgia, whence it spread throughout the warmer regions of North America. In the 17th century, thanks to the French botanist Jean Robin who brought it to Paris in 1601, it also spread throughout Europe and made its way to northern Africa, East Asia, the Middle East and New Zealand.

False acacia will grow on poor and dry soils. It is used for erosion control on banks and sand dunes because its wide-spreading root system holds the soil. It is this root system that makes it possible for it to grow where other trees would be hard put to do so. The roots extend far from the trunk and are so efficient at taking nutrients from the soil that practically none are left for other plants. This is one reason why nothing else grows in its immediate vicinity. Another is that there is little humus — the leaves shed by the false acacia contain large amounts of tannin and decay very slowly.

The roots aid the tree in still another way. They produce numerous suckers which promote its spread. As a result it is difficult to eradicate this tree.

Even though false acacia has many disadvantages it can be put to good use. In city streets it does not mind the air pollution and is a pleasant sight in June with its hanging racemes of white flowers.

The pleasant fragrance attracts bees which find here an abundance of food and at the same time pollinate the flowers. Apiarists make use of this fact and put their bee-

False acacia

hives in large stands of false acacia.

Its flowers are like those of other members of the pea family. They are made up of several parts: a hairy calyx and white corolla; the keel is recurved and has a yellow-green blotch in the centre. The flowers attract insects not only by their colouring and fragrance but also by the nectar produced in the bottom of the calyx.

Norway maple *Acer platanoides*
The most distinguishing feature of the maple is its fruits. Each one consists of two winged seeds fused together to form what is called a double samara. When ripe, the double samaras break up into two separate sections which are carried by the wind, whirling like helicopter rotors, far from the parent tree.

If a seed comes to rest in a suitable spot it germinates and the seedling begins life with vigour. In time it develops into a tall, bushy maple, 25—30 m (82—98 ft) high. The flowers are arranged in erect cymose panicles. Though rather insignificant and not brightly-coloured (they are yellowish-green), they are quite pretty with their five sepals and five petals with a central disc that produces nectar. Round the margin are eight stamens and in the centre a single carpel. At the base of the stalks supporting the cymes are several striking bracts. The flowers appear before the leaves. These have five to eight sharply pointed lobes. The closely related sycamore has similar leaves, but it is not difficult to distinguish between them. An important identifying feature is the stalk of the young maple leaf which exudes a milky liquid when broken.

Maples do not grow only singly in parks or spacious tree avenues. On the contrary, they are a common component of mixed forests in the temperate zone of the northern hemisphere. And in such a forest light is a scarce commodity. The trees solve this problem by forming a mosaic of their leaves; the leaves with stalks of varying lengths are spread out in a single plane at right

Norway maple

Sycamore

angles to the light so they do not shade one another.

The greyish-white wood of the Norway maple is hard and flexible and is used to make furniture as well as in turnery, but not to the same extent as the wood of the sycamore.

Sycamore
Acer pseudoplatanus
The sycamore is often found in deciduous woods at high elevations, particularly on screes and rocky slopes. It grows slowly, living to an age of 80 to 100 and reaching a height of about 35 m (115 ft). Its greyish-white wood is of very good quality and for that reason it is considered one of the most important of forest trees. The wood is hard and durable and is used to make furniture and tools. That it is truly durable is testified to by the finds of primitive tools of this wood dating from the Stone Age.

The bushy crown of the sycamore affords welcome shade. It is also a pretty sight and for that reason this tree is often planted in parks.

The sycamore bears flowers later than the Norway maple, at the same time as the leaves or immediately after. The small bisexual flowers are coloured yellow-green and arranged in racemes that hang downwards. They are pollinated either by insects or by the wind. The fruits (double samaras) are less widely spread than those of the maple which have wings that form such an acute angle they are almost parallel. Even without their fruits, it is easy to tell the sycamore from the Norway maple by its leaves. Compare the two illustrations and you will see that the leaves of the sycamore are sharply lobed and sharply cleft and their surface is wrinkled. The stalks are often red and the leaves are a darker green on the upper surface than on the lower surface, which is shiny.

Ornamental Shrubs

Man selected from nature not only useful plants but also those that gave him pleasure with their beauty. Plants growing in the wild amidst other vegetation usually have only single flowers that cannot begin to compare with the magnificent beauty of their cultivated relatives — pampered and coddled by man in garden beds. The blossoms of garden forms generally have a greater number of petals. The original single flowers are changed into semi-double or double flowers. However, these extra petals may be obtained at the cost of the stamens and pistils and thus the flowers are often sterile. Parks and gardens frequently include species from distant geographic locations and often it is when such species are cross-bred that gardeners produce the most beautiful specimens of all.

Fuchsia *Fuchsia*
This attractive plant was named after the famous 16th-century German botanist Leonhart Fuchs. Some fuchsia hybrids are grown indoors for room decoration. Often they do not last long and die, as they do not tolerate strong sunlight or drought, or attack by aphids or white fly. There are varieties, however, that can be grown outdoors, and some of these will even survive the winter.

In their native lands — Central and South America and New Zealand — fuchsias grow as shrubs, small trees and sometimes also as climbers in the undergrowth in shaded, moist spots of forests at higher elevations. There are some 60 known species. The flowers of most are coloured red and white. They are long-stalked and hang downwards. They have a long slender corolla tube, coloured four-pointed sepals, and striking protruding stamens and carpels. The fruit is a berry.

Red dogwood *Cornus sanguinea*
The red dogwood is found throughout all of Europe, where it grows in open woodlands, riverine woods, waterside thickets and forest stands. Sometimes it is planted in gardens together with white dog-

Fuchsia

232

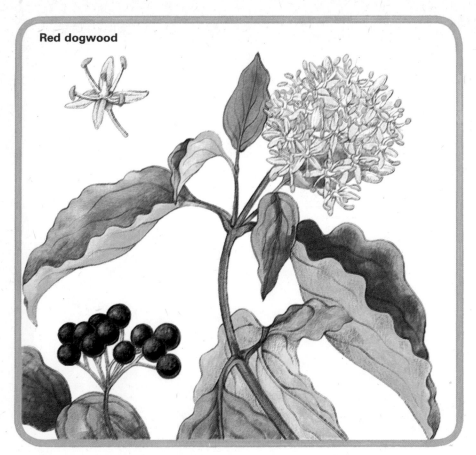

Red dogwood

The leaves appear at the tips of the branches after the flowers have faded, when the ovaries develop into fruits — round, coral-red berries pressed tightly to the branches and decorating the shrub in summer. There are only a few fruits on each twig and they fall soon after they ripen. The fruits, flowers and bark contain the poisonous glycoside daphnin. But this does not harm birds, which feed on the berries and distribute the seeds.

The low shrubs, about 1 m (40 in) high, grow in shaded groves and damp deciduous woods, mainly beech woods, throughout practically all of Europe and part of Asia.

The fact that it flowers in early spring is the reason mezereon is grown for decoration in parks and gardens.

wood, particularly in hedges, even though it is not as attractive as its North American and East Asian relatives such as *C. florida and C. kousa*. Red dogwood, with its straight red stems, is loveliest in autumn when its leaves turn bright red and the twigs bear terminal clusters of blue-black berries, which are inedible. When the shrub has shed its leaves its glossy dark-red shoots make an attractive display in winter.

The firm hard wood of this dogwood is used to make walking sticks and trinkets.

Mezereon *Daphne mezereum*

It is not usual for a plant to attract insect pollinators with its calyx, but so it is with mezereon. Its long, trumpet-shaped calyx is coloured pinkish-red and serves as a substitute for the missing corolla. Not only the colour but also the sweet, almond-like fragrance attracts insects, which feast on the sweet nectar produced at the base of the flower. The small clusters of flowers cover the bare ends of the branches in February and March.

Mezereon

Magnolia *Magnolia*

Comprising 35 species, it includes shrubs and trees native to North and Central America, tropical Asia and the Himalayas. In the warm and temperate regions of Europe they are valuable and extremely decorative features in parks. They are leading ornamentals, not only because of their striking large leaves but first and foremost for their large, solitary flowers. Besides being beautiful they are also of interest from the botanical viewpoint. Their structure is extraordinarily old and primitive: spirally arranged petals, stamens and pistil, large foliaceous stamens, raised receptacle with free carpels.

The magnolias of Europe's parks generally flower in spring. Japanese and Chinese magnolias bear flowers before the leaves (*M. stellata*). North American species flower after putting out leaves (*M. tripetala*). The blossoms are white to pink; those of *M. grandiflora* measure 20—30 cm (3—12 in) across. Besides type species, also grown in gardens are hybrid forms such as the saucer magnolia (*M. soulangeana*).

Magnolia

Service-berry, Snowy mespilus
Amelanchier

Most service-berries are shrubs up to 12 m (39 ft) high or small trees. The 25 species of this genus, which belongs to the rose family, are native to North America except for several south European and Asian species. Service-berries are very good for planting out in parks. They have no special site requirements and make nice shrubs. The white flowers, arranged in loose racemes, appear at the end of April and throughout the month of May in lesser or greater profusion. These are followed by clusters of bluish-black fruits (pomes). They are usually covered with a faint bloom and have a sweetish juicy pulp. Though edible, they take long to ripen and do so in succession.

The species of greatest ornamental value is *A. laevis* which, besides clusters of white flowers, has strikingly coloured leaves. The young tender leaves are brownish-red, turning fiery-red in autumn. *A. canadensis* has scarlet-red foliage in autumn, but the young spring twigs and leaves are silvery. So are those of the common service-berry.

Service-berry

Witch hazel

Witch hazel *Hamamelis*

Witch hazel wakens from its winter sleep much sooner than other plants. Its hardy flowers, coloured yellow, red or orange, open regardless of whether it is cold or not and even when there is snow on the ground. They have wavy petals and are arranged in clusters. Those of Chinese witch hazel (*H. mollis*) are relatively large, whereas those of *H. vernalis* are small, but both have a sweet fragrance. The flowers of *H. japonica* are interesting and are yellow with sepals that are coloured purplish inside.

The genus consists of six species, some native to North America, others to East Asia. They are deciduous shrubs with leaves greatly resembling those of hazel or alder. The foliage turns an attractive orange to scarlet-red or yellow in autumn, depending on the species. The most attractive feature of witch hazels is that all, excepting *H. virginiana,* flower in winter and early spring. *H. virginiana* is a native of eastern North America, where it forms large masses. The small flowers are borne in autumn from October to November, but the capsules do not ripen until the following year. When ripe they suddenly burst open with a pop, scattering their seeds throughout the neighbourhood. The leaves and bark of *H. virginiana* have medicinal properties – both as a tonic and as a mild astringent.

Gifts of the Southern Hemisphere

Growing in the tropics and sub-tropics are unusual herbaceous and woody plants that are not found in lands with cooler climates. The few that are cultivated in greenhouses give us some idea of the exotic plant life of distant lands. Others are no strangers. The fruits of the coconut tree are so long-lasting that they can be shipped great distances without damage. On the other hand, such interesting fruits as those of durian spoil quickly.

Cycad *Cycas*

Cycas is a palm-like tree, but it is not in fact related to the palms. Instead, it is a member of the group called gymnosperms. Cycads and other gymnosperms are distinguished from lower plants, such as ferns, by the fact that the structure of their woody stems is more complicated. The sex-organs also are more efficient. However, they are much simpler than those of flowering plants and the male organs are always separate from the female organs. Both types are generally in cones. The male cones are composed of scales with two pollen sacs; the female cones have scales that bear naked ovules.

In *Cycas,* however, the ovules are not borne in cones. Instead they are borne in two rows on special structures that resemble leaves with fringed blades and stalks. After being pollinated by the wind, the ovules change into red seeds with a hard shell.

Sago is an edible starch, a product of commerce obtained from *Cycas,* from the inner portion of the trunk. The column-like trunks with remnants of old leaves are felled, cut into cross-wise sections and split longitudinally. Inside is a starchy pith which is then processed. Repeated washing and passing through sieves yields grains of edible starch which are dried and then roasted to give the end product: sago. This is used in making hard biscuits and in preparing light, easily digested foods. The two cycads used for this purpose are *Cycas circinalis* of Indochina and *C. revoluta* of southern Japan and China.

Cycad

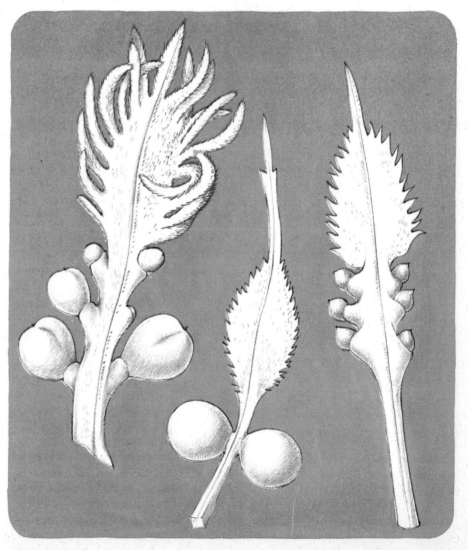

236

Durian *Durio*

Durians, the fruits of *Durio zibethinus,* are the tastiest fruit in the world to some people. Others, however, say they are not fit to eat. The reason for these contrary opinions can be found in the fruit itself. Unripe fruits are green and odourless. As they ripen, they turn yellow and acquire a pronounced scent. This, however, is far from pleasant — it is an odour that may be best described as a combination of rotten eggs, sweaty feet and garlic.

The thick rind of the fruit is rough and covered densely with conical spines. Inside is a buttery pulp coloured yellow to pinkish. The fruit is a five-chambered capsule with several large seeds in each chamber. When roasted the seeds taste like roasted chestnuts. Each seed is softly cushioned inside the capsule in a fleshy aril which developed from the stalk of the ovule. Those who do not mind the odour can partake of a tasty treat.

Durian belongs to the bombax

family and is native to tropical Asia. It is evergreen with rather small leaves covered with red scales on the underside. The flowers grow on thick branches as well as directly from the trunk.

Bottle-brush tree *Callistemon*

The flowering branches of this tree look exactly like brushes for cleaning bottles. The flowers are a lovely red, sometimes pink or even yellow. The cylindrical spike is composed of many tightly-clustered flowers. Protruding from each one are numerous stamens with red filaments. The flowers are arranged like the bristles of a bottle-brush on a slender stalk which extends beyond the flower spike and carries leaves. The flowers are replaced by round, woody capsules that remain on the tree for a long time, interspersing the foliage on the twig like a close-set necklace.

These decorative plants, generally twiggy shrubs or trees, are native to Australia. Their wood, like that of several other genera of the myrtle family, is very hard.

Because of its interesting appearance and unusual flowers, the bottle-brush tree is grown as an ornamental. In congenial climates, on the Riviera or the Caribbean coast, it is also grown outdoors. Elsewhere it is a rewarding plant of cool greenhouses. The species *C. lanceolatus* may also be grown as a house plant.

Durian

Coconut palm *Cocos nucifera*

The coconut palm or tree is one of the most interesting palm trees in the world. It is a very important tree of commerce. The fruits — coconuts — are hard-shelled and can be shipped long distances without spoiling, which is why they are well-known throughout the world.

The coconut palm is probably native to the Polynesian coast, but nowadays these trees border shores throughout the tropics. The tall, slender, flexible trunks, each topped with a bunch of long leaves, are often the first indicators of land on the horizon to ships at sea. Between the leaves emerge panicles of small flowers which, after fertilization, produce some of the largest seeds in the world. Each fruit (drupe) contains a single seed. The smooth leathery membrane on the surface is removed after harvesting to reveal the stone — a brown fibrous husk enclosing a thin hard shell, which is the form in which coconuts are delivered to the market. Inside the stone is the seed composed of a thin, brown seed coat and a two-centimetre thick layer of white edible 'meat'. The centre of the seed is hollow and is filled with a sweet milky fluid called coconut milk (as much as 0.5 litre (0.9 pt) in an unripe nut).

Another fluid obtained from the tree is sweet palm sap from the unopened flowers. It is drunk fresh or fermented into palm wine. Distilled from the fermented sap is an alcoholic drink called arrack, and evaporation of the sap yields a syrup and a coarse, dark brown sugar called jaggery. Coconut meat is rich in fats (it contains up to 70 per cent) and the dried meat of ripe coconuts, called copra, is either boiled or pressed to yield coconut oil used mainly for making high-quality vegetable fats. Shredded coconut is used in confectionery. Pressed copra, from which the oil has been extracted, is used as fodder for livestock or as a fertilizer.

The shell of the coconut is both firm and fragile at the same time. It is used to make various ornamental objects and carved buttons. The fruit, which is dispersed by water, is kept afloat by its thin fibrous husk. The prepared fibres, obtained best of all from unripe fruits, are called coir. They are very coarse and are used to make rugs, mats, sacking, brushes and ropes for use on ships for they stand up well to salt water. Besides the fruits and flowers of the coconut palm, use is also made of the firm, flexible wood of the trunk in the construction of dwellings, bridges and ships; the leaves are used to make matting and fences; germinating plants are also eaten as a vegetable by local people.

Coconut palm

Eucalyptus

Eucalyptus *Eucalyptus*

This interesting and useful genus, comprising many hundreds of species, is characteristic of the Australian region. The eucalyptus is a giant and the tallest of the broad-leaved trees — in 1880 a specimen of *E. amygdalina* (now *E. regnans*) was recorded as being 114.3 m (375 ft) high. Eucalyptus wood is very dense, heavy and durable. It is difficult to process, but that is the very reason why it is an excellent construction material, used also for building ships and making furniture. The colour of the wood varies, depending on the species, and may be light brown, dark brown or red.

Eucalyptus grows very rapidly and absorbs large quantities of water from the soil. For this reason it is planted out in warm, swampy regions where it limits the spread of mosquitoes by drying up the swamps.

Before opening, the bristly flowers (often with coloured filaments) are covered with a cap, which later falls off. The ovary ripens and changes into a woody capsule, increasing only slightly in size. An adult tree has two types of leaves. One kind, coloured dark green, have stalks, are narrow and crooked and grow alternatively on older branches. They are not the least like the broad, leathery, greyish-green leaves of young shoots, which furthermore do not have stalks and are usually pressed tightly to the branch. It is hard to believe that they belong to the same tree. However, a closer look at branches of various ages will reveal that one type of leaf gradually changes into the second type. The leaves are distinguished by still another interesting characteristic: they turn the flat side of the blade away from the burning rays of the sun. The stalks turn the blades downward, edgewise to the sun.

In Australia eucalyptus trees form large stands. In the dry interior regions they form vast impenetrable thickets together with wattles; such thickets are called scrub.

Besides yielding timber and drying up swamps, the leaves of many species yield an essential oil used in perfumery and medicine. On a hot day the air above a eucalyptus forest becomes thick and heavy with this oil. Others yield a useful gum-resin known as Australian kino. Yet another product is a sweet manna formed by the thickening of the sugary juices exuded by cut branches.

Though eucalyptus trees are now native to the Australian region, numerous fossil finds of their leaves indicate that they were also part of the vegetation of central Europe 50 million years ago.

Wattle *Acacia*

Several hundred species of wattle, members of the large pea family, grow in the tropical and subtropical regions throughout the world. Together with their companions — the eucalyptus trees — they form impenetrable thickets (scrub) in Australia. Wattles and eucalyptus trees have other things in common as well. For instance they both have extremely hard wood. It is so heavy that it even sinks in water. Such wood is called 'ironwood'. The reason why it does not float is that the ducts are filled with thick, rigid resin. In eastern Australia such 'ironwood' is provided by the species *Acacia excelsa*.

Wattles, like eucalyptus trees, bear different kinds of leaves when they are young and when they are older. Young plants have pinnate leaves; in older plants the leaves are reduced to scales and photosynthesis is carried on by the green leaf-stalks. In some species the leaf-blades are absent altogether and the stalks are widened and leaf-like. These changes are adaptations to the harsh, dry habitats where wattles grow. In Central American species (e. g. *A. sphaerocephala*) the stipules are transformed into large, hollow spines which are inhabited by ants. These wattles provide more than just a dwelling place for the ants — they also provide them with food in the form of oily white particles located at the tips of the leaves. In return the ants protect the wattles against herbivores.

Some Australian and Indian wattles yield a brown gum called wattle-gum. Of far better quality, however, is the clear or yellow gum known as gum arabic obtained from *A. senegal* of northern tropical Africa. It was known to the ancient Egyptians who used it in ink and in paints. Gum arabic is an exudation such as is found also in many fruit trees. It is a healing, sticky substance exuded from the trunk at an injured spot which dries and hardens into a resin-like mass on exposure to the air. It is readily soluble in water.

Grown in southern Europe is the ornamental species of wattle *A. farnesiana* of the West Indies. Its fragrant flowers are sold throughout Europe as 'mimosa'.

Wattle

Protea *Protea*

The Proteaceae is a very old family which is documented by fossil finds dating from over 25 million years ago. At that time it was apparently at the peak of its development. Now there are about 56 genera with some 1100 species found only in the southern hemisphere. Most species are found only in South Africa and Australia, but a few species made their way to Japan, tropical America and New Caledonia. The Proteaceae are considered the loveliest shrubs and trees of South Africa where they are widely grown in gardens.

The genus *Protea* includes shrubs and small trees with stiff, narrow leaves. Their native habitats are regions with damp winters and dry summers. They are beautiful and striking plants with leathery evergreen leaves. The small flowers are arranged in striking inflorescences in the form of racemes, spikes or heads. The inflorescence has brightly coloured bracts. The fruits are one or two seeded whiskered nuts.

One of the loveliest is *Protea mellifera,* a small tree with white or reddish flowers in capitula; the flowers provide food for bees.

Australian Christmas tree
Nuytsia floribunda

The Australian Christmas tree bears a profusion of yellow-orange flowers at Christmas-time which is how it came to be so called. It grows only in Nuyt's Land in Western Australia, where the Dutch navigator Pieter Nuyts was the first to land in 1627, and was named *Nuytsia floribunda* in his honour by the botanist Robert Brown. It is the only member of this genus.

The large, 10-12 m (33-39 ft) high tree is, surprisingly enough, a parasite. It appears to grow on its own, the same as other trees. Its roots, however, feed on the underground parts of the surrounding herbaceous plants. Even grasses are a source of food for this giant, whose method of obtaining nourishment and whose stiff leathery leaves class it as a member of the mistletoe family.

Protea

Australian Christmas tree

241

Mulla-mulla

Mangrove *Rhizophora*

On the muddy shores of the tropics, in shallows where there are no breakers, grow the strange trees called mangroves. They form thick impenetrable woodlands which extend in a broad belt from shallow, thin mud to thicker mud to firm ground. The vast tangled mass of roots and branches catches the alluvium and sediment in sea water. Whereas the roots hold and anchor the muddy soil, the bushy crowns of the trees intertwine and provide a shelter for sea birds.

Mangroves are the dominant trees in such woodlands mostly because of their effective method of propagation. A mangrove seed does not fall but remains hanging from the branch and the young seedlings grow from the seed on the parent tree to a good size — 30-100 cm (12-40 in) long. Only then do they fall into the muddy water, where they rapidly take root, grow and produce new offspring after several years. A single such seedling carried by the water to a congenial spot is enough to establish a whole wood.

The good stability of mangroves in the soft substrate is made possible by the arching, stilt roots. The height of these indicates the difference in water level at low and high tides. And because there is not enough oxygen in the thick mud, the buried roots furthermore send up special breathing roots called pneumatophores, which grow upright into the air and provide access to the atmosphere. They are provided with tiny openings at the tips, through which gases can pass.

The leaves of *Rhizophora* and other mangroves are slightly fleshy or leathery in order to limit the evaporation of water.

Some species of mangrove have practical uses. For example *Rhizophora mucronata* is used in making tannin and red dye. The wood of *Rhizophora mangle* is hard and coloured brownish-red. It is used to make boats and poles as well as charcoal. Some parts of the tree have medicinal properties: the bark is used to treat tonsillitis and tuberculosis of the lungs.

Mulla-mulla *Ptilotus*

The unusual appearance of this Australian plant of the amaranth family is best expressed by its scientific name *Ptilotus,* derived from the Greek word *ptilotos* meaning feathery. This refers to the floral parts, which are covered with whitish or greyish hairs. In *Ptilotus obovatus* the hairs are star-like.

The various species of *Ptilotus* growing in Australia are either herbs, sub-shrubs or shrubs. Some, like *Ptilotus manglesii,* which attains a height of about 30 cm (12 in), are quite small and good for room decoration. Particularly attractive are the flowerheads. The flowers are borne singly in the axils of glossy, translucent bracts. They form a spike 5 cm (2 in) across.

Of the 60 species of this lovely genus, illustrated is the Pink mulla-mulla, *Ptilotus exaltatus,* with pink flowers and part of the white-flowering *Ptilotus spathulatus.*

Mangrove

Vegetables from the Sea

Plants growing on dry land are mostly green, but there is very little green in the sea. Instead brown and red algae, or seaweeds, predominate. Marine algae are extremely diverse in shape and the differences in their size are even greater, ranging from small microscopic species, which make up a large part of the free-floating plankton of the sea, to seaweeds measuring several metres. Some giant kelps measure over 100 m (328 ft) in length.

Unicellular as well as multicellular green algae are green because of the chlorophyll they contain.

Europe's coastal regions. It has broadly expanded lobes with wavy margins and resembles lettuce. It is common on seashores, where it attaches itself to rocks and piers. The wavy, flat 'leaves' of this seaweed form thick, bright green masses. Certain other species, such as *Ulva latissima*, reach a length of 10-60 cm (4-24 in). In the Mediterranean region oysters, mussels and cockles are laid out on the fresh green blades in the fish market.

The flat, leaf-like plant body is composed of two layers of cells and is produced by division of the cells in three planes.

Enteromorpha *Enteromorpha*

Other edible seaweeds are the green algae of the genus *Enteromorpha*. They are a favourite food of the natives of Australia, where they grow in estuaries. The members of this large genus may occasionally be found also in fresh and brackish water.

The thallus of *Enteromorpha* is composed of branching tubes with only one layer of cells in the walls. It is initially attached to the substrate, such as a rock, by the narrowed end which resembles a leaf stalk, but later it may become detached and float freely in the water.

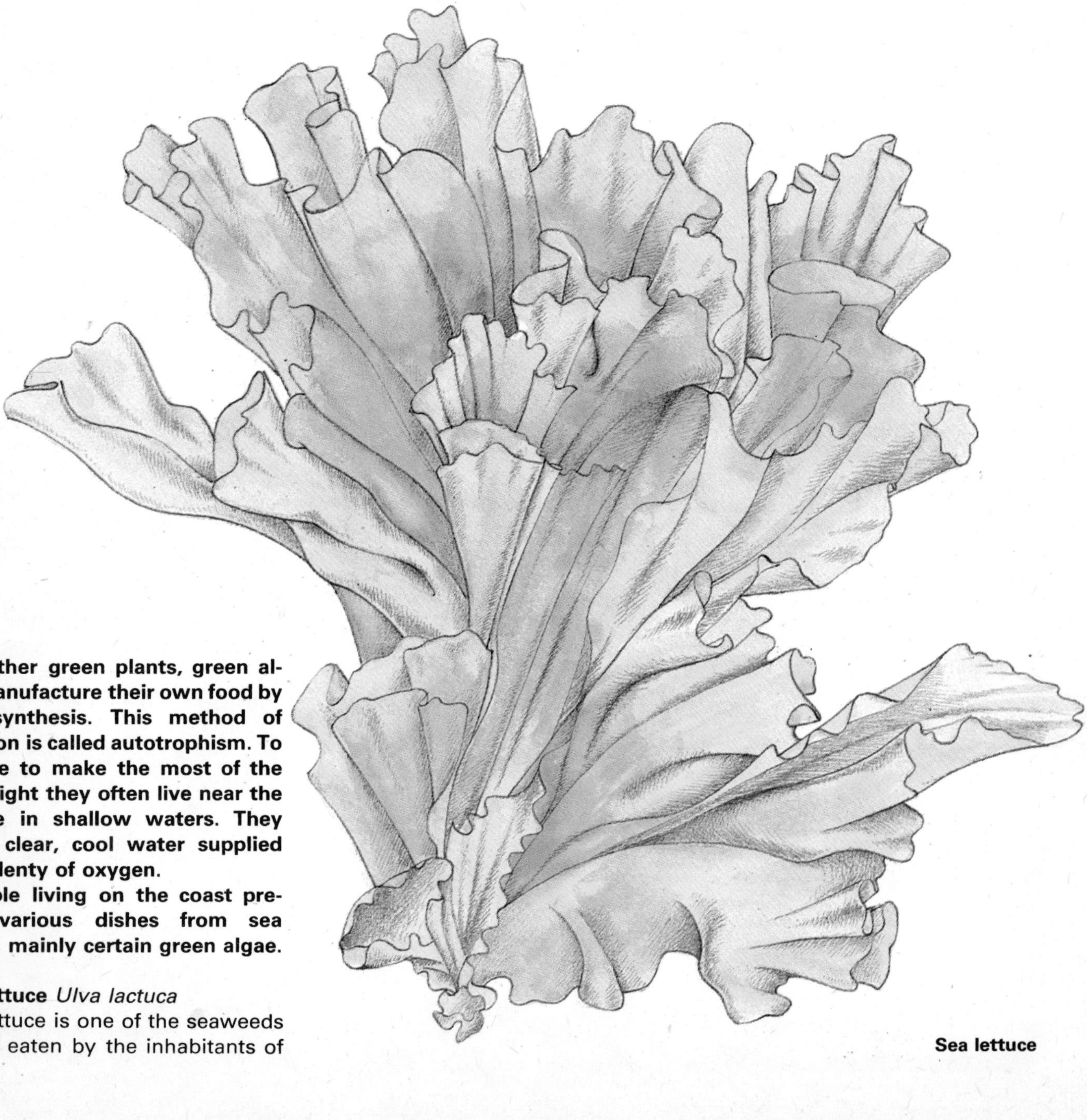

Like other green plants, green algae manufacture their own food by photosynthesis. This method of nutrition is called autotrophism. To be able to make the most of the sun's light they often live near the surface in shallow waters. They prefer clear, cool water supplied with plenty of oxygen.

People living on the coast prepare various dishes from sea plants, mainly certain green algae.

Sea lettuce *Ulva lactuca*

Sea lettuce is one of the seaweeds that is eaten by the inhabitants of

Sea lettuce

244

Enteromorpha

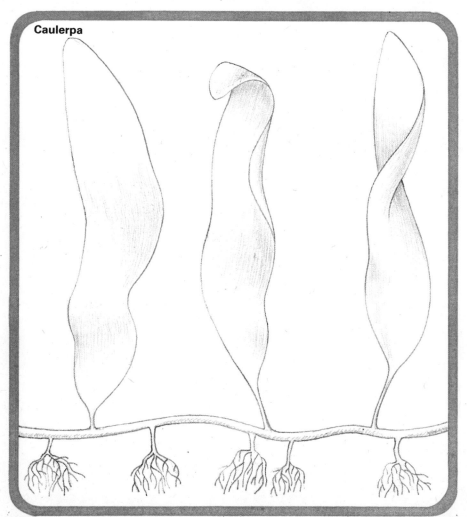

Caulerpa

Caulerpa *Caulerpa prolifera*

Caulerpa is an extraordinary green seaweed 10-20 cm (4-8 in) long. It is coenocytic, which means that its nuclei are not separated by cell walls. What look like colourless roots, a colourless creeping stem and green leafy shoots are all parts of a single mass of protoplasm containing several nuclei and green chloroplasts. The chloroplasts enable the algae to carry on photosynthesis in the same way as in higher plants and, as usual, the end product of this process is starch. The rigidity of the plant is increased by the formation of numerous transverse and longitudinal stiffening rods made of cellulose.

During the process of reproduction the entire living mass is transformed, dividing up into reproductive cells, or gametes. Most of the thallus disintegrates, releasing the gametes, which part and form new individuals.

Caulerpa greatly resembles the multicellular higher plants, even though its structure is so simple. It attaches itself to the sandy bottom with its root-like appendages and forms thick, spreading masses in the warm, shallow seas.

Ocean Meadows

Except for a few species, the brown seaweeds are distributed only in the seas and oceans. Like the green algae, they contain chlorophyll, but the green colour is masked by the large quantity of brown pigments, mainly fucoxanthin. Carbohydrate formation in brown algae is comparable to that in sugar-storing vascular plants rather than that in starch-storing ones. In *Laminaria* it is the sugar known as laminarin which, together with the alcohol mannitol and fats, forms the plant's food reserves.

Brown algae grow at greater depths than green algae and often attain vast dimensions. Their number includes the common bladder wrack *(Fucus vesiculosus)* and serrated wrack (*F. serratus),* the floating seaweeds (genus *Sargassum*) and the giant kelps (genera *Nereocystis* and *Macrocystis*). They frequently occur also on coasts with heavy surf. They are able to withstand the pounding of the surf thanks to their tough thalli with firm cell walls. In addition to this their surface is slippery because of the presence of the gelatinous sugars algin and fucoidin in the cell walls. This characteristic also helps the algae to survive in rugged conditions.

Bull or Ribbon kelp
Nereocystis luetkeana
The ribbon kelp measures 20-25 m (66-82 ft). Of this the stipe accounts for 10-13 m (33-43 ft) and the blades 3-5 m (10-16 ft). The rope-like stipe terminates in an air bladder 15 cm (6 in) long which keeps the plant floating in the water. Attached to the upper end of the bladder are several ribbon-like blades that extend up to the surface, where they float. This seaweed grows in very deep water off the Pacific coast of America.

Oar weed *Laminaria*
The surf zone of the cold northern seas is inhabited by seaweeds built to withstand the pounding of the waves. This type of multicellular brown alga is differentiated into three distinct parts like the bodies of higher plants. They are attached

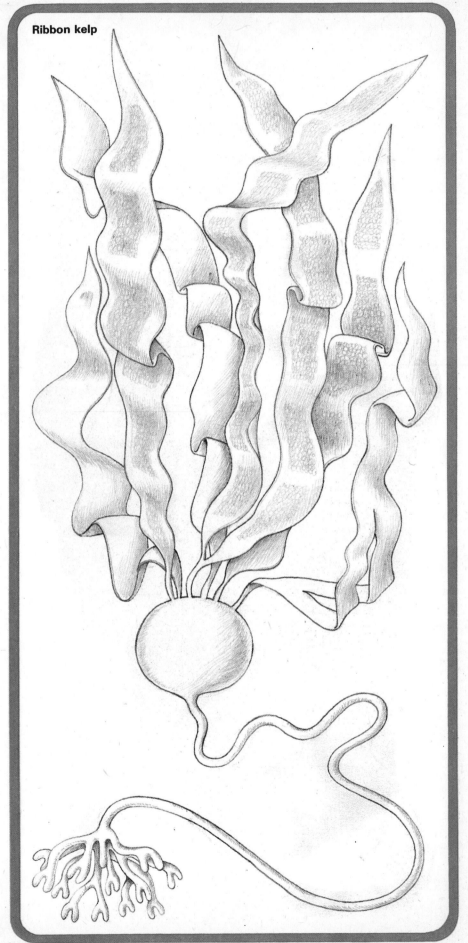

Ribbon kelp

Gulfweed

to rocks, stones or underwater constructions by means of the root-like organ called the holdfast. From this rises the thick, persistent stem-like region (stipe) terminated by the leaf-like portion (blade) which is much longer than the stipe. The blade may be single and narrow with undivided margin, like a whip e. g. sugar wrack, or split into longitudinal segments, e. g. tangle and other species. The blade, be it single or split, is annual; the stipe is perennial. The transition zone between the stipe and the blade is the growing region containing cells that lengthen and strengthen the stipe and, what is most important, annually replace the old blade with a new one.

Oar weeds are well-adapted to the rugged conditions of their habitat, with their flexible bodies and gelatinous surface. Despite this, the pounding surf continually tears loose masses of these seaweeds and sweeps them up in piles on the beach. Such cast-up seaweeds are used as fertilizer, mainly to enrich the sandy soil of coastal fields. Besides being a source of nitrogen they also help retain moisture in the soil. Oar weeds have other uses as well. Sodium bicarbonate and iodine may be obtained from their ash. The dried stipes, particularly those of tangle, swell enormously when moistened and were therefore used formerly in surgery to dilate narrow passages.

Gulfweed *Sargassum*
This brown seaweed bearing the generic name *Sargassum* is known chiefly for its association with the Sargasso Sea or Sargasso Meadows. These 'meadows' cover vast expanses of the open sea between the West Indies and the Azores.

The formation of these floating 'meadows' of gulfweeds begins on the rocky coasts of the Antilles where they grow. In stormy weather they are torn free by the pounding waves and carried by the current far into the open sea. There, in calmer patches, they form large drifting masses, themselves causing the surface of the sea to become even calmer. The floating seaweeds continue to grow at the tip for a while, but at the same time they slowly die from the bottom upwards.

Sargasso meadows have been the subject of many tales and legends since time immemorial. The danger these masses of gulfweed pose for navigators was often exaggerated in the telling of such tales. This natural obstacle was discovered by Columbus, and Darwin wrote about the 'Sargasso meadows', as did the Greek philosopher Aristotle.

The gulfweeds of the warm seas have a multicellular thallus differentiated into a stipe or stem-like region and a blade or leaf-like portion. The thallus looks like a fruit-laden branch because it is covered with numerous berry-like air bladders. These bladders are formed by the modification of the lateral 'twigs' shaped like leaves with a toothed margin. Most brown algae multiply by sexual means, one exception being *Sargassum natans* which apparently multiplies only by vegetative means — by fragmentation of the thallus.

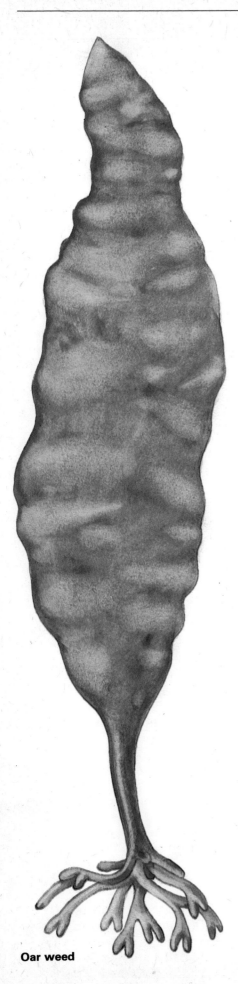

Oar weed

Salt-loving Communities

Where the sea is shallow and the shore flat and sheltered from the surf, small particles of mud are deposited. The first plants to grow on the newly-formed layer of greyish-black mud are salt-loving annuals, such as glasswort and seablite. As soon as the layer of mud extends beyond the high tide level, further species join the community, such as sea sweet-grass and sea arrow-grass. These are followed by grasses that tolerate salty soil, such as cord grass, red fescue and creeping bent. Their roots hold the soil and as the soil layer thickens it is covered by a spreading expanse of tufts. Gradually, the area turns into a pastureland.

Salt-loving plants in many ways resemble desert plants. Many have fleshy stems and leaves; others have leaves which are generally very small and thin, sometimes almost negligible. These adaptations greatly reduce the evaporation of water from the bodies of such plants. Halophytes also grow very slowly. Being typical plants of the seashore they are light-loving and do not tolerate shade. In less salty soils, salt-loving species give way to the usual meadow plants.

Sea spurrey *Spergularia salina*
This and other species of spurrey prefer salty soils. Small plants, 10-20 cm (4-8 in) high, of the pink family, they are slightly less attractive than the salt-loving plants of the goosefoot family. The red-flowered red spurrey is a field weed. Less striking pink or white flowers are produced by the annual, sometimes also perennial, sea spurrey. Their petals are shorter than the sepals, but the fruits (capsules) are one-and-a-half times as long as the sepals. Fleshy, linear leaflets with inconspicuous stipules grow from the point of union of the flower-stalks.

Sea spurrey flowers from May until September on seashores throughout the northern hemisphere and on inland salty soils.

Herbaceous seablite
Suaeda maritima
Herbaceous seablite is a member of the goosefoot family, but unlike common glasswort and saltwort it does not bear its flowers singly. They are borne in scanty clusters of three to five tiny, five-petalled flowers in the axils of the fleshy, linear leaves. However, even the clusters are not particularly conspicuous because they are coloured green. The fruits are black achenes. Herbaceous seablite flowers from July till September.

This small, 5-30 cm (2-12 in) high annual branches profusely from the base. It makes low, spreading, herbaceous shrublets that join to form large blue-green masses. It is an insignificant plant, but has spread throughout the world; it is plentiful on seashores and occasionally grows on inland salty soils.

Saltwort *Salsola kali*
This fleshy plant up to 1 m (40 in) high, is not as closely tied to the seashore or salt-marshes as other salt-loving species. In Europe, Asia and North Africa it grows on waste ground and as an annual weed of cultivated land as well as in the sandy and gravelly soils of the surf zone. It made its way to North America, where it became established in the prairies. It has also

Sea spurrey

Saltwort

become established in New Zealand.

The greyish-green stems of saltwort branch from the base. The branches grow either upright or along the ground and are covered with spine-tipped leaves. The small flowers are borne singly in the axils of the upper leaves. The unattractive green perianth is composed of five segments, three of them wide and two narrow. All have a membranous appendage on the upper surface. The perianth is not shed but remains on the plant when the fruit ripens, the appendages increasing markedly in size and spreading out to form a ring round the fruit. Saltwort flowers from July till September.

Common glasswort
Salicornia herbacea

The annual glasswort of the goosefoot family is a pioneer plant of salty soils. It tolerates greater concentrations of salt in both soil and water than any of the other halophytes and survives temporary submersion at high tide without difficulty. It is therefore not surprising that the common glasswort contains so much salt in its tissues. In maritime countries, for instance in France, sodium was at one time obtained from the plant. In addition, young stems were eaten as a salad.

Common glasswort grows in every saline soil; both by the sea and elsewhere it seeks out spots with the greatest concentration of salt. Frequently it is found in the muddy ruts of roads. The fleshy, seemingly leafless stem is 20-30 cm (8-12 in) high and forms opposite branches from the base. The stem as well as the branches are composed of segments that narrow towards the base where they carry two tiny, reduced leaflets resembling horns. The flowers are barely visible, concealed in hollows at the base of the upper segments. In the autumn, when the fruits (achenes) are formed, the whole stem turns red.

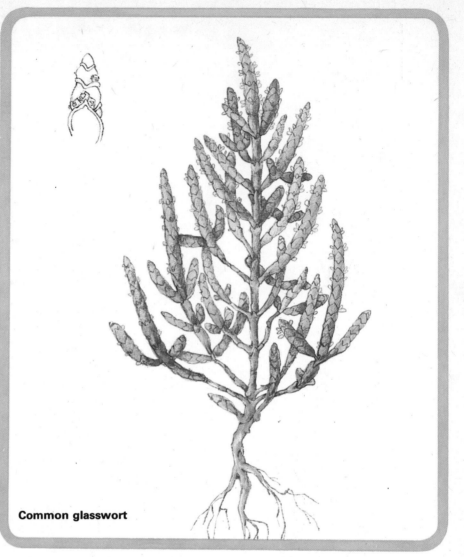

Common glasswort

Sea milkwort

Sea milkwort
Glaux maritima

This small, 5-20 cm (2-8 in) high, fleshy-leaved member of the primrose family is called sea milkwort. It is a greyish-green, shrubby, creeping plant that forms dense masses. The stems are thickly covered with oval leaves — opposite at the base, alternate at the top. Located in the leaf axils are single pink to reddish bells. These tiny flowers do not have petals; their purpose is filled by the coloured sepals. The fruit is a capsule with seeds.

Sea milkwort is an important salt-loving species that tolerates high concentrations of salt, submersion in sea water and being buried in sand. Its presence calls attention to the fact that there are high concentrations of salt in the soil. It grows in salty mud by the sea and occasionally in inland salt-marshes.

Buck's-horn plantain
Plantago coronopus

Plants that grow on sandy seashores, in salt-marshes and on inland waysides also include members of the large plantain family. The various species have the same type of inflorescence and most have the same leaf shape. The flowers of plantains are generally arranged in long or short spikes. Buck's-horn plantain, an annual species, bears white flowers in short, dense, cylindrical spikes. The fruit is a four-seeded capsule. However, the leaves are slightly different from those of other plantains — the basal leaves forming the rosette are divided into leaflets, each of which has a single vein. From June until September several leafless, downy stems bearing fleshy spikes of tiny flowers rise from the centre of the rosette. In some regions this plantain was formerly used in folk medicine.

Buck's-horn plantain is widespread in central and southern Europe, North Africa and the Middle East. It has been introduced into Australia and North America.

Townsend's cord-grass
Spartina townsendii

This grass grows in grassy and muddy places and helps stabilize salt-marshes.

Townsend's cord-grass is a vigorous and rapid-growing hybrid, a cross between *Spartina maritima* of Europe and *Spartina alterniflora*, native to America. The latter, however, also occurs on the coast of Europe, chiefly in Spain and France. Townsend's cord-grass, which is an unusual hybrid in that it is fertile, flowers from July until August. Hidden deep in the ground are long, creeping rhizomes whereby it spreads rapidly and forms large masses. It also grows on sand banks, being able to tolerate submersion by the tide. Such an uncertain location even has its advantages, because the water supplies the grass with food in the form of detritus which catches in the tufts.

Townsend's cord-grass, almost half a metre (20 in) high, has stiff, smooth stems and rough, pale-green leaves. The stout spike, measuring up to 20 cm (6 in), is strikingly large.

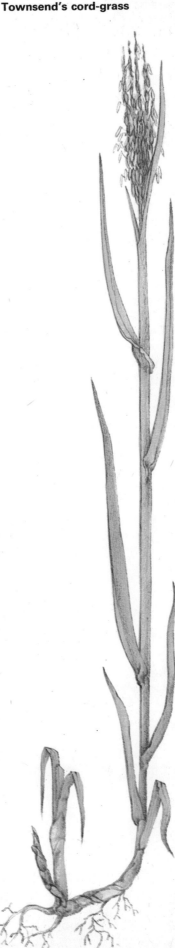

Townsend's cord-grass

Buck's-horn plantain

Sand Dunes

Sand dunes are widespread on seacoasts throughout the world and are also found inland, not only in deserts but to some degree even in the temperate regions. Dunes are generally 6-8 m (20-26 ft) high and are formed gradually, often quite unnoticeably. They are built up from dust and sand carried by the wind and caught by some obstacle such as grass or thickets. Further layers are simply heaped on top of this. Dunes are continually shifted by the wind, with sand on the windward side moving to the leeward as long as the colonizing plants allow this. The colonization of a new dune, however, does not take long. First it is settled by pioneer annual species, followed by perennial species. These are mostly grasses with a tangled network of roots that holds the upper layer of sand and slowly prepares the ground for the growth of other plants. Sand-loving plants may grow even in pure sand and are very hardy. Even when covered with sand it does not take long for them to surface again.

Sea holly *Eryngium maritimum*
Most of the 220 species of *Eryngium* look more like thistles than plants of the parsley family. Sea holly is one that inhabits seashores. The top parts of the plant have an attractive violet tinge. The greyish-blue leaves are tinted a delicate blue-violet, the flowers a deeper shade. The flower is so modified that at first glance one would think it was the inflorescence of a composite plant or even more likely a single flower. In actual fact,

however, it is an elongate capitulum of densely clustered, tiny, pale-blue flowers made spiny by the tapering tips of the sepals. The petals are shorter than the sepals. The whole inflorescence is enclosed and made more striking by the large coloured bracts.

Sea holly is a warmth-loving plant found only on seaside dunes. It grows in southern Europe round the Mediterranean Sea and on the shores of North Africa, western Europe and the North, Baltic and Black Seas. With its extremely long root it is a pioneer plant of sand dunes, helping to hold the shifting sand.

Lyme-grass *Elymus arenarius*
Lyme-grass is another tough grass with greatly-branching rootstock. It is not only effective in binding sand dunes, but is also very decorative with its striking, broad, grey-blue leaves. It is often grown as an ornamental grass in the garden. However, it has a vigorous tendency to spread and become a troublesome weed. In the wild it grows in pure sand, but elsewhere it appreciates

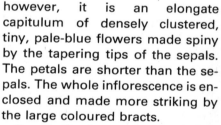

Sea holly

a light soil and sunny situation. Before flowering, the plant is about 70 cm (28 in) high. Flowering stems appear from June onwards and reach a height of 120 cm (47 in). The rough, flat leaves furl in dry weather and terminate in a prickly point. The spikes of this grass are also interesting. They are stout, measure 20-30 cm (8-12 in) in length, and are composed of two rows of large, three-flowered spikelets. These are awnless and attached directly to the main stem. The grains were formerly used to make bread in Iceland.

Lyme-grass grows wild on the coastal sands of the Baltic and North Seas. However, except in the Mediterranean, it is often planted out to bind sands on other European coasts as well. It is also found in Siberia and North America and is planted out on sandy soils in the interior, where in places it has become naturalized.

Lyme-grass

Sand sedge *Carex arenaria*

Sand sedge is so like grass that even its flowers leave one puzzled at first glance. In most sedges the spikelets are spaced out and more or less stalked. In sand sedge, however, all 6 to 15 spikelets are arranged in a spike-like panicle at the tip of the leafy stem — the same arrangement as is found in many grasses. The terminal spikelets are male, the lower ones female and the middle ones are male at the top and female at the bottom. However, the fact that this is a sedge can be deduced from its stem and by its fruits. Those of sand sedge are ovoid, winged at the sides and located in the lower part of the inflorescence.

Sand sedge flowers from spring until autumn. It is most widely distributed on the sand dunes of the Atlantic and Baltic seacoasts and in North America. It forms extensive masses and with its branching, far-creeping rhizomes helps bind the shifting sands. It is also found in inland 'mini-deserts', heaths and pine woods.

Marram grass *Ammophila arenaria*
The scientific name of this plant indicates its association with sand, for *Ammophila* is derived from the Greek word meaning sand-loving and *arenaria* from the Latin word meaning sandy. And because it is a robust grass up to 1 m (40 in) high it is also sometimes called sand reed. The flowering stems, which rise above the inrolled leaves, are upright, firm and terminated by a dense cylindrical panicle, up to 15 cm (6 in) long, composed of yellowish spikelets. The flowering period is in June and July.

The thick clumps of this beach grass rapidly spread to form a dense greyish-green cover. It is a perennial grass with stout, branching rhizomes that bind the shifting sands of coastal dunes. It does better on dunes that are farther from the sea where the sand is not so salty. It is naturally a part of the coastal vegetation of North America, the Mediterranean region and practically all European seacoasts. In places where it does not grow naturally it is often planted by man.

Marram grass

Sand cat's-tail
Phleum arenarium
Sand cat's-tail is an annual grass that forms scattered tufts on the shifting sands of seashores. From the west European coast it spread as far as southern Scandinavia and is also found in the Mediterranean region.

Sand cat's-tail is usually only a few centimetres high. The stems are erect, reddish and slightly swollen at the top. The blades of the upper leaves are short and rough on the underside and the sheaths are swollen. The grass flowers from May until June.

Sand cat's-tail

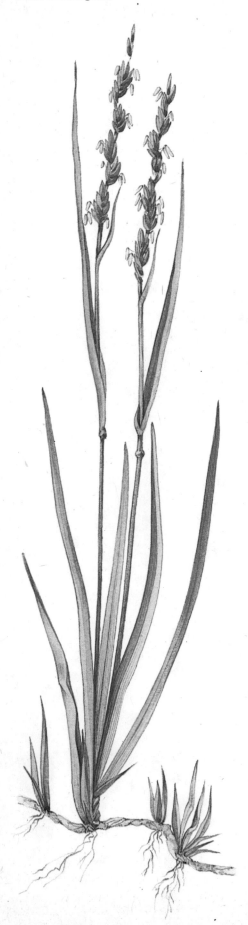

Sand couch grass

Sand couch grass
Elytrigia juncea

Couch grass is a troublesome weed and one of the most widespread of all grasses. It is found on all types of soils and one species — sand couch grass — also grows on sands. Couch grass is notorious for its spreading rootstock and sand couch grass is also one of the plants that is effective in binding shifting sands and incipient dunes. It is about half a metre (20 in) high and has greyish-green flat leaves that furl after a time, particularly in dry weather. The flat inflorescence is composed of five to eight-flowered spikelets arranged in two rows. The main axis of the spike is very brittle.

Common salt-marsh grass, Sea poa
Puccinellia maritima

Another grass of seashores is the common salt-marsh grass, or sea poa, which resembles meadow-grass, hair-grass and bent, particularly when it flowers. It is extremely fond of salt water. If the salty soils of seashores do not satisfy its needs, it frequently spreads into shallow water. The 50-centimetre (20 in) high, yellow-green plants form loose masses. When it invades water, it is effective in strengthening the bottoms of shallows, which then begin silting up more readily — the first step in the process of natural land reclamation.

This perennial grass generally produces sterile individuals, but often multiplies readily by vegetative means. From June until October it bears five to nine-flowered spikelets arranged in a narrow and slightly one-sided panicle with spreading to drooping branches. The spikelets are often purplish. Common salt-marsh grass is distributed on the shores of several seas.

Common salt-marsh grass

Moulds and Mildews

The most familiar fungi are the mushrooms and toadstools that grow in fields and woods. However, there are thousands of other types that grow in a variety of places. Some grow on decaying food. Others are parasites; they feed and grow on living plants and animals.

White rust *Albugo candida*
Both plants of commercial value and weeds are affected by fungal pests. For example, white rust is parasitic on shepherd's purse and certain plants of the cabbage family. This fungus is very plentiful in spring. The plants it infects have distorted stems and leaves that appear to be sprayed with white lacquer.

Downy mildew of grape
Plasmopara viticola
The upper sides of the leaves of a grape vine infected with this parasite are covered with yellow, so-called 'oily spots' that later turn brown. The underside of the leaves is covered with a white furry coating made up of tufts of spore-containing structures. Inside the leaves the fungal threads form a dense mass. The spore-containers are generally shed whole and carried

Downy mildew of grape

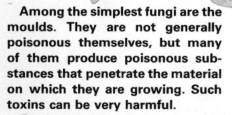

White rust

Among the simplest fungi are the moulds. They are not generally poisonous themselves, but many of them produce poisonous substances that penetrate the material on which they are growing. Such toxins can be very harmful.

On the other hand, some moulds are beneficial. *Penicillium,* when grown on certain materials, produces penicillin, an antibiotic that is not too toxic for man, but destroys germs.

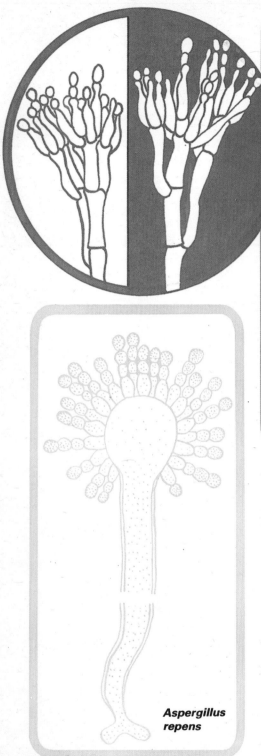

Penicillium crustaceum

Penicillium italianum

Aspergillus repens

from dead plants. They are harmful to man in that they grow on various foods, which are thus spoiled.

A typical member of this genus, *Mucor mucedo,* is common on bread and on other organic substances. It thrives in damp. The related *Mucor piriformis,* which coats decaying apples and pears, even smells nice. Many species, however, are important in industry in that together with yeast they ferment starches and sugars to form alcohol.

Aspergillus repens

In damp places all kinds of foods are soon covered with the greyish-green to brown, furry coat of the fungus *Aspergillus repens.* It generally reproduces asexually, but the yellow fruiting bodies which are formed very occasionally contain structures called asci, which indicates that this fungus belongs to the higher group of fungi known as Ascomycetes.

Penicillium expansum

Penicillium expansum forms a yel-

low-green to brown furry coat on all kinds of food left in damp places. Several species are useful. These include *P. roquefortii, P. camembertii* and *P. gorgonzola.* They have the ability of breaking down fats and proteins and act together with other factors in ripening certain kinds of cheese. Others, for instance *P. notatum,* produce important antibiotic drugs.

Apple and pear mould

by the wind throughout the vineyard to the leaves of other, healthy plants. There, the spores attack not only the leaves but the flowers and berries as well.

Apple and pear mould
Mucor piriformis

Moulds of the genus *Mucor* are very common. They are saprophytic organisms which take their food

Cup Fungi and Their Relatives

The group of fungi known as Ascomycetes include a large number of different types. There are tiny organisms visible only through the microscope, but which are of great economic importance, and larger types, some of which are edible, found in woodlands. The microscopic species, such as yeasts, generally multiply by vegetative means. But the fruiting bodies of the larger species produce spores inside sac-like structures called asci (singular: ascus). In this Ascomycetes differ from the Basidiomycetes, which include the familiar mushrooms and toadstools and

the formation of gas (yeast breaks down sugars to form alcohol and carbon dioxide). Baking destroys the yeast cells and evaporates the alcohol.

Brewer's yeast is a cultivated strain used in brewing beer and in the manufacture of alcohol from potatoes. 'Wild' species of yeast, however, cause damage in breweries by clouding the beer and giving it an unpleasant taste and aroma. Because it is rich in vitamin B (as well as enzymes and proteins), yeast is also used by the pharmaceutical industry in the preparation of various drugs.

mainly rye. Minute spores appear on the surface of this mycelium together with a sticky, sweetish liquid known as honeydew. This attracts insects which collect the secretion and carry the spores to other flowering heads of rye. The plant ovary, in which the mycelium develops, becomes a hard and horny body, coloured violet-red to black. It then protrudes from the spike and falls to the ground when the grain is harvested. Overwintering ergots develop rosy-purple, flask-shaped fruiting bodies, inside which asci and spores are formed. Mature spores are distributed by the wind.

Ergots contain a great many alkaloids, most of which are poisonous. The fungus is collected for the pharmaceutical industry.

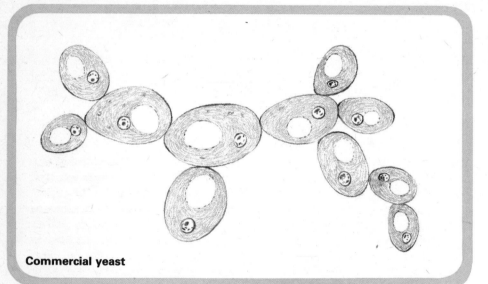

Commercial yeast

which produce spores externally on club-shaped structures called basidia.

Commercial yeast
Saccaromyces cerevisiae
A yeast plant consists of a single cell. As a rule it does not produce a mycelium or asci, but multiplies very rapidly by vegetative means – by budding. The newly-formed cells do not detach themselves immediately and chain-like colonies are formed.

Commercial yeast consists of a large number of yeast cells mixed with starch and sugars and pressed into cakes. For baking the packaged yeast is first dissolved in lukewarm water, in which the yeast cells begin to multiply rapidly. When added to the dough and kneaded, the yeast causes the dough to rise by

Ergot of rye *Claviceps purpurea*
This parasitic fungus is best known in the stage of its development when it forms dark-coloured compacted masses of tissue called ergots. These are formed in spring in the ovaries of flowering grasses,

Ergot of rye

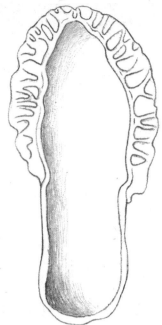

Morel

Peziza badia

When ripe, the open fruiting bodies of this cup fungus look like wide cups or bowls. They are coloured brown and measure 5-6 cm (2-2.4 in) across. The spore-bearing layer is spread over the surface of the fruiting body thus forming a lining to the cup.

Peziza badia is generally found in sandy moist soil in forests.

Morel *Morchella esculenta*

Morels can be found in the spring. This edible ascomycete generally grows in shady places — in grass in meadows, woodlands, gardens and ditches throughout the temperate regions of the northern hemisphere. It is also found in Australia.

The morel has a fleshy fruiting body composed of a cap and stipe. The cap is usually club-shaped with firm branched ribs forming deep, wide pits lined with a spore-bearing layer.

The morel is a tasty edible mushroom. However, it must be scalded before eating or better still simmered or fried.

Truffle *Tuber melanosporum*

The fruiting bodies of this fungus are not evident because they grow below the surface of the soil. They are tuberous, measure approximately 10 cm (4 in) in diameter, and have a warty black surface. The

Peziza badia

dark violet flesh is interlaced with paler tunnels lined with a spore-producing layer of round asci. They open into the tough outer portion known as the cortex. Truffles have a very characteristic odour and man trains dogs or pigs to locate them by scent. The mycelium is symbiotic with the roots of oaks, but only in the limestone region of southern France. Here truffles do so well that they are cultivated in oak stands (truffle farms). Attempts to grow truffles in similar places elsewhere in Europe, however, have not met with success.

Truffle

Fungi that Grow on Wood

Mushrooms and toadstools can be found in a wide variety of places. Many are woodland species and some of these can be found growing on tree trunks, logs and stumps.

Those that grow on dead wood are saprophytes and take their food from the decaying wood. Others, however, are parasites and grow on living trees. Most of these do their hosts little or no harm, but some, such as the honey fungus, eventually kill their hosts.

A few of these fungi are edible, such as the oyster mushroom, *Pleurotus ostreatus* and the changing pholiota, *Pholiota mutabilis.* But many are inedible and are best avoided. Never eat any toadstool that has not been identified.

Some have medicinal properties. *Lentinus edodes* is cultivated in Japan, not only for its excellent taste, but also for a drug that is used to treat arteriosclerosis.

Tinder fungus, Surgeon's agaric
Fomes fomentarius
Tinder fungus is a parasite of hard-wood trees, mostly beech. It has hoof-shaped bodies that grow like shelves from the trunks of trees and measure up to half a metre (20 in) in diameter. The huge mycelium, whose filaments penetrate and interlace the wood of beech, forms a thin, leathery layer and causes what is known as 'white rot' in wood. The tree slowly dies until it is finally knocked down by the wind. The fungus continues to grow even on the felled tree, in which case the fruiting bodies turn their spore-bearing layer downward or else the fungus produces new fruiting bodies at right angles to the old ones.

The surface of the fungus is concentrically furrowed. These furrows, however, are not 'annual rings' as in trees but irregular growth layers formed under congenial atmospheric moisture conditions.

Formerly, the thick felt-like flesh was macerated to make tinder for starting a fire from a spark made by flint and steel struck together. The fruiting bodies were also ham-mered out into fine leathery pieces and used to sew warm vests and caps. The macerated flesh was also used at one time in surgery to hasten the clotting of blood.

Many-zoned polypore
Trametes versicolor
This bracket fungus often adorns stumps or dead, fallen trunks like some dainty, coloured lace. It thrives on deciduous trees.

The annual fruiting bodies grow in clusters above each other. They are thin, velvety above, leathery, and striped grey, red, olive, orange and yellow. The tubes are very short. In this it differs from tinder fungus, which has persistent, long-lived, small tubes arranged in several layers one above the other.

Many-zoned polypore

Tinder fungus

Honey fungus

poisonous properties and its unpleasant astringent taste disappear when it is cooked.

Although this is a popular edible fungus, foresters rightfully view it as a pest for it infests most kinds of otherwise healthy trees. As the mycelium grows and spreads it causes the wood to rot. The tree then dies and is readily toppled.

Honey fungus is well equipped for vigorous growth. It has two mycelia. One is a delicate white that glows at night on decaying stumps. This phenomenon has led to stories of ghostly lights and burning men. The second mycelium is blackish-brown and ribbon-like; the ribbons, sometimes several metres long, penetrate the bark of the tree and generally spread immediately beneath it. They serve as a food store from which the fungus takes nutrients chiefly for the production of fruiting bodies. The fruiting bodies are not formed on living wood, only on rotting stumps.

Staghorn fungus *Calocera viscosa*
This small, bright-orange fungus readily catches the eye of the mushroom picker but is sought by no one, for it is tough, rubbery and inedible.

The horned fruiting bodies are only 3-6 cm (1.2-2.4 in) high, but their root-like base penetrates as much as 20 cm (8 in) deep into decaying stumps and tree roots.

Honey fungus *Armillaria mellea*
The honey fungus begins to grow in September. Unlike many toadstools, it produces not one or two fruiting bodies but whole clusters — a great advantage for the mushroom picker. However, it is recommended that only the caps of young fruiting bodies be collected. These must be cooked thoroughly (scalding is not enough) because the raw fungus is slightly poisonous and can cause indigestion. Its

Staghorn fungus

Boleti

For many people boleti are mushrooms with excellent qualities. And not only the edible boletus. The rough-stem boletus, orange-cap boletus, and elegant boletus are all equally popular with mushroom pickers.

Nearly all mushrooms and toadstools have particular places in which they grow. Many woodland mushrooms grow near particular species of tree and experienced mushroom pickers know exactly where to look. This phenomenon is due to the fact that many fungi form symbiotic associations with certain trees. The fungal threads, or hyphae, of the fungus form a tangled network, or mycorrhiza, with the roots of the tree.

Satan's boletus *Boletus satanas*

The brightly coloured Satan's boletus is one of the prettiest of toadstools. However, although it is related to the tasty edible boletus and looks very similar, it is poisonous. The two boleti often grow in similar places, such as forest clearings and grass under oaks. Satan's boletus can be readily distinguished from edible species by the pores on the underside of the cap which are coloured deep carmine-red in young specimens; this turns a paler shade in older caps. The stem is very stout and has red and

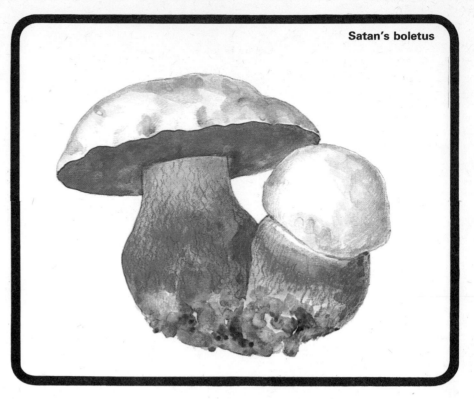

Satan's boletus

yellow colouring. The cap is greyish-white, sometimes tinged olive green.

It is very important that you be able to identify this mushroom because, even though it is fleshy and looks good to eat, a small piece can cause persistent vomiting. Do not be misled by the taste which is quite pleasant — like that of hazelnuts. Another, very similar inedible toadstool, the scarlet-stemmed boletus, *(B. calopus)* has an unpleasant bitter taste. It is often confused with Satan's boletus, but its pores are lemon yellow, not red.

Rough-stem boletus
Boletus scaber
Boletus rugosus

The slender-stalked rough-stem boletus, with its semi-spherical, greyish-brown caps can often be found on the ground beneath white birches. *Boletus rugosus* is found under hornbeams and other deciduous trees and is even more common in summer than *Boletus scaber,* which grows mostly in the autumn. When cut, *Boletus rugosus* changes colour more than *Boletus scaber,* whose white flesh remains practically unchanged. The young fruiting bodies of both species are edible and very tasty.

Scarlet-stemmed boletus

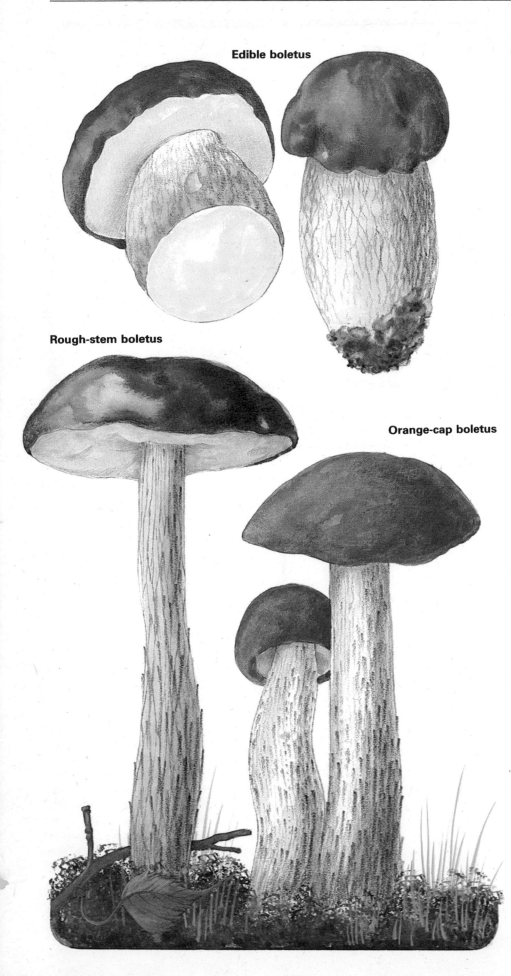

Edible boletus

Rough-stem boletus

Orange-cap boletus

Edible boletus *Boletus edulis*

The edible boletus, also known as the cep or penny-bun boletus, is one of the most familiar woodland mushrooms. In a good season large numbers can be found, mostly in broad-leaved woods and pinewoods. The edible boletus is a sturdy mushroom with a brown cap and stout brownish stem (stipe) patterned with a delicate network. It is a very tasty mushroom, but unfortunately it is often riddled with small maggots. Its firm white flesh can be prepared in a number of ways.

When the mushroom is cut in two the flesh remains white. In this the edible boletus differs from other boleti such as *Boletus luridus, Boletus erythropus,* the scarlet-stemmed boletus and Satan's boletus, which turn blue, green or some other colour when cut. The edible boletus is an excellent and popular mushroom.

It is unfortunate that it cannot be grown like a field mushroom. All the many attempts have failed. The reason is that the edible boletus is an extremely complex organism living in symbiosis with the roots of small trees and it is very difficult to grow the mycelium in a laboratory. Nevertheless, after several years of experimentation, attempts to grow the mycelium under laboratory conditions finally met with success but that was all — the mycelium grew but did not produce fruiting bodies.

Orange-cap boletus
Boletus aurantiacus

The orange-cap boletus is closely related to the rough-stem boletus. However, it is much more colourful and has firmer flesh. The semi-spherical, rusty-orange to brownish-red cap is perched on top of a solid, firm stem covered with brownish, fibrous scales. This excellent edible mushroom is a great favourite with mushroom pickers. It turns black when it is sliced and dried, but this does not detract from its taste.

Orange-cap boletus is found in deciduous and mixed woodlands, mostly where aspens grow.

Not All Toadstools Are Poisonous

Toadstools all have certain striking characteristics in common. The young fruiting bodies are covered with a protecting veil. As they grow, the veil ruptures and only fragments remain on the adult toadstool. The lower part, shaped like a cup, remains around the base of the stalk and the upper part sometimes forms scaly patches on the cap which may be washed off by rain. The veil may also join the edge of the cap to the stalk leaving a ring on the stalk when it ruptures.

The toadstools in this section are distributed throughout the temperate zone of the northern hemisphere, where they grow in deciduous and coniferous forests. They may often be found growing elsewhere as well. The most widely distributed of the species shown here is the fly agaric, which is also found in northern Asia, North Africa and Australia.

Fly agaric

Fly agaric *Amanita muscaria*

This pretty toadstool is the one most commonly illustrated in children's fairy tales. Its slender white stalk is topped by a wide, bright-red cap covered with white flakes and the ring round the stem makes a short, full skirt.

However, although it is attractive in its woodland setting, it must never be eaten for it is poisonous! It contains several substances that cause not only unpleasant digestive upsets but also damage to the nervous system. That it is not deadly, however, is confirmed by the fact that fly agaric is used by some people to make narcotics. The deadliest part of the whole toadstool is the skin of the cap. It is used to kill flies. After removing the stalk, the cap is placed on a dish red side up and sprinkled with sugar. Flies suck up the sweet liquid together with the alkaloid muscarine and are killed or at least stunned. That is how fly agaric got its name.

Blusher *Amanita rubescens*

This toadstool is called the blusher because the flesh turns a rosy colour, particularly when it is bruised. Also the tunnels made by larvae are redder than the surrounding flesh. And a characteristic feature of the blusher is that the bulbous base and stalk are often riddled with such tunnels.

Otherwise this excellent and tasty mushroom is a typical toadstool. Though it may be confused, at first glance, with the poisonous panther cap, it is readily identified on close examination. A reliable distinguishing feature of the blusher is its furrowed ring and the reddening of the flesh which is otherwise white. Also characteristic is the bulbous base which is cracked and covered with warts. The blusher is best used fresh in soups.

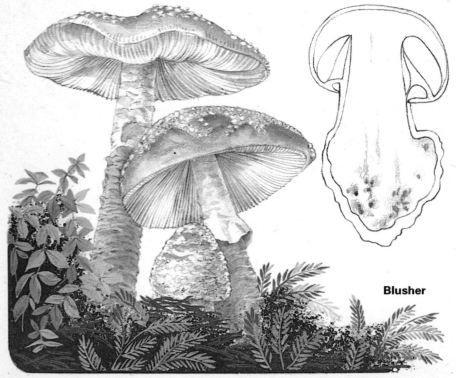

Blusher

Panther cap *Amanita pantherina*

This extremely poisonous toadstool is really dangerous! Firstly it should be pointed out that it does not always have all the characteristics of other species of *Amanita*. The white remnants of the veil on the greyish-brown cap are often washed off by rain; also the smooth, limp-ring is sometimes absent and the stalk may be only slightly bulbous at the base. The flesh is white, does not change colour, and is misleadingly tasty. It even has a pleasant aroma, whereas the edible fly agaric has an unpleasant, astringent taste when eaten raw.

This poisonous toadstool is all the more dangerous in that it can be mistaken for the excellent blusher or even more often for the edible though not very good *Amanita excelsa*. Unlike the death cap, poisoning caused by eating the panther cap proceeds very rapidly. The first symptoms often occur within half an hour. The mycoatropine it contains causes brain damage, the nature of the poisoning being very similar to that caused by excessive consumption of alcohol. Luckily, however, such poisoning is rarely fatal.

Death cap

Death cap *Amanita phalloides*

The bright red cap of the poisonous fly agaric is like a flag of warning. However, the death cap, which is deadly poisonous, is a rather nondescript mushroom. Its appearance gives little indication that it is the most poisonous mushroom known.

The slightly convex cap is coloured greyish-olive-green, the stalk a lighter hue. The related fool's mushroom and avenging angel, which are coloured white, are also deadly poisonous. Inexperienced or unwary mushroom pickers can easily confuse these with field mushrooms. A sign of warning is the bulbous base of the stalk enclosed in a membranous cup (volva) called the 'death cup'. However, it may not be well developed. Sometimes the mushroom may be pulled up without it and incorrect identification of this toadstool results in dozens of fatal poisonings every year.

It is therefore surprising that the death cap also contains medicinal substances. However, these cannot be used for no one has yet found a way of separating them from the poisonous substances, which far outweigh the medicinal ones in quantity.

The death cap is found from July until October in deciduous and mixed woods. However, it sometimes appears earlier and beyond the margin of the forest.

Panther cap

A Tasty Foursome

The parasol mushroom, orange milk cap, chanterelle and the toadstool *Russula vesca* are all edible and very good. They can be readily identified even by the amateur. The parasol mushroom is a tall, striking mushroom of open woodlands; the egg-yellow chanterelle is much smaller but often plentiful; the orange-green orange milk cap oozes a bright orange milk, and *Russula vesca* can be identified by its taste. The rule of thumb for russulas is that those with a pleasant taste are edible and those with an acrid taste are not.

The mushrooms in this section are distributed throughout the temperate regions of the northern hemisphere and in Australia; the parasol mushroom is also found in Africa and India.

Parasol mushroom

Parasol mushroom *Lepiota procera* This mushroom bears a strong resemblance to a parasol. The wide cap of adult specimens perches on top of a rather slender, hollow stem. The ring can slide up and down the stem. The parasol looks rather flimsy and in fact it is. When the mushroom is picked the cap usually breaks off and if you take a look you will see that the flesh does not change colour. If it turns reddish, then it is not the parasol mushroom but its relative the shaggy parasol, which is also edible, but only the caps should be eaten.

The parasol mushroom is edible and aromatic, though not very fleshy, and a favourite of mushroom pickers. It is collected only for immediate consumption, being unsuitable for drying.

Russula vesca

autumn, usually in groups. If you find one or two fully-grown specimens there are sure to be several more close by, at various stages of growth.

The chanterelle is edible and prized by mushroom pickers even though its flesh is rather tough. As a rule it is not riddled with maggots and does not spoil quickly, so it is sometimes transported to distant markets. It is also tinned. The flesh is a lighter colour than the cap — almost white — and has a faint but pleasant fruit-like aroma and a delicate peppery flavour.

Russula vesca

Unlike their relatives, the milk caps (*Lactarius* spp), russulas do not ooze milk when bruised nor do they change colour — only rusty patches appear here and there. Many species of the genus *Russula* are brightly coloured, but *Russula vesca* is rather drab. The cap may be tinged brown, pink, brownish-green or pale violet. The gills, flesh and stem are white. However, it is tender, very tasty (it tastes like almonds or hazelnuts) and is an excellent edible mushroom.

Saffron milk cap

Chanterelle *Cantharellus cibarius*

Chanterelles are easily spotted, for their egg-yellow colour stands out in sharp contrast to the browns and greens of the forest vegetation. They grow from summer until

Chanterelle

Saffron milk cap
Lactarius deliciosus

Milk-like saps may be found in herbs, spurges and bell-flowers. One group of fungi also has this characteristic. These are the milk caps, such as the saffron milk cap, and *Lactarius volemus*. The milk is often white, but may be coloured. In the case of the saffron milk cap it rapidly turns bright orange. This is also the colour of the gills, the stalk and the cap, which is convex, later funnel-shaped and marked with dingy-green concentric rings.

If you come across the saffron milk cap (it grows mostly in autumn under young spruce trees) you should pick it, for it is a very good and popular mushroom. It is tender and has a pleasant spicy flavour, most pronounced in caps pickled in vinegar.

Two in One House

A lichen is a combination of two organisms — an alga and a fungus — which together make up a single plant body. This association is beneficial to both. The alga, by photosynthesis, makes the necessary organic food and the inorganic substances are provided by the fungus, which absorbs water containing dissolved minerals from the soil and gives the alga protection.

Lichens often multiply by vegetative means, by the development of minute bud-like outgrowths on the upper surface of the thallus. These consist of one or more algal cells enclosed by a few fungal filaments. When they break away from the thallus they are dispersed by the wind throughout the neighbourhood. If they fall in a suitable spot they take hold and develop into a new lichen. Sometimes the fungus produces spores by sexual reproduction. These develop into a new lichen only if the fungal threads that develop from the spores find a suitable alga.

Lichens are divided according to the shape of the thallus into three basic types: crustose (encrusting), foliose (flat and leaf-like), and fruticose (upright and branched). They play an important role in nature as pioneers. They grow in places that other plants find unsuitable, such as bare rocks, disrupting the surface both chemically by means of the acids they produce and mechanically by growing into the substrate.

Lichens are indicators of air pollution, particularly those that grow on the bark of trees. The slightest contamination of the atmosphere by smoke fumes causes them to die. Many lichens serve as food for herbivores in northern regions and deserts.

Peltigera canina

This lichen was formerly used to treat bites by rabid dogs and this fact is indicated by the specific name *canina* (the Latin word *canis* means dog).

Peltigera canina is one of the largest of the ground lichens, with a thallus that measures up to 20 cm (8 in) across. It grows in moist soil in shady places and is common alongside forest rides, where it makes large masses measuring several hundred square centimetres. In some places it grows together with certain mosses such as *Hylocomium splendens* and *Pleurozium schreberi*. In some cases the stalk of the moss may even grow through the lobed thallus of the lichen. In damp conditions *Peltigera canina* is coloured greyish-green; when dry it is brownish and finely felted. It is attached to the substrate by means of numerous root-like outgrowths.

Peltigera canina is very fertile. Its chestnut-brown fruiting bodies are relatively large. They are inrolled like tubes and located at the edges of upright lobes.

Peltigera canina

Iceland lichen *Cetraria islandica*

Iceland lichen is very plentiful in northern and central Europe. It does not form spreading masses but occurs only in groups or singly between mosses. It prefers acid soils in mountain regions, only occasionally being found in the lowlands in heaths, peat moors and coniferous woods in dry, sandy soil. It often grows in the same places as reindeer moss, and the moss *Hypnum cupressiforme*. The thallus is coloured greyish-brown and the lobes are often furled into horn-like shapes and are covered with bristles along their margins.

Unlike other lichens, the thalli of

Iceland lichen

Iceland lichen are of practical use for they contain cetrarine, which has medicinal properties. It promotes digestion and is also used to treat lung diseases and sea-sickness. This lichen was used as early as the 16th century by those who wished to lose weight and to treat chronic diarrhoea. It has a penetrating bitter taste and in Iceland it is taken in milk. In times of want it was also added to flour. An edible jelly can be made by soaking the lichen first, to remove the bitterness, and then boiling it.

Cetraria glauca

Members of the genus *Cetraria* are

interesting examples of transitional forms between foliose lichens and fruticose lichens.

Whereas the thalli of Iceland lichen generally grow on the ground and are upright, the closely related *Cetraria glauca* is a common epiphytic species that grows on tree trunks. The foliose thallus of this lichen is only shallowly lobed.

Cetraria glauca

Inflated lichen
Parmelia physodes
The inflated lichen is a very common species. Though it generally grows on the bark of forest trees from lowland to mountain elevations it may also occur on other substrates, such as boulders, rocks, barren soil and old dead wood and stumps. It does best, however, on the trunks and branches of trees, which it covers with its greyish-green lobed thalli. It is most abundant in shady, moist places with few air currents, where it coats not only the trunks but even the twigs of young as well as older trees. The foliose thallus is so firmly attached to the substrate that it is almost impossible to pull off. Only the narrowly lobed edges are loose and slightly inflated. Unlike *Peltigera* this lichen is generally sterile.

A vertical section of this lichen

Inflated lichen

shows the typical structure of foliose lichens. At the top and bottom is a dense mass of fungal threads (hyphae) called the upper cortex and lower cortex. Beneath the upper cortex is the algal layer consisting of loosely interwoven hyphae intermingled with green algal cells. Beneath the algal layer is a layer composed of very loosely interwoven hyphae which is called the medulla.

Reindeer moss
Cladonia rangiferina
In the far northern regions there is very little in the way of vegetation. Only a few species of plants are able to survive in the harsh habitat of the tundra. The ones that do are the principal food of the northern herbivores — reindeer. These are chiefly lichens and mosses, and of the lichens mainly reindeer moss, followed by *Cladonia sylvestris* and Iceland lichen. Lichen pastures, however, take 20 to 30 years to grow again and thus large areas are needed to support the herds of reindeer.

Reindeer moss belongs to the genus *Cladonia* which is marked by great diversity and changeability of the species and includes the greatest number of ground lichens. Unlike most *Cladonia* species, reindeer moss has no scaly ground thallus, only greatly developed stalks (podetia) which are hollow and up to 15 cm (6 in) high. They are coloured greenish-greyish-white and are bush-shaped. In the reindeer moss the branches are bent to one side with the fruit located at the tips.

This lichen is a xerophyte, in other words a plant that stands up well to drought. Reindeer moss is able to survive even long periods of drought during which it dries up so that it disintegrates into a white powder when touched. However, the loss of water is rapidly replaced when it rains; even an increase in atmospheric moisture is sufficient. In the forests of the temperate zone these ground lichens are found only in the poorest soils.

Reindeer moss

The Forest's Green Carpet

The thick carpets of moss in forests are composed of tiny individual plants. The stem of each one bears small green leaves and is anchored in the ground by rhizoids formed at the base. In the life cycle of mosses there is an alternation of two generations — one sexual, the other asexual. The former, called the gametophyte (plant that bears sex cells), produces simple sexual organs at the tip of the stem. These may be either male (antheridia) or female (archegonia). They are generally arranged in separate groups either on the same plant or on separate plants. The male cells swim through the water on the plant to the female cells and fertilize them.

White fork-moss

A fertilized female cell gives rise to the asexual generation, called the sporophyte (spore-bearing plant). This is not green and is dependent on the gametophyte from which it grows for water and nourishment. It is composed of a stalk (seta) bearing a capsule containing spores and topped by a cap. When ripe, the capsule opens by the breaking away of a lid to release the tiny spores. These germinate in the moist woodland soil to produce a green, filamentous, alga-like structure called the protonema from which grow the new leafy plants. Mosses also multiply by vegetative means.

Today the importance of mosses

for man's use is negligible. However, in former times moss was used by country folk as thermal insulation. And moss is a very important element in the ecology of all forests.

Dicranum scoparium

This moss makes soft, gleaming green carpets 'brushed' to one side. The five to ten-centimetre-long gametophytes are covered with long, thin leaflets and their tips curve in an arch. The long red setae are topped by long-beaked capsules.

Dicranum scoparium is found on stones and logs, but generally grows on poor, rather dry soil in forests, forming spreading masses particularly in sandy areas.

Dicranum scoparium

White fork-moss
Leucobryum glaucum
The sparse undergrowth of coniferous forests on poor soils often includes the roundish, bluish-green cushions of white fork-moss. The cushions often join to form large spreading masses. When dry they are whitish-green and gleam like silver on the needles covering the forest floor. They are attached to the substrate only lightly because the stems have very few rhizoids at the base. The stems die from the base and grow at the top. Their large, empty dead cells fill with water when it rains. That is why the thick cushions are soaked with water like a sponge while the soil beneath is

dry. The water is held inside the cells and only that which runs over the surface finds its way to the soil.

The presence of white fork-moss is an indication of a poor forest habitat. In places where it forms spreading masses it damages forest stands even more, because it prevents the germination of fallen seeds. The seeds remain on the tops of the cushions and cannot develop into new plants.

Pleurozium schreberi
This moss forms loose spreading

Pleurozium schreberi

masses, often covering large expanses of ground. It is a typical moss of coniferous and mixed woodlands, doing best on sandy soil in pine woods. It does not tolerate deep shade and in sun-dappled places makes a fluffy carpet even on logs and rocks. Its thick cushions are very absorbent and hold water from melting spring snow and rainfalls for a long time.

The stems of this moss are upright, regularly branched in two rows and coloured a striking red. The tips of the branches are recurved. The leaves covering the stem are oval, convex and arranged like tiles on a roof. Female specimens are readily identified by the long, thread-like, crooked setae (capsule-stalks), which are coloured red, and by the crooked capsules.

leaflets that form a rosette at the top. This moss is usually very fertile.

Common hair-moss
Polytrichum commune
The common hair-moss is frequently described in school textbooks. This is because it is one of the largest mosses with the most perfect structure. Its upright stem generally reaches a height of about 25 cm (10 in). It has a well-developed central cylinder, which, like the vascular bundles of higher plants, conducts water and dissolved minerals up the plant. Its surface is thickly covered with narrow dark-green leaflets with sharply toothed margins. In damp weather these leaflets stand away from the stem; in dry weather they are pressed tightly to the stem, thus protecting it from drying up. The large, angular capsules, markedly swollen at the base, are borne on long, stout stalks coloured yellowish-red. The capsules are upright at first, later horizontal, and entirely covered by a cap.

Hylocomium splendens

Mnium punctatum

Common hair-moss

Hylocomium splendens
This moss makes loose, gleaming clumps, often forming a soft spreading carpet. The upright, red stem is covered with small oval leaves and is regularly branched, giving the plant a plume-like appearance. The seta (capsule-stalk) is red, stout, up to 4 cm (1.6 in) high and bears a brown ovoid capsule at the top.

Hylocomium splendens is found from lowland to high mountain elevations where it grows both in pure stands of spruce as well as in mixed woods on richer soils.

Mnium punctatum
This moss can often be found in damp places in the forest and alongside woodland streams. The short, erect stem of the gametophyte is covered with hairs from the base and higher up by transparent

Witnesses of the Distant Past

Clubmosses and horsetails
Clubmosses are the simplest and oldest group of pteridophytes. Today clubmosses and horsetails are only insignificant components of the forest undergrowth, but in the distant past the extinct relatives of present-day forms were as large as trees and formed large forests. Their

primarily in shady mountain spruce woods where it occurs up to the upper forest limit. It has a cosmopolitan distribution and is found even in arctic regions. In such rugged conditions, where the growth period is only very brief, it multiplies more by vegetative than by sexual means.

ulcers. The powder also has a marked antiseptic effect, and the extract obtained by boiling it was used as a gargle and applied to inflamed gums. In pharmacies it was used to coat pills to keep them from sticking together. The green parts of the plants contain certain alkaloids and are poisonous.

Fir clubmoss

dead trunks eventually formed part of today's coal.

Horsetails have leaves arranged in characteristic whorls on hollow, grooved stems. Their spore-bearing cones, or strobili, are more perfect than those of clubmosses and conceal a greater number of sporangia (spore-containers).

Fir clubmoss *Lycopodium selago*
The fir clubmoss is found in the most deeply shaded damp places in forests. It makes dense, green tufts of erect stems up to 25 cm (10 in) high and it does not produce cones. The sporangia are located in the middle section of the stem on the upper side of the leaves, thereby putting no limit on the stem's growth. It may continue to grow even after the sporangia are ripe.

In lowlands the fir clubmoss grows in deep valleys where light does not penetrate. However, it is found

Stag's-horn clubmoss
Lycopodium clavatum
Stag's-horn clubmoss, a creeping, evergreen species sometimes up to 1 m (40 in) long, grows in shady, mainly coniferous forests from lowland to mountain regions. Ascending from the thickly-leaved ground stems are long, leafless, yellow-green stalks with forked tips bearing a single spore-bearing cone on each branch of the fork.

Behind each of the leaves that make up the cone is a kidney-shaped sporangium containing a large number of spores. The spores germinate when they fall on the ground and develop into a gametophyte, which produces male and female sex cells. After fertilization a new spore-producing plant grows.

The fine powder formed by the spores is not easily wetted and for this reason was applied in folk medicine to wounds, rashes and

Stag's-horn clubmoss

Wood horsetail
Equisetum sylvaticum

A delicate web of thick, fresh-green whorls adorns the stem of the wood horsetail in summer. In spring the stem is entirely different. It is erect, pinkish-brown and bears a cone at its tip. The cone falls after the spores are shed. The stem then turns green and forms branches in whorls at the nodes. Each whorl is composed of ten to 16 arching branches.

Wood horsetail is partial to shady, damp places, particularly in forest wetlands. Wherever it occurs in masses, water is sure to be close beneath the surface of the soil.

Wood horsetail

Field horsetail *Equisetum arvense*
A stubborn weed, field horsetail is found in large numbers on sandy and loamy soils. Unlike the wood horsetail, it produces two stems — a fertile spring stem and a sterile summer stem. The first is yellowish (it contains no chlorophyll), simple and stout with a spore-bearing cone at the tip. In dry weather the spores are shed like clouds of white dust. They germinate in moist conditions to produce gametophytes.

After the spores are shed the fertile stems die and are replaced in summer by green, sterile stems that grow from the same rhizome. These stems with their whorls of branches are the same size as the spring stems — about half a metre (20 in) high. The green stems carry out photosynthesis and store a supply of food for the growth of the fertile spring stems the next year.

Field horsetail is collected as a medicinal plant because it contains large amounts of silicon dioxide. The extract obtained by boiling this plant promotes the flow of urine. This extract also kills aphids.

Field horsetail

273

Ferns

The great period of horsetails, club-mosses and ferns, jointly called pteridophytes, is long past. Their greatest development took place in the Palaeozoic era, when they attained the size of trees, growing in extensive bogs and swamps where they formed forests. Old trunks fell into the mud. There, covered by silt and water, and in the absence of oxygen, they did not rot. Instead they formed peat, which later turned into coal. In this form the remains of the pteridophytes have survived to this day, providing us with much-needed fuel. However, they have also survived in the form of living descendants, which, though they exhibit a marked diversity of shape and distribution, do not attain such large dimensions. This is true, at any rate, of the ferns growing in temperate regions. They are mostly shade-loving plants with persistent underground stems. In the tropics there are still tree-ferns that resemble their ancient ancestors. In the tropics you will also find ferns growing as climbers (e. g. *Lygodium*) and epiphytes, (e. g. *Asplenium nidus*).

Pteridophytes and most bryophytes (mosses and liverworts) have a more complex structure than algae — their bodies are divided into root, stem and leaves. However, they do not produce flowers and their reproduction is by means of spores. For a long time people believed that ferns produced miraculous flowers on certain nights. These flowers were said to make a person invisible and enable him to understand the speech of animals and trees. Also fern 'roots' were believed to protect man from magic spells.

Male fern

Dryopteris filix-mas

The male fern is a common fern of various types of forests. It is a large plant reaching a height of 1 m (40 in). The fronds form a funnel-shaped rosette. They grow from a short, creeping rhizome covered with rusty hairs. The young, spring fronds are tightly coiled into spirals that slowly unwind and expand into longish blades, which continue to grow at the tip for a long time. Some of the fronds bear spores. The sporangia (spore-containers) are clustered in groups, called sori, on the underside of the leaf and covered by a special scale-like outgrowth called the indusium. The evaporation of water from special cells in the wall of the sporangium causes the sporangium to tear open and release the spores, which are then distributed by the wind. When a spore lands in a suitable spot, it germinates and develops.

The male fern is widespread throughout most of the world and in some places grows even at alpine elevations. Sometimes the rhizomes are collected as a drug used to get rid of tapeworms. However, it is a dangerous medicine because the whole plant, particularly the rhizome, is poisonous.

Male fern

Water fern

274

Bracken

Hard fern *Blechnum spicant*

A characteristic of the hard fern is that it has two types of fronds. The outer, sterile fronds spread sideways and have a number of short lobes arranged like the teeth of a comb. The inner fronds grow upright and have narrower, more distantly-spaced lobes. These are the fertile fronds and they bear sori, which are covered by the inrolled edges of the lobes. The two types of fronds also differ in colour. The sterile fronds with short stalks are dark green; the spore-bearing fronds with long stalks are brown.

The hard fern stands up well to rugged climates with frequent rainfall. In the north temperate zone it is the only representative of this attractive genus. It grows on heaths and moors, but not on chalk or limestone soils. The other 60 species are distributed in the temperate zone and tropical mountains of the southern hemisphere. Some are grown in greenhouses.

Hard fern

Water fern *Salvinia natans*

The water fern is an annual measuring up to 20 cm (8 in) and is well-adapted to life in water. It floats on the surface and its single or branching stem bears leaves in whorls of three, one of which is finely split and dangles in the water. Another leaf is modified into a marble-like structure located close to the stem. This bears sori containing sporangia with spores. Some of the sori contain larger sporangia with a single large spore, others contain smaller sporangia with a large number of tiny spores.

Bracken *Pteridium aquilinum*

Bracken is a large, coarse fern found in all parts of the world except South America and Antarctica. It grows in poor, lowland pine woods, open woodlands, and on heathland right up to the Arctic Circle. Its rhizomes spread vigorously in all directions both in gravelly and sandy soils, forming spreading masses covering up to several square kilometres. It is difficult to eradicate and in some places it is a serious pest.

Each plant has only one frond. This frond, however, is large – 1-2 m (40-80 in) high. It is composed of a stout, rigid stalk and a much-divided blade of triangular outline. The sporangia are arranged around the margins on the underside of the lobes and covered by the folded-under edges of the lobes.

Bracken is one of the most widely distributed ferns in the world. With its branching rhizomes, it covers and binds nonfertile soils, loosens soil, and its large fronds help to prevent the soil from drying out.

Common spleenwort
Asplenium trichomanes

The common spleenwort is a small fern with a stout rhizome which is widespread in the northern hemisphere. Its stiff fronds generally grow as pioneers from rock crevices and cracks in old walls. The reddish-brown stalks are thickly covered with two rows of tiny, ovate leaflets that fall off in the autumn. The largest leaflets grow in the middle section of the oval blade. The sori are long and thin and are located on the lower surface of the leaflets. They are covered with papery scales that fall off when the spores are ripe.

When rubbed between the fingers the fronds have an aromatic scent. The dried fronds were used in folk medicine to treat the spleen and the liver. In medieval days it was believed that the spleenwort had magic powers and protected people from evil spirits.

Common spleenwort

Wall rue

Wall rue
Asplenium ruta-muraria

Wall rue is another small fern that grows on shaded as well as sunny rocks, in rock crevices and walls from lowland to mountain elevations. It is quite widespread and finds even the lime-mortar in the brick walls of old castles and ruins sufficient for its needs.

Wall rue makes thick, low tufts (only about 10 cm (4 in) high) of dark-green fronds which differ from those of the common spleenwort not only in the way in which they are divided but also in the colour of the leaf-stalk. In wall rue the stalk is green, changing to dark brown at the base. The oval sori are covered with thin, papery scales.

Hart's-tongue fern
Phyllitis scolopendrium

Hart's-tongue fern is distributed throughout most of the world, but is particularly noteworthy in Europe for it is the only European fern with undivided, entire fronds. The glossy, leathery fronds are arranged in a rosette and remain green even in winter. They are tongue-shaped, (heart-shaped at the base) and slightly wavy along the margin. The stalks are reddish-brown and covered with rusty hairs. On the underside of the fronds are striking long sori covered by linear, scale-like growths (indusia).

This decorative fern is also frequently planted in shady rock gardens. Its natural habitats include the shade of deep mountain ravines and the moist rocks and sides of waterfalls, as well as woodlands and hedgerows.

Common polypody
Polypodium vulgare

This polypody is common throughout the northern hemisphere. In the north its range extends as far as the Arctic Circle and in the south to

Common polypody

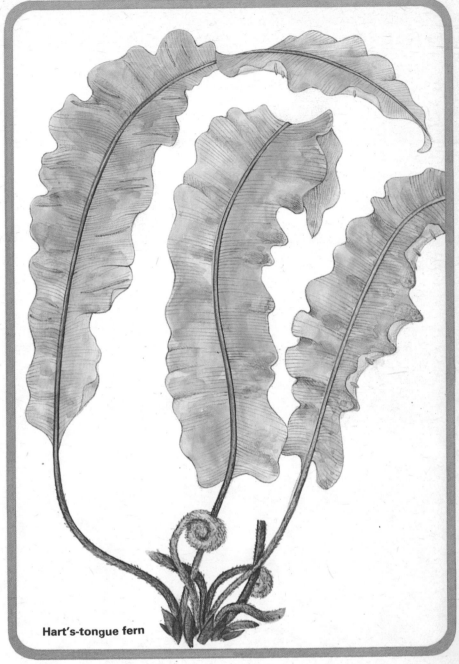

Hart's-tongue fern

Africa. It generally grows on moist rocks, moss-covered boulders and in shady forests. It likes both limestone and sandstone. Although it prefers rocks, it may also be found on rotting tree trunks.

The long, thin creeping rhizome is very sweet and slightly bitter. It branches repeatedly and is thickly covered with brown scales which fall off as the rhizome gets older. On the underside it produces roots; on the upper side it produces widely-spaced, leathery fronds up to 40 cm (16 in) long. They remain throughout the winter. Young fronds are coiled

in a spiral and mature fronds coil in dry weather to limit the evaporation of water. The older, hind portion of the rhizome gradually dies and decays, while the younger, forward end continues to grow and produce new fronds. On the underside of the leaflets are two rows of round sori.

Formerly, the rhizomes of the common polypody were widely collected and used in folk medicine. Nowadays it is still an important drug plant. It contains a bitter substance (which has not been investigated as yet), mucilages, saponins and essential oils.

Mineralogy

After studying the living world of nature, the Earth, on which plants and animals live, may seem static and unchanging. The Earth is the inorganic, or non-living part of nature. But even stones are 'born' and 'live'. This is because minerals and rocks, like living organisms, can remain stable only in an environment that suits their needs. When conditions change, they disintegrate or are altered into new, secondary minerals and rocks. For example, some minerals and rocks are changed when they come into contact with air or water. Other changes result from great heat or pressure. Such changes form part of a continuous process, called the rock cycle. But as the rock cycle is usually measured in millions of years, it is hardly noticeable in one person's lifetime.

Composition of the Earth

The Earth's crust is composed of rocks, which are themselves made up of minerals. A few minerals occur as pure or nearly pure elements, and are called native elements. But most minerals are compounds of two or more elements. The elements are the same as the ones that make up the living world but their arrangement is different.

The most common element in the top 16 km (10 miles) of the Earth's crust is oxygen. It occurs in combination with other elements to form a group of minerals called oxides. Oxygen comprises 46.60 per cent of the mass of the Earth's crust. Silicon, the second most common element, makes up another 27.72 per cent. Combinations of oxygen and silicon, usually with one or more other elements, form a group of minerals called silicates, which include amphiboles, feldspars, micas, olivines, pyroxenes and quartz. These minerals are described on pages 286-287. The other major elements in the crust are aluminium, iron, calcium, sodium, potassium and magnesium. Together, these six elements make up another 24.27 per cent of the crustal mass. Eighty-four more elements occur naturally in the Earth's crust,

these forming only 1.41 per cent of the total.

The Earth is composed of three main zones. The thin crust is made up mostly of light silicates. Beneath the crust is the mantle, which can be divided broadly into two main parts. The outer mantle contains heavier silicates, with a greater proportion of magnesium. The inner mantle is denser (heavier by volume), containing oxides and sulphides of various metals. The mantle encloses the extremely dense core, which probably consists mostly of iron, with some nickel.

The solid outer layer of the Earth, including loose stones, soil, sand and mud, consists of rocks, which, in turn, are composed of grains of minerals. Minerals are thus the basic constituents of rocks. Rocks, however, are heterogeneous, because the proportions of minerals in them vary from one sample to another. Minerals, on the other hand, are homogeneous (the same throughout). They have specific chemical formulae and special properties. The formulae and properties of the minerals described in the following pages are given at the end of each section. Rocks, being heterogeneous, do not have chemical formulae or precise properties, such as specific gravity.

In chemical formulae, each element is represented by a symbol. For example, the symbol for the element gold is Au. This symbol also applies to gold which occurs as a native element in the Earth's crust (see pages 292-293). But, as we have seen, most minerals are combinations of two or more elements. For example, O is the symbol for oxygen and H stands for hydrogen. When one atom of oxygen is combined with two atoms of hydrogen, they form water, which is a mineral with the chemical formula H_2O. Other formulae are made up in a similar way. The formula for halite, or rock salt, is NaCl (see pages 294-295). It contains one atom of sodium, which has the symbol Na, for every atom of chlorine (Cl). Similarly, the common mineral quartz has the formula SiO_2 (see pages 285-286). It contains one

Mineralogy

atom of silicon (Si) for every two of oxygen (O). The chemical symbols for all the elements referred to in the following chapters are listed below.

Many chemical formulae are more complicated. In some of them, you will find two or more elements surrounded by brackets and separated by commas. The chemical formula for olivine is $(Mg,Fe)_2SiO_4$ (see page 287). The reference to $(Mg,Fe)_2$ in the formula means that atoms of magnesium (Mg) can substitute for atoms of iron (Fe). Hence, some kinds of olivine are rich in magnesium, while others are rich in iron. Olivine is, therefore, the name for a group of minerals, rather than a single mineral. For example, a magnesium-rich variety of olivine is called forsterite, while an iron-rich variety is named fayalite.

Mineral crystals

In most minerals, the atoms are arranged in a regular and repeating pattern, called a lattice. The lattice formed by the atoms determines the shape of crystals formed by the minerals. For example, we are all familiar with cubic crystals of salt. Water, too, forms crystals when it solidifies into ice. Snowflakes are essentially composed of tiny crystals, grouped in shapeless masses.

Nearly all minerals form crystals, providing that there is space for them to grow. However, in nature, well-formed crystals are generally rare, because suitable conditions for their develop-

Some minerals do not appear to be crystalline. Some are cryptocrystalline, which means that their crystal structure is hidden. For example, the mineral chalcedony (see page 313) does not look crystalline. But, under a microscope, you will see that it is composed of tiny crystals invisible to the naked eye. However, amorphous minerals such as opal (page 311), do not have any regular internal arrangement and are non-crystalline.

Characteristics of crystals

Well-formed crystals, which may be found in cavities or veins, are prized by collectors. They are usually found in groups. Sometimes, pairs of crystals develop in contact with each other or they may be intergrown. Such crystals are called twins. Each crystal is bounded by flat planes (surfaces), with edges of various lengths, arranged at definite angles to each other. As we have seen, the shapes are determined by the angles and distances between the atoms in the lattice. Crystals are classified according to the system of symmetry to which they belong.

Symmetry occurs in many familiar objects. Think of a building block, used by children in play. Many of these blocks are cubes, which are solids bounded by six equal, square faces. If you cut one of these cubes down the middle, one half will be a mirror image of the other half. In fact, a cube can be cut in nine different ways—horizontally, vertic-

CHEMICAL SYMBOLS USED IN THIS SECTION

Ag	Silver	F	Fluorine	O	Oxygen
Al	Aluminium	Fe	Iron	P	Phosphorus
As	Arsenic	H	Hydrogen	Pb	Lead
Au	Gold	K	Potassium	S	Sulphur
B	Boron	Li	Lithium	Si	Silicon
Be	Beryllium	Mg	Magnesium	Sn	Tin
C	Carbon	Mn	Manganese	Th	Thorium
Ca	Calcium	N	Nitrogen	Ti	Titanium
Cl	Chlorine	Na	Sodium	U	Uranium
Cu	Copper	Ni	Nickel	V	Vanadium
				Zn	Zinc

ment seldom occur. For example, the rock granite forms when molten material cools and hardens (see page 284). The minerals in the molten material, such as quartz, feldspar and others, generally do not have the space in which to develop well-formed crystals. As a result, the minerals crystallize into small, irregular and interlocking grains. Minerals may also form in aggregates, which are masses of imperfectly-formed crystals. For example, reniform aggregates are rounded and kidney-shaped, while botryoidal aggregates are shaped like a bunch of grapes. Other aggregates may have a branching, thread-like or wiry form.

ally and diagonally — so that one half is a mirror image of the other. The surfaces along which the cube is cut are called planes of symmetry. A cube can also be described according to the symmetry around its axes, which are imaginary lines joining the centres of the opposite faces. For example, the vertical axis is an imaginary line linking the centres of the top and bottom faces. If you rotate the cube around the vertical axis, it will look the same four times during one complete revolution, because the four side faces are the same. We call this a *four-fold axis of symmetry*. The cube has two other axes, which join the centres of the opposite side faces. If the cube is rotated around these axes, it will appear the same four times during each revolution. Hence, we can say that the cube has three four-fold axes of symmetry. The three axes are equal in length and at right angles to each other. Any crystal with these features belongs to the cubic system of symmetry.

Apart from the cubic system, there are six other crystal systems: tetragonal, orthorhombic, monoclinic, triclinic, hexagonal and trigonal. The hexagonal and trigonal systems are similar. Both have four axes, three of which are horizontal and the other vertical. But they differ in their symmetry. The hexagonal system has one, vertical *six*-fold axis of symmetry. This means that it appears the same six times when it is rotated around its vertical axis. But the trigonal system has one, vertical *three*-fold axis of symmetry. In the other systems, the lengths of the axes and the angles between them vary (see diagrams). However, all crystals belong to one of these seven basic systems.

Other features of minerals

The arrangement of atoms determines not only the external form and symmetry of crystals, but also other important properties of minerals. For example, carbon occurs in nature in the form of soft, black graphite or as glittering diamond, the hardest of all natural substances. This phenomenon is called *polymorphism*. It is caused by a different arrangement of atoms in the lattice.

Hence, hardness depends on the internal structure. It is measured on the Mohs' scale, which is numbered from 1 to 10. At one end of the scale is talc, which has a hardness of 1 and which can be crushed with a fingernail. At the top end of the scale (10) is diamond. Between talc and diamond, in order of hardness, are gypsum (2), calcite (3), fluorite (4), apatite (5), orthoclase (6), quartz (7), topaz (8) and corundum (9). Simple hardness tests are useful guides in identifying minerals. For example, a penknife, which has a hardness of about 5 1/2, will scratch apatite but not orthoclase. Quartz will scratch glass, but you would need a special steel file to scratch quartz.

Another feature of minerals is the ratio between the weight of a mineral and an equal volume of water. This is referred to as specific gravity. A lump of pure gold feels heavy. It is 19.3 times as heavy as an equal volume of water and so gold, therefore, is said to have a specific gravity of 19.3. The average specific gravity of all minerals is 2.6. However, the specific gravity of individual minerals ranges between 1 and about 23.

Colour is sometimes a help in identifying min-

CRYSTAL SYSTEMS

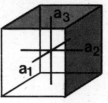

Cubic

The three axes are equal in length and at right angles to each other. Rotating the crystals around the axes shows the three four-fold axes of symmetry.

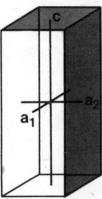

Tetragonal

The three axes are at 90° to each other. The vertical can be longer or shorter than the two horizontal axes which are both equal in length. There is one four-fold axis of symmetry.

Orthorhombic

The three axes are at 90° to each other and each axis is of a different length. There are three two-fold axes of symmetry.

Monoclinic

The three axes are all different in length. The two horizontal axes are at right angles to each other and the vertical axis is tilted at more than 90°. There is a single two-fold axis of symmetry.

281

Mineralogy

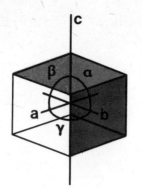

Triclinic

The three axes are all different in length and none of the angles between them is at right angles. There can be one or no axis of symmetry.

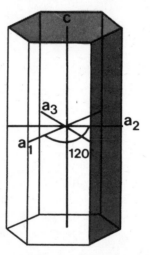

Hexagonal and Trigonal

Both systems have four axes. The three horizontal axes are at 120° to each other and of equal length; the vertical axis is either longer or shorter. The Hexagonal has one six-fold axis of symmetry. The Trigonal has one three-fold axis of symmetry.

erals. But it can be misleading, because many minerals occur in a wide range of colours. Another colour test can be done by crushing a bit of a mineral into a powder, which is called the streak. The colour of the streak often differs from the colour of the mineral. For instance, it is possible to confuse the dark-looking minerals haematite and magnetite. But if you crush haematite, you will find that it has a red streak, while the streak of magnetite is black. However, the streak test is not always as useful. For example, the streak of nearly all the minerals in the large silicate group is white.

Another feature of minerals is the way in which they behave when light strikes them. Some transmit light easily and are transparent. Translucent minerals are those that transmit some light, but you cannot see through them. Opaque minerals do not transmit light. Instead, they reflect or absorb light rays. The lustre of minerals is another guide to identification. Lustre is the surface sheen or gloss of a mineral and it is determined by the nature of the surface and the extent to which it reflects light. The lustre may be metallic, adamantine (brilliant, like a diamond), vitreous (like broken glass), resinous (like amber), pearly, silky, greasy, dull or earthy.

Cleavage occurs in some minerals. This means that, when they are struck, they split along one or more regular planes. For example, mica minerals, such as biotite (page 286) and muscovite (pages 300-301), often occur in blocks which split into thin sheets. Minerals which split readily along regular planes are said to have good cleavage. Some minerals have two or three cleavages, while others have none. Minerals without cleavage may fracture in a typical way. For example, the fractured surface of some minerals, such as quartz, is conchoidal or looks like sea shells.

Other features of minerals include elasticity, brittleness, malleability, electrical and magnetic properties, radioactivity, their reaction to various acids, and so on. Chemical analysis of minerals is complicated to do and so most mineralogists rely on their knowledge of the properties of minerals in order to identify them.

How minerals are formed

Most minerals were formed, and continue to be formed, below the surface of the Earth, where they are exposed to immense pressures and extremely high temperatures. At great depths there are pockets of molten rock called magma. The magma consists essentially of silicates—combinations of silicon, oxygen and other of the more common elements. Magma also includes smaller amounts of the less common elements. The magma rises upwards, under pressure, into cooler layers. There it solidifies slowly to form igneous rocks, which are one of the three main types of rock (see page 284). In some places, magma reaches the surface and may spill out over the land through fissures (long cracks). It may also emerge, during eruptions, through the vents of volcanoes as blazing-hot lava or as fine volcanic dust and ash. Magma which hardens on the surface also forms igneous rocks.

Magma beneath the surface usually cools slowly, while magma on the surface cools rapidly. Underground, as the magma cools, crystals develop around minute cores, a process that continues until all the magma solidifies. But some minerals cool at higher temperatures than others. In granitic magma, the common silicate minerals, like quartz and feldspar, are the first to crystallize.

INTERNAL STRUCTURE OF THE EARTH

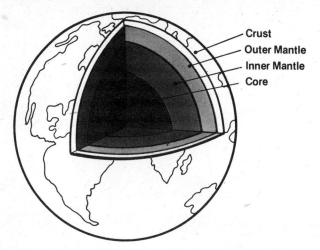

Crust
Outer Mantle
Inner Mantle
Core

MAIN ELEMENTS IN THE EARTH'S CRUST

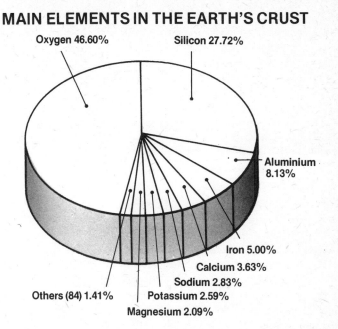

Oxygen 46.60%
Silicon 27.72%
Aluminium 8.13%
Iron 5.00%
Calcium 3.63%
Sodium 2.83%
Potassium 2.59%
Magnesium 2.09%
Others (84) 1.41%

Beneath the Earth's crust is the mantle, divided into two main parts, and the core (above). The crust is composed of 92 elements. The 8 major elements make up 98.59% of the total (right).

In the remaining magma, the gas and water content increases and the magma becomes thinner. This thin magma often contains many of the less common elements, which crystallize into minerals at lower temperatures than the silicates. This remaining magma often flows towards the edges of the original magma chamber, where it forms rocks called pegmatites during the final stage of crystallization. These rocks often contain large crystals of less common minerals, which form during the slow cooling.

Some gases and water vapour may remain enclosed in the rocks, where they may form cavities. Later, the gases and water escape and the empty cavities may be filled with crystals of other minerals, precipitated from hot or cold solutions. Hot liquids often cool in joints and fissures in rocks. There, they crystallize to form hydrothermal veins, which range in thickness between a few centimetres and several metres. Because these liquids are often rich in less common elements, such veins often contain valuable mineral ores.

Minerals formed by weathering
On the Earth's surface, rocks and minerals are affected by the weather and erosion. As a result, they are slowly but constantly changed. Rocks are disintegrated by changes of temperature during the course of the day and the seasons.

Chemical changes are caused chiefly by oxygen and carbon dioxide in the air, and by water. Solid rocks and their mineral components are gradually decomposed by chemical action. They disintegrate into smaller grains, are dissolved and are altered. For example, one kind of feldspar, called orthoclase, disintegrates and changes into kaolin, the clay used to make porcelain, and soluble potassium salts which are important for the growth of plants. Bauxite, the ore from which we get the metal aluminium, is another rock formed by chemical action in warm, wet climates (see page 291). Limestone is dissolved by water containing carbon dioxide, because the presence of the dissolved gas turns the water into a weak acid. However, in limestone caves, some of the dissolved calcite may be precipitated from dripping water to form stalactites and stalagmites. Other minerals, such as halite, anhydrite and gypsum, are precipitated from enclosed seas when the water is evaporated.

Rocks of biological origin
Living organisms also form new rocks. For example, tiny marine animals, called coral polyps, build hard cups of limestone around themselves. Many coral polyps live in colonies and, over millions of years, they build up layer upon layer of these cups to form thick beds of rock. Other limestones consist largely of the shells and skeletons of sea organisms. Phosphorite is another rock formed from the remains of dead organisms.

The fossil fuels, coal, peat, oil and natural gas, are all of biological origin and so, sometimes, are sulphur, saltpetre and pyrite. Materials composed of once-living matter are not minerals, although coal and oil are often called 'mineral resources'. However, by definition, minerals are inorganic (lifeless) substances.

Rocks

There are three main kinds of rocks. Igneous rocks are formed from magma which has cooled and crystallized. If this has occurred beneath the surface, the rocks are called intrusive igneous rocks. Those that form on the surface are extrusive igneous rocks. Granite and basalt are two of the commonest igneous rocks. Chemically, igneous rocks are classified according to the proportion of silica that they contain. Acid igneous rocks are rich in silica, while basic igneous rocks contain much less silica.

Sedimentary rocks include rocks formed from worn fragments of the land which have piled up, usually in water. Some sedimentary rocks are made from once-living matter and others result from chemical processes.

The third group of rocks, metamorphic rocks, is formed by metamorphism and include gneisses, schists, marble and slate.

Granite

Granite comprises much of the crustal rocks in the continents. This hard, resistant rock is used to make paving stones and also for building. Granites are usually medium-grained because the magma cooled slowly underground and crystals (grains) had time to grow. Chemically, granite is an acid igneous rock. Its main constituents are orthoclase feldspar, quartz and often some mica and amphibole. These are essential, or rock-forming minerals. Accessory, or non-essential minerals occur occasionally in small quantities. In granite, they include apatite, beryl, rutile and topaz. The colour of granite varies: White, grey, pink and red are most common, but it may also be greenish, yellowish or bluish. Granite is easy to quarry because it has rectangular jointing in three directions.

Granite pegmatites are related, coarse-grained rocks. These acid igneous rocks contain other accessory minerals such as corundum, tourmaline, fluorite and spodumene, left over after the main granite mass has solidified.

Basalt

Basalt is an extrusive, basic igneous rock. It is fine-grained to compact, having cooled quickly on or near the surface. Some parts of the world are covered by vast, thick sheets of basalt and some fills fissures in the crust.

Basalt is corrosion-proof and has great resistance to heat. Blocks of

Snowflake

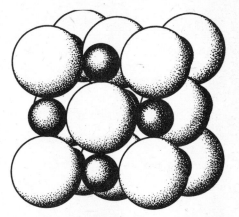

Rock salt

Two illustrations showing the internal structure of crystals

basalt are often made by melting the rock in furnaces and pouring the melted rock into moulds. As it cools, it hardens into a finely crystalline substance called fused or cast basalt. This has the same and sometimes even better properties than the original rock. It is used to make tunnel vaulting or troughs and pipes that resist abrasion and the effects of various acids.

In nature, basalt may shrink on cooling and split into long hexagonal or pentagonal columns in parallel or fanshaped groups. One famous example of this is the Devil's Tower in Wyoming.

The main minerals in basalt are plagioclase feldspar, nepheline, augite, amphiboles and olivine. Accessory minerals include apatite, biotite, ilmenite and magnetite. Basalts may contain oval cavities, which may later be filled with such minerals as calcite or zeolites.

Cross-section of granite   **Granite**

Conglomerates

Conglomerates are coarse-grained sedimentary rocks formed from rounded pebbles of quartz, quartzite, lydite and other minerals and rocks, cemented together by other minerals, usually clay minerals, limonite, calcite and haematite. The strongest is quartz cement. The colour varies, depending on the main constituents and the binding agent.

Other sedimentary rocks, including rock salt, phosphorite, coal, limestone, travertine and sandstone, are described later.

Gneiss

Gneiss is a metamorphic rock, formed when high pressures and temperatures deep within the Earth alter sedimentary or igneous rocks. Gneiss is common and is used as a building material and as gravel. In mineral and chemical composition, gneiss corresponds to granite. Its main constituents are quartz, orthoclase feldspar, biotite, muscovite, pyroxene and amphibole. Accessory minerals include apatite, garnet, rutile and zircon. This medium to fine-grained rock has a parallel banded structure. The colour may be grey, greyish-white, yellowish, brown, reddish-brown or greyish-black.

Rock-forming Minerals

Mineralogists have identified nearly 3000 minerals. They are composed of the 92 elements which occur naturally in the crust. A few minerals are native elements, but most of them are compounds of two or more elements. However, of the known minerals, only aobut 40 to 50 are widespread. The best-known are quartz, feldspars, micas, pyroxenes, amphiboles and olivines. These minerals are the major constituents of many rocks and are therefore called rock-forming minerals. All these minerals are silicates. Quartz is a simple compound of silicon and oxygen, but the others are much more complex, as can be seen from their chemical formulae. The only major rock-forming minerals which are not silicates are the carbonates, especially calcite, which makes up most of the widespread rock limestone.

Quartz

Quartz is one of the commonest and most widely distributed minerals on Earth. It occurs in igneous, sedimentary and metamorphic rocks. Some rocks, such as *quartzite* and some *sandstones,* are made up mostly of quartz. Large, almost perfectly developed crystals of quartz are often found in cavities and in granite pegmatites. They are usually six-sided, elongated and terminated by six faces. Quartz also occurs as tiny, glass-like grains, and it may occur in veins with various metals. Quartz is relatively hard and very resistant.

Besides being used in jewellery and ornaments, quartz has many other uses. It is used to grind and polish other materials. Quartz sand is the basic raw material used to make glass for windowpanes. And, at high temperatures, quartz is fused into a clear glass that stands up well to sudden changes in temperature, and also to the effects of chemicals. For such reasons, it is used in mercury-vapour lamps (sun lamps) and optical instruments.

Quartz is important in modern telecommunications. If a small

Basalt

Conglomerate

Gneiss

plate is cut from a pure quartz crystal in a certain direction, and pressure is applied to the surface of the plate, an electrical voltage develops across the surface. This is called the piezoelectric effect, and it is used in radar equipment, radio transmitters, radio receivers and in some modern timepieces. However, most quartz crystals used in such instruments are made synthetically.

Quartz (silicon dioxide), SiO_2; Crystal system: trigonal; Hardness: 7; Specific gravity: 2.65; Colour and transparency: colourless or white, but it also occurs in a wide range of colours, transparent to opaque; Streak: white; Lustre: vitreous (like broken glass); Cleavage: none, but quartz has a conchoidal (shell-like) fracture.

Quartz

Orthoclase

The commonest minerals next to quartz are a group of related minerals, called *feldspars*. They occur in many igneous and metamorphic rocks. Chemically, they are potassium, sodium or calcium aluminosilicates. Orthoclase is a potassium feldspar. It occurs chiefly in acid igneous rocks, such as granite.

Orthoclase is decomposed fairly readily by chemical weathering. Under suitable conditions, the weathered remains of feldspars form kaolin deposits.

Feldspars are an important raw material in the glass and ceramic industries. They are used to make glazes and enamels. The greatest quantities are mined in the United States, Canada, the USSR, Sweden,

Norway and France.

Orthoclase (potassium aluminosilicate), $KAISi_3O_8$; Crystal system: monoclinic; Hardness: 6; Specific gravity; 2.5-2.6; Colour and transparency: white, grey, brownish-yellow; flesh pink, reddish, translucent to opaque; Streak: white; Lustre: vitreous; Cleavage: two perfect cleavages.

Biotite

Biotite is one of a group of common silicate minerals called micas. Micas may be found as scattered flakes in rocks, in blocks, or as crystals. Blocks of mica have excellent cleavage, because they readily separate into paper-thin, flexible and elastic sheets. Large biotite crystals are found in Canada, the USA, the USSR (the Urals), Sweden, Greenland and Italy (on Mt Vesuvius).

Biotite, a dark mica, may be black, brownish-black or dark green. It is a major constituent of some granites, granite pegmatites and basalts. It is also common in some metamorphic rocks. An important light-coloured mica, muscovite, is described in the chapter on 'Heat-Resistant Minerals'.

Biotite (complex aluminosilicate of magnesium, potassium and iron), $K(Mg,Fe)_3AlSi_3O_{10}(OH)_2$; Crystal system: monoclinic; Hardness: 2.5-3; Specific gravity: 2.8-3.2; Colour and transparency: black, dark brown, dark green, transparent to translucent; Streak: white; Lustre: vitreous or pearly; Cleavage: perfect.

Orthoclase

Augite

Augite is the commonest mineral of the rock-forming pyroxene group. Augite crystals are often embedded in basic igneous rocks such as basalt and tuff (a rock composed of volcanic ash). In Italy they occur in the material ejected by the volcanoes Etna, Stromboli and Vesuvius.

Augite may sometimes be confused with amphiboles. But, while augite has good cleavage in two directions which are almost at right angles to each other, the angle between the two cleavages in amphiboles is about 120°. Chemically, augite is extremely complex. In addition to the elements shown in the formula, it may also contain some manganese and titanium.

Augite (complex silicate of calcium, magnesium, iron, sodium and

Biotite

*aluminium),   (Ca,Na)(Mg,Fe,Al)
(SiAl)₂O₆; Crystal system: monoclinic; Hardness: 5-6; Specific gravity: 3.2.-3.6; Colour and transparency: dark green to black, opaque;
Streak: greyish-green; Lustre: vitreous; Cleavage: distinct.*

Augite

Amphiboles

Amphiboles are a group of silicates
that are closely related to pyroxenes. One variety is nephrite,
a form of jade. However, much
more common are basalt amphibole (hornblende) and common
amphibole, which is similar but
contains less iron. Amphiboles are
common in dark basaltic-type
rocks. A metamorphic rock, am-
phibolite, consists essentially of
common amphibole. Amphibole
crystals are found in tuff, volcanic
dust and weathered basalt. They
may resemble those of augite but
they differ in their cleavage. Amphiboles probably crystallize from
magma at lower temperatures than
pyroxenes.

*Amphibole (basalt amphibole, or
hornblende, and common amphibole – highly complex aluminosilicate of calcium, magnesium,
potassium, iron and sodium, with
fluorine),   (Ca,Na,K)₂₋₃(Mg,Fe,Al)₅
(Si,Al)₂Si₆O₂₂(OH,F)₂;   Crystal
system: monoclinic; Hardness:
5-6; Specific gravity: 2.9-3.6; Colour and transparency: black,
brown, brownish-green to green,*
*translucent to opaque; Streak:
greyish-brown,   greyish-green;
Lustre: vitreous; Cleavage: good.*

Olivine

Olivines form another group of
common rock-forming minerals.
The name derives from its olive-
green colour. Transparent varieties,
called *peridot* and *chrysolite*, are
used in jewellery.

Olivine is found in various basic
igneous rocks, but well-developed
crystals are fairly rare. Occasionally, olivine is found in meteorites.
*Olivine (magnesium iron silicate),
(Mg,Fe)₂SiO₄; Crystal system: orthorhombic;   Hardness: 6.5-7;
Specific gravity: 3.3-3.7; Colour
and transparency: shades of green,
transparent to translucent; Streak:
white or grey; Lustre: vitreous;
Cleavage: imperfect.*

Amphibole

Olivine

Metallic Minerals

Metals are the backbone of manufacturing industries. Many minerals contain metals, but relatively few are economic ores — that is mineral ores with a high enough metal content to make their extraction worthwhile. Below are descriptions of some metallic minerals.

Haematite

and in veins. Major producers include Sweden, the USSR, the USA, Canada, Brazil and India.

This mineral sometimes forms superb black crystals with a metallic gloss. Some deposits are spherical or reniform (kidney-shaped) but haematite often forms earthy or compact masses composed of tiny spheres. This mineral's name comes from the Greek word *haima,* meaning blood, because of the colour of its streak.

Haematite (ferric oxide), Fe_2O_3; Crystal system: trigonal; Hardness: 5.5-6.5; Specific gravity: 5.2; Colour and transparency: reddish-

Iron

The Iron Age, which began about 3,300 years ago, marks the start of modern civilization. Without this tough metal modern farming would not exist, no industrial goods would be manufactured, and no modern transport would be possible.

Although iron is today commonplace, there were times when it was rare and precious. For example, the coffin of the Egyptian pharaoh Tutankhamen contained a dagger made of iron. This was an object of great value in those days.

Haematite

Haematite (or hematite) is the most important and commonest iron ore. By weight, it contains up to 70 per cent iron. It is widely distributed in igneous and metamorphic rocks

brown to grey and black, opaque:
Streak: cherry-red; Lustre: metallic
to dull; Cleavage: none.

Magnetite

Magnetite is an iron ore containing
up to 72 per cent iron. It occurs in
various rocks but only in negligible
amounts. In some places, however,
there are huge magnetite deposits
in igneous rocks, with ore bodies
up to 80 metres (264 feet) thick.
Magnetite also collects in coastal
sands.

The mineral forms black crystals,
but it is generally granular or massive. Its highly magnetic property,
which causes it to attract iron and
affect the needle of a compass, is
an aid to identification. You can
pick out grains of magnetite from
sands with a magnet.

The richest deposits are in north-
ern Sweden, around Kiruna and
Gällivare beyond the Arctic Circle.
The USSR and the USA also have
large deposits.
Magnetite (iron tetraoxide), Fe_3O_4;
Crystal system: cubic; Hardness: 6;
Specific gravity: 5.2; Colour and
transparency: black, opaque; Streak:
black; Lustre: metallic to dull;
Cleavage: none.

Manganese

The black powder used in pocket
flashlight batteries is called manganese oxide. It occurs in nature
as the manganese ore pyrolusite.
Manganese is essential in the refin-
ing of iron. Steel made with manganese is resistant even to sea
water which is highly corrosive. The
chemical and glass industries also
use many manganese compounds.
Manganese is essential to life. Both
plants and animals need small
amounts for good growth.

Manganite

Manganite is one of several manganese ores. It forms columnar
crystals and fibrous masses. It is
often found with other manganese
ores in veins. When manganite decomposes it generally changes into
pyrolusite. Superb manganite crystals come from the Harz Mountains
and Thuringia (East Germany). It

Magnetite also occurs in Cornwall, in Britain,
in Michigan, in the USA, and in
Nova Scotia, Canada.
Manganite (hydrous manganese
oxide), MnO(OH); Crystal system:
monoclinic; Hardness: 4; Specific
gravity: 4.3; Colour and transparency: grey to black, opaque;
Streak: dark brown; Lustre: metallic; Cleavage: perfect.

Nickel

Nickel, like chrome, is used to protect car headlights, the handle-bars
of bicycles and many other articles
against corrosion. It is a component
of corrosion-resistant steels and
nickel alloys are also used in aircraft, rockets and atomic resistors,
as well as in coins.

Because of its pale, copper red
colour, people once believed that
the mineral niccolite (or nickeline)
was an ore of copper. Later, however, it was found to contain a new
metal, nickel. Niccolite occurs in
ore veins as a compact, massive
mineral. Crystals are rare. In damp

Niccolite

conditions the mineral becomes
coated with a pale green crust.
Niccolite (nickel arsenide), NiAs;
Crystal system: hexagonal; Hardness: 5-5.5; Specific gravity: 7.8;
Colour and transparency: pale copper-red, opaque; Streak: brownish-
black; Lustre: metallic; Cleavage:
none.

Cassiterite

Cassiterite (tin dioxide), SnO$_2$; Crystal system: tetragonal; Hardness: 6-7; Specific gravity: 6.8-7.1; Colour and transparency: brown to

Twinned-crystals

black, nearly transparent to opaque; Streak: white; Lustre: adamantine (brilliant) to submetallic; Cleavage: none, brittle.

Tin

Tin first became important following the invention of bronze (an alloy of copper and tin) about 5,000 years ago. The invention of bronze, which is harder than copper, a metal already in use at that time, marked the start of the Bronze Age, a major advance in the development of civilization. Pewter is another alloy using mainly steel. Beautiful cups and candlesticks were made with it up until the 19th century. Tin is still used today to plate other metals.

The chief tin ore is *cassiterite*. Cassiterite crystals are often twinned (in contact with each other, or intergrown) in an elbow-shape and called 'visor tin'. Generally, however, the mineral occurs in granular or massive forms, or as pebbles and sand in alluvial deposits. It is the richest tin ore, containing 78 per cent of the metal. It is quite common in small quantities, but only a few deposits are suitable for mining. Tin mining was an ancient industry in Cornwall, Britain, and in the Ore Mountains in Czechoslovakia. Today, most cassiterite is obtained from alluvial deposits in Malaysia, Burma, Thailand, Indonesia and China.

Lead

In the hanging gardens of Babylon, one of the seven wonders of the ancient world, the urns that graced the terraces were made of lead, another metal used since early times. Today, world consumption of lead is steadily increasing. It is used especially in pipes, cable sheaths, batteries and shields against radio-activity, and its alloys are made into type used by printers and solders. Lead compounds are also used in making dyes.

The richest source of lead is the mineral galena. Lead forms nearly 87 per cent of the weight of the mineral. There may also be up to 0.5 per cent silver in the ore and it is often extracted as a by-product. Austria, Britain, Czechoslovakia, Germany, Poland and Rumania are among European countries with galena deposits. The largest de-

Galena

posits are in the USA, Australia and Mexico.

In the USA, the chief district is in Missouri, Kansas and Oklahoma, centring on Joplin, Missouri. The Leadville district in Illinois, Iowa, Idaho and Colorado is also important.

Galena crystals are generally cubic and they have perfect cleavage in directions parallel to the planes of the cube, so that they cleave into smaller cubes. The galena that is extracted from mines, however, is either granular, compact or massive. Large, beautiful crystals are often found in *druses* (cavities in which crystals extend outwards from the sides in roughly parallel bands).

Galena (lead sulphide), PbS; Crystal system: cubic; Hardness: 2.5; Specific gravity: 7.5; Colour and transparency: lead-grey, opaque; Streak: grey; Lustre: metallic; Cleavage: perfect.

Zinc

The quality of a picture on television screens is determined by the fact that the layer of the screen on which the picture appears is covered by a fine coat of zinc sulphide. That is also the composition of the chief ore of zinc, which is called *sphalerite, zinc blende* or, simply, *blende*. Zinc has many other uses. It is applied as a protective coating on iron and steel and is used in several alloys, such as brass. It is used in the chemical, electrical engineering and pharmaceutical industries. It is also a pigment in paints.

Sphalerite contains up to 67 per cent zinc. Iron is often present and there may also be some cadmium, indium and germanium. These elements are extracted as by-products and used to manufacture transistors. Sphalerite is a common ore, usually found in hydrothermal veins with galena, and also in limestones with pyrite and magnetite. Although crystals are often found, some deposits occur only in granular or massive forms. The mineral shows a wide range of colours, while sphalerite containing a fair amount of iron is black. Be-

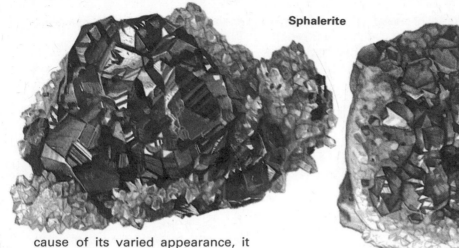

Sphalerite

cause of its varied appearance, it may be hard to identify. Its name comes from a Greek word, *sphaleros,* which means deceitful.

Deposits of sphalerite are widely distributed. One of the purest varieties is mined in Spain. Sphalerite crystals from Spain have a high gloss and are used in jewellery. Other deposits are found in Czechoslovakia, Germany, Poland, Rumania and the USSR. Major ore bodies occur in Australia, Canada, Japan, Mexico and Peru. However, the largest deposits are in the USA. *Sphalerite (zinc sulphide), ZnS; Crystal system: cubic; Hardness: 3.5-4; Specific gravity: 4; Colour and transparency: yellow, green, red, brown to black, transparent to opaque; Streak: brown, yellow, white or grey; Lustre: greasy, adamantine to submetallic; Cleavage: perfect, brittle. Other features: sphalerite luminesces when rubbed in the dark.*

Aluminium

Aluminium utensils are used in almost every home. Aluminium is also a vital metal in transport. Ships, aircraft and railway wagons are made from light aluminium alloys. It is also important in building construction and is a substitute for copper in conductors. Its oxides are excellent abrasives, used for grinding and polishing. Because of its many uses, world consumption of aluminium is rapidly increasing.

Aluminium is the third commonest element in the Earth's crust, after oxygen and silicon, making up 8.13 per cent of its total mass. It occurs in many minerals, the aluminosilicates. The pure metal

was first obtained in 1825, but large-scale production did not begin until the early 1900s.

The only substantial ore of aluminium is *bauxite.* Bauxite is not a mineral, but a sedimentary rock. It consists chiefly of three minerals, boehmite, diaspore and gibbsite. Each mineral has its own properties. Bauxite was formed by the weathering of rocks containing aluminosilicates in tropical and subtropical regions. Heavy rains leach (wash) out the silicates from surface layers, but the aluminium remains as hydrous aluminium oxides. The term 'hydrous' indicates that water is combined chemically in the minerals. Bauxite may also contain oxides of iron and manganese, opal and other minerals. It was named after the town of Les Baux, near Arles in southern France, where there are rich deposits. A similar rock with a greater iron content is found in tropical regions.

Bauxite

It is called laterite. Bauxite may be red, yellow, brown or grey. It is usually compact or earthy.

In Europe, important bauxite deposits are found in France, Hungary, Rumania, the USSR and Yugoslavia. There are also deposits in Africa, notably Guinea, Australia, Guyana, India, Indonesia, Jamaica, Oceania, Surinam and the USA.

Titanium

It was not until 1795 that the metal titanium was discovered in the mineral rutile. Today this strong, silver-grey metal has many uses, particularly in making a steel alloy which is used in the aircraft industry and in the construction of rockets and space ships. It is important in many other fields as well. Ask a painter what gives the clouds in his pictures their unusual, transparent-like whiteness. He will show you a tube

Rutile

Native Metals

of paint called titanium white.

Rutile is the commonest mineral of titanium and the metal makes up about 60 per cent of the ore. Rutile forms columnar crystals, which are often twinned in a knee- or elbow-shape. It also occurs as slender needles embedded in quartz, and this variety is called 'arrows of love' or the 'hair of Venus'. Rutile is found in small quantities in numerous rocks. Commercially, it is generally extracted from sands and alluvial deposits. It is mined in Australia, Brazil, Norway and the USSR. In the USA, it is found in Arkansas, Georgia and Virginia, and it is also obtained from sands in northeastern Florida. Rutile is now produced artificially by the Verneuil process, which is discussed in the chapter on 'Gems and Precious Stones'.

Rutile (titanium oxide), TiO_2; Crystal system: tetragonal; Hardness: 6-6.5; Specific gravity: 4.2-4.4; Colour and transparency: red, reddish-brown to black, transparent (when thin) to nearly opaque; Streak: light brown; Lustre: adamantine to submetallic; Cleavage: poor.

NATIVE METALS

Only 22 elements are found in a pure state in the Earth's crust, although some occur in more than one form. For example, pure carbon takes the form of graphite and diamond. These substances are called native or free elements and they are mostly rare. Some metals, including gold, silver and copper, occur as native elements.

Panning for gold in a medieval mine

Gold

Gold

Gold is the most prized of all metals. It is easy to work and shape. It does not tarnish in air and does not dissolve in acids. These are some reasons why gold has been used to make ornaments and jewellery since the early Stone Age. It has been found in the tombs of ancient Egyptian pharaohs and in the treasure of Mycenae, which dates back to the second millennium B.C., now in the National Archaeological Museum in Athens. Gold is a symbol of power and wealth. It was once used for coins, but most of it now serves as an international monetary metal. It is also used in dentistry and electrical engineering. It is even used in the glass industry to colour glass red.

In nature, gold occurs in veins together with quartz, and worn fragments accumulate in stream beds. Crystals of gold are very rare and the metal usually takes the form of flakes, grains, pebbles or nuggets. The leading gold producer is South Africa, where it is mined in Transvaal and Orange Free State.

Gold (Au); Crystal system: cubic; Hardness: 2.5-3; Specific gravity: 19.3, it is less if it contains traces of other metals; Colour and transparency: yellow, opaque; Streak: gold-yellow; Lustre: metallic; Cleavage: none, hackly (rough) fracture; Other features: ductile and malleable.

Copper

Silver

Silver, a beautiful, rare white metal, has always been rated second to gold. However, an object made of silver often outshines one of gold. Ancient Egyptian silver beads date back 6,000 years and the ancient Greeks made superb silver filigree (ornamental metallic lacework). Even today, silver has lost none of its popularity. In jeweller's shops you will see that gold goes well with some precious stones. But others look better when offset by the white gleam of silver. Silver is used in dentistry and electrical engineering. It is also used for plating other metals and for film emulsions in the photographic industry.

In nature, native silver occurs in ore veins in the form of wiry, scaly, branching or irregular masses. Crystals are rare. Because native silver is seldom found, most silver comes from mineral ores. Many silver deposits that were famous in the past have been exhausted. The major producers today are the USSR, Mexico, Canada, the USA, Peru and Poland.

Silver (Ag); Crystal system: cubic; Hardness: 2.5-3; Specific gravity: 10-12; Colour and transparency: silver-white, often tarnishes black in the air, opaque; Streak: silver-white; Lustre: metallic; Cleavage: none, hackly fracture; Other features: ductile, malleable.

Copper

Copper was the first metal used by prehistoric Man for other than ornamental purposes. It came into use in the Stone Age, about 10,000 years ago. Though it was also made into jewellery, copper tools and weapons were much more important. At first, pure copper was used, but copper is fairly soft. About 5,000 years ago, however, a hard alloy of copper and tin, bronze, was invented. This invention marked the end of the Stone Age and the start of the Bronze Age. Today, copper is the most widely used metal next to iron. It is used chiefly in the electrical engineering industry to make conductors and wires, and also to make bronze and brass.

In nature, native copper occurs as flakes or in branching, wiry or irregular masses. But native copper is rare and most copper comes from mineral ores. The leading producers are the USA, the USSR, Chile, Canada, Zambia, Zaire, Peru, Poland, the Philippines and Australia.

Copper (Cu); Crystal system: cubic; Hardness: 2.5-3; Specific gravity: 8.9; Colour and transparency: copper-red, opaque; Streak: copper-red; Lustre: metallic; Cleavage: none, hackly fracture; Other features: malleable, ductile.

Minerals in the Chemical Industry

The Earth's crust is a gigantic natural chemical factory and a storehouse of treasures. Today, some minerals are indispensable and, in recent years, there is hardly one that has not been used in some way in the chemical industry.

Sulphur

Making a fire by rubbing two sticks together is not easy. A box of

matches is far more convenient. Besides other things, matches contain sulphur, an element which occurs in mineral ores and, occasionally, in a pure state. It has a bright yellow colour. It is used in treating rubber and in making sulphuric acid, a major industrial chemical, numerous dyes, medicaments, wood pulp and cellulose, artificial silk, plastics and fertilizers. It is an ingredient in explosives, as well as in matches and insecticides. Sulphur is also used to make countless other products in daily use.

Sulphur is often of volcanic origin, but vast deposits were also formed by the action of bacteria on sulphates in water. In some places, such as the Yellowstone National Park, Wyoming, in the USA, it is deposited by hot springs. It may also be produced when coal heaps are burned. In nature, sulphur often forms lovely, pyramidal crystals. Generally, however, it is found in granular, massive or compact deposits.

Important sulphur deposits are

situated in Sicily. It is also mined from recently discovered deposits in Japan, Mexico and Poland. The USA is the world's main producer. *Sulphur (S); Crystal system: orthorhombic; Hardness: 1.5-2.5; Specific gravity: 2-2.1; Colour and transparency: yellow to brown, transparent to translucent; Streak: white to yellow; Lustre: greasy to adamantine; Cleavage: none, con-*

Sulphur

choidal fracture—sulphur is brittle and often cracks when held in the hand (at body heat); Other features: sulphur is a poor conductor of heat and electricity; it melts at 119°C (246°F).

Pyrite

A lump of coal may sometimes contain mineral grains which have a golden glint. However, these grains are not gold but pyrite, or 'fool's gold'. Unlike gold, pyrite is

Pyrite

brittle and has a blackish streak. The name pyrite comes from a Greek word *pyrites* meaning 'striking fire', because it throws off sparks when struck with steel.

In ancient Greece, people considered that pyrite had healing powers and they often wore pyrite amulets. The mineral is now used in making sulphuric acid, sulphates, pigments and polishes. Pyrite is sometimes used as a low-grade iron ore, and copper and gold may be obtained as by-products from some pyrite. Pyrite is also used as an ornamental stone.

Italy, Japan, Portugal, Scandinavia, Spain and the USA all have large pyrite deposits. In the USA, important deposits are found in Massachusetts, New York and Virginia.

Pyrite (iron disulphide), FeS_2; Crystal system: cubic; Hardness: 6-6.5; Specific gravity: 4.9-5.2; Colour and transparency: brass-yellow, opaque; Streak: brownish-black, greenish-black; Lustre: metallic; Cleavage: none, conchoidal to uneven fracture.

Halite (Rock salt)

Salt is essential for health and it is the only edible mineral we use daily. Each of us consumes 6 to 7 kilograms (13-15 lb) a year.

Even though rock salt, which mineralogists call *halite,* lacks the valued properties of gold or precious stones, it was an article of international trade in ancient times.

Rock salt is also vital as a food preservative and it is used in many industries. For example, the chemical industry uses more than one-third of the amount mined.

Chemically, rock salt is sodium chloride. It contains about 40 per cent sodium and 60 per cent chlorine. It crystallizes in the form of cubes, but generally forms granular or fibrous, variously coloured masses. Salt is soluble in water. Sea water contains an average of 3.5 grams of salt per litre. The oceans cover more than 70 per cent of the Earth's surface and so sea water represents a vast 'reservoir' of salt.

Huge deposits of rock salt were

Halite

formed when sea water evaporated in hot or dry climates. As the water evaporated, layer upon layer of rock salt accumulated on the shores and beds of closed sea basins. Many of these deposits were later covered by layers of sands and clays. The pressure of the overlying strata has squeezed some deeply buried salt layers, causing the salt to flow upwards through the overlying rocks to form large salt plugs, or salt domes. Salt domes are sometimes traps for oil which has seeped through the surrounding rocks and accumulated around the top of the dome.

Salt is readily obtained by evaporating sea water, or brine. It is also extracted by mining. The largest rock salt deposits are situated in the USA, especially in Michigan, Ohio and Texas, the USSR and China. In Europe, there are notable deposits in Austria, Britain, France, Germany and Poland. *Halite, or rock salt (sodium chloride), NaCl; Crystal system: cubic; Hardness: 2; Specific gravity: 2.1; Colour and transparency: colourless or white, also yellow, red, blue, violet and brown, transparent to translucent; Streak: white; Lustre: vitreous; Cleavage: perfect − the crystals split into smaller cubes; Other features: dissolves in water, distinctive taste.*

Fluorite

Today, most homes contain a re-frigerator. However, few people know that the major component of the refrigerant *Freon* is fluorine, an element obtained from the mineral *fluorite.* Fluorite is used in metallurgy to aid in the smelting of ores, because it becomes liquid much more readily than other metallic minerals. Its name comes from the Latin word *fluere,* meaning 'to flow'. Fluorite is the basic raw material for the production of fluorine compounds. One compound, hydrofluoric acid, is used in etching glass. Fluorine is also present in non-stick coatings used in cooking utensils.

In nature, fluorite generally crystallizes from hydrothermal solutions in ore veins. It forms cubic crystals, but is also found in granular or massive deposits. When heated, fluorite becomes strongly fluorescent in darkness. Crystals also fluoresce in daylight, changing colour as they are rotated.

Beautiful banded varieties of fluorite, called Blue John, are found in Derbyshire, England. There are important fluorite deposits in East Germany, France, Italy, the USSR, and the USA.

Fluorite (calcium fluoride), CaF₂; Crystal system: cubic; Hardness: 4; Specific gravity: 3.1; Colour and transparency: colourless, yellow, green, pink, red, violet and brown, transparent to translucent; Streak: white; Lustre: vitreous; Cleavage: perfect, brittle.

Zeolites

Zeolites are a group of minerals that are valuable aids in the chemical industry. Chemically, they are hydrous aluminosilicates. They contain water, which is loosely bonded in the pores of their internal framework. When heated, zeolites lose water readily and boil. Their name comes from two Greek words, *zeo,* meaning 'to boil' and *lithos,* 'stone'. When zeolites cool, they regain the lost water or they may bind and release other substances, such as ammonium and various hydrocarbons. These processes are called ion exchanges.

Ions in the zeolite structure may be exchanged for other ions dissolved in water. For example, hard water is softened by passing it through a suitable zeolite filter. What happens chemically is that the calcium ions in the hard water are replaced by sodium ions from the zeolite.

Zeolites occur in cavities and fissures in basic igneous rocks. They crystallize from hot solutions and form light-coloured, needle-like, plate-like or tabular crystals with broad, flat faces. The picture shows Icelandic stilbite and heulandite, two zeolites which often occur together. Both minerals have similar properties.

Stilbite, heulandite (both are hydrous calcium aluminosilicates), CaAl₂Si₇O₁₈·7H₂O (stilbite), CaAl₂Si₇O₁₈·6H₂O (heulandite); Crystal system: monoclinic; Hardness: 3.5-4; Specific gravity: 2.2; Colour and transparency: white, yellow, reddish, transparent to translucent; Streak: white Lustre: vitreous, pearly; Cleavage: perfect.

Fluorite

Mineral Fertilizers and Plant Foods

Some minerals play an important part in what is known as the nutrient cycle. They are a major component of the diet of plants and animals and are stored in their bodies. After the living organisms have died and their bodies have decomposed, the elements are returned to the soil.

Apatite

Apatite is a fairly common and widely distributed mineral. It occurs in small quantities in many kinds of rocks. It is also a vital component in the diet of plants and animals. Apatite makes up about 60 per cent of the bones and about 90 per cent of the tooth enamel of vertebrates (animals with backbones). Hence, it is a truly vital substance.

In nature, apatite often forms crystals which have the symmetry of the hexagonal system. However, the shapes of the crystals are very varied. They may be column-like, needle-like, elongated or thickly tabular in shape. Small crystals occur as an accessory (non-essential) mineral in a wide range of igneous rocks. Large crystals are a feature of granite pegmatites and some hydrothermal veins. Some crystals found in Canada measured more than three metres (ten feet) in length. Although crystals are common, apatite is also found in the form of finely granular, massive or compact deposits. Apatite is found in some metamorphic and sedimentary rocks. In sedimentary

rocks, it is the chief component of fossil bones and other forms of once-living matter. Apatite is the most important source of phosphorus for the manufactured fertilizer known as superphosphate.

Transparent and attractively coloured crystals, such as pink and yellow varieties from Switzerland, and from Durango in Mexico, are sometimes ground and polished to produce ornamental stones. Another popular semiprecious variety has a yellowish-green colour and is aptly called 'asparagus stone'. But apatite is too soft for jewellery.

The name apatite comes from the Greek word *apate*, which means 'deceit'. This is because the shapes of apatite crystals vary greatly and also because it occurs in a wide range of colours, although it is most commonly some shade of green. Hence, it has often been mis-

Apatite crystal form

Zeolite taken for other minerals, such as amethyst, aquamarine (a bluish-green form of beryl) and tourmaline. Such mistakes were made especially in the days when people thought that each mineral had one characteristic colour. However, apatite is easily distinguished from these other minerals because it is much softer than any of them. Apatite was finally established as a mineral in its own right when its chemical composition was determined in the late 18th century by the great German mineralogist Abraham Gottlob Werner (1750-1817).

Chemically, apatite is a fairly complex mineral. It is a phosphate of calcium, but it also contains some fluorine (F), chlorine (Cl) and hydroxyl (OH). These three substances can substitute for each other and the amounts of them vary from sample to sample. It is this variation in the proportions of fluorine, chlorine and hydroxyl in apatite that accounts mainly for the differing appearance of specimens.

The largest deposits of finely granular apatite are found on the Kola peninsula in the USSR.

Apatite (phosphate of calcium, with fluorine, chlorine and hydroxyl), $Ca_5(PO_4)_3(F,Cl,OH)$; Crystal system: hexagonal; Hardness: 5; Specific gravity: 3.2; Colour and transparency: colourless (when pure), green, bluish-green, red, violet and brown, transparent to opaque; Streak: white; Lustre: vitreous to greasy; Cleavage: poor, brittle, conchoidal to uneven fracture; Other features: it dissolves in hy-

Apatite

drochloric acid and most kinds of apatite fluoresce in ultraviolet light.

Phosphorite

As we have seen, apatite is vital to life. But after marine animals die and their bodies decompose, the apatite in their bodies makes its way to the sea bottom. There it often accumulates in layers of compact or finely fibrous sedimentary rocks, which are grey or brown to black in colour. The essential components in this phosphate rock, which is called *phosphorite,* are various calcium phosphates related to apatite. The rock also contains the remains of various organisms and carbonaceous substances. Phosphorite may also occur as nodules (lumps) in rocks.

Like apatite, phosphorite is not soluble in water. It is made into a fertilizer, superphosphate, by treatment with sulphuric acid. This fertilizer dissolves more readily in the weak acids that are present in the soil. It can therefore be absorbed more easily by plants than calcium phosphate. The current consumption of phosphorite around the world is generally increasing. This is the result of the expansion of large-scale and intensive farming.

Deposits of phosphorite of commercial quantity are found in North Africa, especially Algeria, Morocco and Tunisia, Belgium, France, Spain and the USSR, notably in the

Chile saltpetre

Ukraine and Kazakhstan. The USA, the world's largest single producer, has deposits of high-quality phosphorite in Idaho, Tennessee and Wyoming, and deposits of pebble phosphates are scattered along the Atlantic coast from North Carolina to Florida. On some Pacific islands, phosphate deposits are constantly building up from the droppings of sea birds. This is called *guano* and it, too, is used as a fertilizer.

Chile saltpetre

Chile saltpetre, a mineral which is also called nitratine or soda nitre, was an essential component of the gunpowder used by musketeers in the Middle Ages. It is also valuable

Phosphorite

as a fertilizer. Its beneficial effect on plants was known to the ancient Egyptians who used it to enrich the soil in their gardens. Chile saltpetre is still mined today and it is used mostly as a nitrogen fertilizer.

Saltpetre is found in the soil, where it is produced naturally by the action of bacteria on the remains of dead plants and animals. It readily dissolves in water and deposits are precipitated from water, generally in massive forms but also as thin crusts. These deposits are found only in the driest regions of the world. The largest and richest deposits are along the coast of northern Chile. Smaller deposits occur in neighbouring Bolivia, Peru, North Africa, the USSR and the USA, in California and Nevada.

Chile saltpetre is often found together with many other salts and rock fragments, including gypsum and halite (rock salt). Impure sodium nitrate is called *calich.* After being quarried, the impurities are dissolved out and the substance is then used to make pure nitrates.

Chile saltpetre (sodium nitrate), $NaNO_3$; Crystal system: trigonal, but crystals are rare; Hardness: 1-2; Specific gravity: 2.3; Colour and transparency: colourless, white, grey, yellow and brown, transparent to translucent; Streak: white; Lustre: vitreous; Cleavage: perfect; Other features: readily soluble in water, cool taste.

Fossil Fuels

Substances found in the Earth's crust that were formed from the remains of once-living organisms and are now used as fuels are called *fossil fuels*. They include peat, various types of coal, along with crude oil and natural gas. Coal was formed from the remains of huge forests which grew millions of years ago. There are four stages in the formation of coal. Each stage is represented by a distinctive substance: peat, brown coal, black coal and anthracite.

Peat

The first stage in coal formation, peat is formed when plants grow in fresh water or coastal swamps. The remains of dead plants accumulate in the water and are rapidly submerged as new vegetation grows above them. The rotting remains are cut off from oxygen in the air. Hence, the normal process of decay, which reduces all organic matter to water, carbon dioxide and simple, inorganic salts, is halted. The partly rotted remains therefore accumulate, layer upon layer, as they do in peat bogs today. Older fossil peat deposits were formed in much the same way, but from other kinds of plants. Peat is usually light brown in colour in the top layers but, in thick deposits, the bottom layers may be dark brown. It contains about 58 to 60 per cent carbon. It is a porous, light substance, and the plant remains can be seen clearly in any sample. When peat is extracted, up to 90 per cent of its weight may be water. The number of commercially important peat deposits around the world is large. In some places, dried peat is used directly as a fuel. Elsewhere, it is used to make briquettes and charcoal. Because it is porous, it is a useful heat-insulating material in the building industry. It is also used in the manufacture of paper and plastics. Gardeners and farmers use it as a fertilizer on soils that are poor in humus (rotted plant and animal remains), and also in making certain plant feeds. Last but not least, the healing properties of peat are used in medicine in the form of peat packs.

298

Peat

Brown coal

If layers of peat are completely submerged by an appreciable depth of water, plants cease to grow and layers of sand and clay start to accumulate above the peat. This may occur when the sea level rises or the land is depressed by Earth movements. Pressure from these overlying layers slowly squeezes the water out of the peat. Gradually, pressure combined with the higher temperatures below the surface increase the amount of carbon in the deposit.

Brown coal represents this second stage in coal formation. Most brown coal deposits are of comparatively recent origin. They date from the Tertiary period (between 2 and 65 million years ago). Brown coal still shows its vegetable origin and is generally dull, without lustre, compact and relatively soft. It contains between 70 and 73 per cent carbon, but water may still account for up to 50 per cent of its weight when it is first extracted. There are various types of brown coal, which are distinguished by the nature of the plants from which they formed and by their geological age. One of the more recently formed types is *lignite,* which splits readily and which still shows clearly the texture of the original plants. Lignite deposits are widespread and it is mined for use as a fuel. It is also an important raw material in the chemical industry.

Brown coal

Black coal

As long as geological conditions remain the same, so do the properties of brown coal deposits in the ground. If, however, the pressure and temperatures increase, brown coal changes into black coal as a result of a further concentration of carbon in the vegetable matter.

Black coal is generally banded and composed of gleaming and dull layers. It contains 82 per cent carbon and thus gives out more heat than brown coal. Black coal is also called bituminous coal, household coal, gas coal, steam coal, and so on. Most of it was formed during the Carboniferous period (between 280 and 345 million years ago). At that time, many plants flourished in the warm moist climate.

The use of black and brown coal as fuels has become less profitable in recent years because of competition from oil and natural gas. But coal is the raw material that has made possible the development of the modern chemical industry. It is used to make coke, tar, gas, synthetic petrol, medicines, dyes, mineral oils and many other substances in daily use.

Anthracite

Anthracite is the variety of coal which contains the highest percentage of fixed carbon. It represents the final stage in coal formation. All traces of its vegetable origin have been removed during its transformation and the carbon content has increased to 94 per cent. This black,

tremely high pressures and temperatures, which were probably caused by great movements in the Earth's crust. For example, enormous pressures are caused by plate movements in the crust. Plates are huge blocks of crust that move slowly around. The pressure created when two plates push against each other may buckle rocks on the sea bed upwards into mountain ranges, like the Alps. But anthracite is comparatively rare, occurring mainly in the deepest parts of coal beds. It is found in quantity in Wales, in various basins in the USSR, and in western Pennsylvania, in the USA.

This hard coal is used for smelting iron ores in blast furnaces in place of coke and as a source of

Black coal

Deposits of black coal are found throughout the world, including the icy polar regions, which clearly must have had a much warmer climate in the Carboniferous period. Large coal basins occur in Belgium, Britain, France, Germany, Poland and the USSR, the world's leading producer, and also in many places in the other continents. In the USA, the world's second largest producer, black coal is found in quantity on the western side of the Appalachians from Pennsylvania south to Alabama, as well as in Illinois, Iowa and Oklahoma. The world output is about 2,500 million tonnes per year. The USSR and the USA have about three-quarters of the world's known reserves.

shiny rock is clean to handle and has a brilliant, vitreous to metallic lustre. It is hard, compact and brittle, with a conchoidal fracture.

Anthracite was formed under ex-

energy. It is particularly suitable for domestic use because it burns at an even temperature and, of all the varieties of coal, it gives off the greatest heat.

Anthracite

Heat-resistant Minerals

Hundreds of years ago, alchemists tried to find substances that would withstand great heat. These substances are now essential for many industrial processes which are carried out in the extreme temperatures generated in blast furnaces. The few heat-resistant minerals include graphite, asbestos, talc, muscovite and magnesite.

Graphite

An ordinary pencil contains at its centre a stick of 'black lead'. This soft, heat-resistant substance is the mineral *graphite,* which is also called plumbago. The name graphite comes from the Greek word *grafein,* meaning 'to write'. Graphite was first used to make pencils in the mid-1500s. The useful properties of graphite were, however, known long before then. In prehistoric times, the sides of clay vessels were sometimes coated with graphite. The blacking stopped any leaking and the graphite surface could be decorated with patterns.

Only a small proportion of the graphite that is mined is now used in pencils. It is much more important in the manufacture of crucibles for melting various ores and metals and it also has many uses in the electrical engineering industry. For example, it is used to make electrodes for batteries, carbon 'brushes' for motors and in electroplating processes. It is also used as a solid lubricant and in heat-resistant sealing compounds. Chemically, graphite, like the entirely different looking diamond, is a form of the element carbon. In nature, graphite usually occurs in

Asbestos

Graphite

metamorphic rocks and, less frequently, in veins of igneous rocks.

Important graphite deposits in Europe are situated in Austria, Czechoslovakia, West Germany, Italy and the USSR. Commercial sources outside Europe include Canada, Korea, Mexico, Sri Lanka and Madagascar and the USA.
Graphite (carbon), C; Crystal system: hexagonal; Hardness: 1; Specific gravity: 2.2; Colour and transparency: black to steel-grey, opaque; Streak: black; Lustre: metallic to dull; Cleavage: perfect; Other features: flexible but not elastic, it has a greasy feel and it marks fingers and paper.

Asbestos

Asbestos is perhaps the best known of the heat-resistant minerals. It is also used in various special fireproof garments because it provides reliable protection.

Asbestos is the name for fibrous varieties of several minerals with varied chemical compositions. The most important is *chrysotile,* a variety of the mineral serpentine. Chrysotile now accounts for over 90 per cent of the world production of asbestos. Chrysotile got its name from two Greek words: *chrysos,* meaning 'gold', and *tilos,* 'fibre'.

Cloth for fireproof garments is woven from the longest fibres. This is called *fabric asbestos* and it is used to make safety curtains used in theatres. Shorter fibres are used

to manufacture asbestos cement roofing, various pipes and insulating material in electrical engineering. Asbestos is also used for brake linings and clutch facings in cars.

In nature, chrysotile is found in the metamorphic rock serpetinite, which consists chiefly of serpentine minerals. In the fissures, chrysotile forms veins of microscopically fine fibres arranged in parallel formations. The main source of chrysotile asbestos is Quebec, in Canada. Elsewhere, it is found in East Germany, the Ural region of the USSR, Rhodesia (Zimbabwe), Transvaal province in South Africa and the USA.
Chrysotile (hydrous magnesium silicate), $Mg_3Si_2O_5(OH)_4$; *Crystal system: monoclinic; Hardness: variable, 3-5; Specific gravity: 2.4-2.7; Colour and transparency: whitish, grey, greenish-grey, yellow-green and brown, transparent to opaque; Streak: white; Lustre: silky; Cleavage: none; Other features: it contains easily separable fibres, flexible and elastic.*

Muscovite

In Russia, around 300 years ago, large, thin sheets of muscovite were used to make windowpanes. These were exported from Moscow to western Europe as 'Muscovy glass', which explains the origin of the name muscovite. Today, muscovite is used in many branches of modern technology, because of its heat resistance and its excellent properties as an electrical insulator. Ground into a fine powder, it is also used as a glittering 'pigment' in paints.

Muscovite, like the related dark mica mineral biotite which has already been described, is a common rock-forming mineral. Glittering flakes are found in many granites. Muscovite also occurs as blocks, which cleave readily into thin, transparent sheets, and, more rarely, as crystals. Muscovite is also found in various sedimentary and metamorphic rocks.

Large crystals, measuring as much as several metres, are found in Brazil, Canada, Japan, Madagascar and the USSR. The main pro-

Mica

ducers are the USA, India and South Africa.

Muscovite (a mica mineral, silicate of aluminium, potassium and hydrogen, with fluorine), $KAl_2(AlSi_3O_{10})(OH,F)_2$; Crystal system: monoclinic; Hardness: 2-2.5; Specific gravity: 2.7-2.8; Colour and transparency: colourless, yellow, brown and greenish, thin sheets are transparent, thicker layers are translucent; Streak: white; Lustre: vitreous to pearly; Cleavage: perfect, it splits into thin, flexible and elastic sheets.

Talc

When clothes are made, the tailor may measure the cloth and mark it with what looks like a piece of chalk. This substance is really a piece of soapstone, although it is called 'tailor's chalk' or 'french chalk'. Soapstone, or *steatite*, is a variety of talc, the softest common mineral.

Ground talc is well known because it is used as a base for talcum powder, face powders and other cosmetics. Today, it has many other uses because of its special properties. It is heat-resistant, is unaffected by acids and is a good electrical insulator. It is used to make fireproof linings, bricks, burners, crayons and soap. It is also used as a solid lubricant in the textile, rubber and paper industries.

Talc occurs in metamorphic rocks. It is a mineral of secondary origin, which is often formed by the alteration of amphiboles, olivines and pyroxenes. It forms finely crystalline masses, although the tabular crystals are rare. It also occurs as coarsely layered or scaly aggregates. Steatite is the compact, massive variety of talc. Rich deposits of talc are found in Austria, southern France, West Germany and northern Italy and there are many large deposits in the USA.

Talc (hydrous magnesium silicate), $Mg_3Si_4O_{10}(OH)_2$; Crystal system: monoclinic; Hardness: 1; Specific gravity: 2.6-2.8; Colour and transparency: white, grey, yellowish and greenish, transparent to opaque; Streak: white to very pale green;

Talc

Lustre: greasy to pearly; Cleavage: perfect; Other features: sheets of talc are flexible but not elastic (like mica); it has a greasy or soapy feel.

Magnesite

Magnesite is a mineral that serves the needs of modern industry. Blast furnaces could not operate without it, because it is used chiefly to make heat-resistant bricks which line the furnaces. It is used in the building industry for making special cements, in the porcelain and ceramics industries, and as a raw material for making magnesium and its compounds.

In nature, magnesite is formed either by the weathering and decomposition of rocks rich in magnesium, such as serpentinite, or by the alteration of limestone and dolomite. In such rocks, it forms veins or irregular masses. Depending on its origin, magnesite may be coarsely crystalline, although individual crystals are not common, or it may, more usually, be granular or massive, or compact to earthy. It may resemble chalk, but magnesite is a much less common mineral than calcite.

The richest deposits are found in Austria. Important deposits also occur in Czechoslovakia, Greece, Italy, Manchuria and the Ural region of the USSR. In North America, substantial deposits are found along the Pacific coast.

Magnesite (magnesium carbonate), $MgCO_3$; Crystal system: trigonal; Hardness: 3.5-5; Specific gravity: 3-3.2; Colour and transparency: white, grey, yellow and brown, transparent to opaque; Streak: white; Lustre: vitreous to dull; Cleavage: perfect; Other features: it may resemble calcite but, while calcite dissolves readily in cold dilute hydrochloric acid, magnesite is hardly affected.

Magnesite

Minerals in Glass, Ceramics and Buildings

Man has learned much from nature. For instance, he noticed that a glassy substance is formed when lightning strikes sand in deserts. In volcanic regions, he found black volcanic glass. He also saw how insects and birds cement their nests with clay and how these nests have remarkably regular shapes. Observation of such phenomena, with thought and experiment resulted in new ideas and techniques which, led to major developments in ceramics and glass-making.

The basic and essential raw material in the glass-making industry is quartz sand. Quartz has already been described in an earlier chapter, together with feldspars, which are used for glazes.

Kaolinite

Old porcelain is a popular collector's item, especially pieces produced under the Ming dynasty in China, when it was perfected. But porcelain was not introduced into Europe until the 15th century and it was a long time before Europeans succeeded in making it. At first, because it was a luxury item, it was copied, mostly in the Netherlands, from as early as the 15th century in the form of fine majolica (glazed earthenware called *faience*).

In 1709, a German pharmacist, Johann F. Böttger (1682-1719), succeeded in making porcelain. He was also an alchemist and had sought ways to make gold. But, like many other alchemists, he finally

Kaolinite

discovered something quite different. The basic raw material for the production of porcelain is the sedimentary rock *kaolin,* a greyish-white clay whose main constituent is the mineral *kaolinite,* which is a hydrous aluminium silicate. Kaolinite is formed when rocks containing feldspar are weathered. It is a major component of clays and is commonly found in dull, earthy masses. It forms tiny, scaly crystals, which distinguish kaolinite from other clay minerals. But these minute crystals are visible only through an electron microscope.

When moistened, kaolin becomes plastic and it can then be moulded into the desired shapes. After it is fired in a kiln, it changes into a hard, white substance, porcelain. Kaolinite is found in Britain, Czechoslovakia and Germany but the largest deposits are in China.

Kaolinite (hydrous aluminium silicate), $Al_2Si_2O_5(OH)_4$; *Crystal system: triclinic; Hardness: 1-2; Specific gravity: 2.6; Colour and transparency: white, grey, translucent to opaque; Streak: white; Lustre: dull, earthy; Cleavage: perfect.*

Calcite

Calcite is one of the most widely distributed and commonest carbonate minerals. It is the main rock-forming mineral which is not a silicate and it forms most of the common sedimentary rock *limestone.*Limestone regions have a special character, because the rock is worn away by water containing carbon dioxide from the air or the soil. Limestone surfaces are often bare rock, worn into blocks. Pits in the surface lead down to tunnels and caves, where you may find hanging stalactites and pillar-like stalagmites. These are formed from finely crystalline calcite, precipitated from drops of water.

Calcite

Limestones formed millions of years ago as sediments on the sea floor were compressed and cemented together. This explains why limestone often contains fossils of marine organisms, such as molluscs and corals. From the study of these fossils, scientists have learned much about ancient life on Earth. Limestone is the basic raw material for making cement and lime for the building industry and it is used in making glass, ceramics and glazes. It is also a major raw material in the chemical industry. In farming, acidic soils are treated with limestone. In the building industry, varieties of compact limestones are used for building and decorative purposes. Marble, another building stone described in a later chapter, is metamorphosed limestone.

Calcite crystals occur in many forms, ranging from thin, leaf-like or tack-like shapes to long columns or needles. Transparent, colourless crystals of a variety of calcite, called Iceland spar, have an unusual property, double light refraction. Hence if you placed a piece of Iceland spar on this page, you would see each line of type twice.

Calcite (calcium carbonate), $CaCO_3$; Crystal system: trigonal; Hardness: 3; Specific gravity: 2.6-2.8; Colour and transparency: colourless, white, grey, yellow and brown, transparent to opaque; Streak: white; Lustre: vitreous; Cleavage: perfect; Other features: calcite readily dissolves and effervesces in cold dilute hydrochloric acid.

Gypsum

The practical uses of the soft mineral gypsum have been known since ancient times. When heated to 120° − 180°C (248° − 356°F), part of the water that is chemically combined in the structure of the mineral evaporates. The gypsum then changes into the powder known as Plaster of Paris. The ancient Egyptians used it as we do today − for wall decorations and casts of statues. It is also used in the building industry to make light walls, plasters and various moulds. Doctors use it for surgical casts.

Gypsum is added to some cements and is also used in farming and in the chemical industry. The finely granular, massive variety, called *alabaster,* is used for statues. *Satin spar* is a fibrous variety with a silky lustre and *selenite* is a colourless, transparent variety.

Colourless, transparent and perfectly shaped crystals of gypsum are common. The crystals are often joined to form so-called swallow-tail twins. Gypsum is an abundant mineral. It is found in sedimentary rocks with clays and limestones and it often forms thick layers. It is precipitated from sea water when

the water is evaporated. It is, therefore, often found with salt deposits. It is common in ore veins and it occurs in soil as a secondary mineral formed by the weathering of sulphides, such as pyrite.

Gypsum (hydrated calcium sulphate), $CaSO_4 \cdot 2H_2O$; Crystal system: monoclinic; Hardness: 1.5-2; Specific gravity: 2.3; Colour and transparency: colourless, white, grey, yellow, brown and reddish, transparent to translucent; Streak: white; Lustre: vitreous, pearly, silky; Cleavage: perfect.

Anhydrite

Rock salt deposits normally contain gypsum and anhydrite, because these three minerals are commonly associated with each other. Anhydrite is often called the 'dry brother of gypsum'. Both have the same chemical composition but anhydrite does not contain water.

Anhydrite is mostly massive or granular and, less often, coarsely crystalline. It may be formed by precipitation from sea water, or it may be a secondary mineral

Anhydrite

Gypsum

formed when gypsum loses its water. It is found in Austria, Canada, Germany, Italy and Poland.

Anhydrite (calcium sulphate), $CaSO_4$; Crystal system: orthorhombic; Hardness: 3-3.5; Specific gravity: 2.9; Colour and transparency: colourless, white, pink, red, bluish, violet, brown and grey, transparent to translucent; Streak: white; Cleavage: three good cleavages.

Minerals in the Nuclear Age

Uraninite

Scientific progress and technological advances have recently been influenced by a new source of power, atomic energy. The basic raw materials for its production are minerals that contain radioactive elements, particularly uranium and thorium. There are about 100 minerals that contain uranium or thorium. However, most of them are of interest only to mineralogists. The best minerals for the atomic power industry are *uraninite,* a group of secondary minerals.

Uraninite

In the past, uraninite was a useless mineral. Miners cast it on to waste heaps that stood beside silver mines. But, today, it is an extremely valuable raw material, the chief ore of uranium and the source of nuclear energy. The element uranium was first discovered in 1789. About a hundred years ago, uraninite, which also usually contains some radium and polonium, was being used to make a vivid, yellow paint.

In 1898, the scientist Marie Sklodowska Curie (1867–1934) discovered the new element radium in the waste matter from the manufacture of these paints. At first, it was used mainly in medicine. But radium and other radioactive elements are now the basis of a modern branch of science. They provide a vast source of energy and the minerals from which they are obtained have, since World War II, acquired great strategic importance.

Uraninite occurs in granite, where it occasionally forms cubic crystals. However, it is far more often found in ore veins as *pitchblende,* the name given to massive forms of uraninite. It takes the form

of black, gleaming layers and reniform (kidney-shaped) masses. It also occurs as veinlets (tiny veins) in other rocks and is sometimes found in river gravels. Uraninite is strongly radioactive. If a piece of rock containing veinlets of uraninite is placed on a photographic plate, in two days time the developed plate will show the exact pattern of the veinlets. This is known as an *autoradiograph.*

This important mineral is found in quantity in Africa, notably South Africa and Zaire, and in Canada, in Ontario and Saskatchewan pro-

vinces. Well-known European deposits are those of Britain and Czechoslovakia. In the USA, crystallized uraninite is found in Connecticut and it is mined chiefly in Arizona, New Mexico and Utah.

Uraninite (uranium dioxide), UO_2; Crystal system: cubic; Hardness: 5.5; Specific gravity: 7.5-9.5, the crystals having a higher specific gravity than massive forms; Colour and transparency: brownish-black to black, opaque; Streak: brownish-black to grey; Lustre: resinous, greasy; Cleavage: none, uraninite is brittle and has a conchoidal to uneven fracture.

Uraninite veinlets and exposed photographic plate

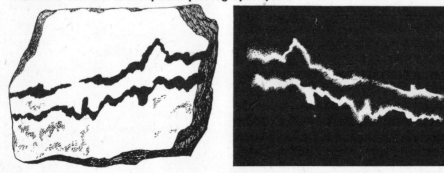

Torbernite and Carnotite

For a long time, the sprawling wilderness of northern Canada was accessible only by boat. But, in 1930, an enterprising mineral collector took off in a hydroplane. Over the Great Bear Lake, he noticed that the ground on one of the islands was coloured bright yellow. The yellow band extended to

Thorite

the mainland and so the collector landed to take a closer look. He found that the yellow colour was caused by small, flaky crystals around veins of uraninite. He had discovered one of the largest deposits of uranium, cobalt, nickel and silver.

Because uraninite is not a stable mineral, it changes into a number of secondary minerals when it is in an oxidized zone, that is, when it comes into contact with atmospheric oxygen. These secondary minerals are leaf-like and scaly. They have good cleavage and a pearly lustre and they are usually yellow or green. They belong to the *uranium mica group*. One member of this group is *torbernite*. It forms in veins containing uraninite and copper minerals. It is emerald green and forms flat, square-shaped, tabular crystals. It also forms clusters of crystals that fill rock crevices. Another member of the group is *carnotite*. It is bright yellow or yellow green and forms a finely crystalline or powdery coating. It is deposited from water

which has been in contact with uranium or vanadium minerals.

Torbernite and related minerals are good quality sources of uranium. They are mined in Britain, Czechoslovakia, France, Germany, Portugal and Spain. Other rich sources are in Africa, notably Zaire, Australia and the USA.

Torbernite (hydrated phosphate of uranium and copper), $Cu(UO_2)_2(PO_4)_2.8\text{-}12H_2O$; Crystal system: tetragonal; Hardness: 2-2.5; Specific gravity: 3.4-3.6; Colour and transparency: emerald-green, transparent to translucent; Streak: pale green; Lustre: vitreous to pearly; Cleavage: perfect.

Carnotite (hydrous vanadate of uranium and potassium), $K_2(UO_2)_2(VO_4)_2.3H_2O$; Crystal sys-

tem: monoclinic; Hardness: 4; Specific gravity: 4.4; Colour and transparency: yellow, yellow-green, crystals transparent; Streak: light yellow; Lustre: pearly, dull; Cleavage: perfect.

Thorite

Of all the radioactive elements, only uranium and thorium form the essential components of separate minerals. The others are present in minerals only as trace elements. Thorium minerals are far fewer in number than uranium minerals. One of them is silicate of thorium, or *thorite*. Thorium is used not only as a nuclear fuel, but also to make electronic equipment and special alloys.

Unlike uraninite, thorite is a very stable mineral and is not as readily altered by weathering. As a result, when rocks containing thorite are weathered, the mineral is washed down into alluvial deposits, where it can be easily extracted. Thorite forms columnar or pyramidal crystals in granite pegmatites. It also occurs in massive forms or as grains in rocks or river gravels. It generally contains rare earth elements (elements with atomic numbers ranging from 57 to 71) and some uranium, sometimes as much as 10 per cent. Thorite crystals are found in Norway.

Thorite (silicate of thorium), $ThSiO_4$; Crystal system: tetragonal; Hardness: 4.5; Specific gravity: 4.4-5.4; Colour and transparency: yellow to orange, brown to black, transparent to opaque; Streak: brown; Lustre: vitreous, greasy to dull; Cleavage: poor.

Uranium mica group

Torbernite

Carnotite

Stone in Sculpture and Architecture

Ornamental stone, used as building material, acquires its attractive appearance or superbly polished surface only in the hands of craftsmen. But the beauty of natural stone has been exploited by builders and artists for thousands of years. In many parts of the world, we can now admire the decorative facades of ancient buildings, statues, relief sculpture, tiles and other great works, large and small.

Stone that has been properly dressed remains the best material for building and decorative purposes. Modern plastics cannot compete with it. Various cultures have used different kinds of stone and the dressing of the stone is closely linked with the styles that were popular in the various periods. For example, the ancient Greeks and Romans made great use of marble and the Romans also prized travertine. Many ornamental stones were used during the Romanesque period and the Gothic period is noted, in particular, for its soaring structures.

Marble

Marble, in its many white and coloured forms, was the most typical ornamental stone used in ancient Greek art and architecture. The ruins of Greek temples are admired to this day and the superb marble statues of antiquity are precious treasures in the world's great museums. Marble was also popular with the Romans and many other peoples. Marble does not age and is not subject to changing fashions.

Marble

Even modern sculptors prefer marble for the final versions of their works, particularly because it is easy to cut and it takes a high polish. Modern architects use it when they want to emphasize the dignity and importance of a building.

Marble is a metamorphic rock, formed when great heat and pressure recrystallized the sedimentary rocks limestone, which consists mostly of the mineral calcite, and dolomite. Calcite and dolomite are fairly soft and marble is easily scratched with a knife. This feature distinguishes it from the much harder metamorphic rock, quartzite, which it may resemble. Marble may be compact or fine- to coarse-grained, when it may be sugary in appearance.

As a rule, marble is white or grey. But black, green, red and yellow varieties also occur. It may be evenly coloured or it may be patchy, streaky, veined or spotted. Sometimes, distorted fossils can be seen in marble. But, at the highest temperatures, all fossils are destroyed.

High-quality marble is found in many places. The famous deposits of white marble at Carrara, in Italy, were used by such great sculptors as Leonardo da Vinci and Michelangelo in the 1400s and 1500s. This snow-white marble is quarried today. Beautiful marble is still also found in Austria, Belgium, Czechoslovakia, France, Greece and Norway. Greenish or golden-yellow marble, sometimes wrongly called onyx, is found in Argentina, Mexico and in California, in the USA.

Travertine

In the building of ancient Rome, then the centre of the civilized world, marble was not the only construction material. It shared the stage with *travertine*. Travertine for the construction of the Colosseum, the Pantheon, temples, palaces and residences was brought from Tibur (now Tivoli). The Romans called the rock *lapis Tiburtinus* (the stone of Tibur). The word travertine comes from a distortion of this Latin name. Like marble and other time-tested building stones, travertine is still used today, mainly for decorative facings on the outsides of buildings and for interior fittings.

Travertine is a sedimentary rock. It is a porous limestone, formed by the deposition of calcium carbonate (calcite) from cold and hot wa-

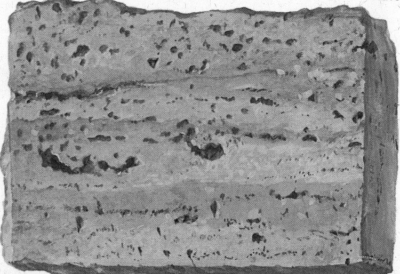

Travertine

ter springs, combined with the action of bacteria and algae. The layers are irregular and the rock is often streaky. It may be compact or it may contain numerous cavities. The colours are generally white, yellow, brown or grey. Well-known deposits are found in Algeria, Czechoslovakia, Germany, Italy, Yugoslavia and the USSR. In the USA, travertine is quarried at Mammoth Hot Springs, in Yellowstone Park, Wyoming. Around the Mammoth Hot Springs, there are beautiful terraces made up of travertine. The terraces resemble a waterfall that has been frozen or turned to

stone. A similar feature is the so-called Pammunale 'Falls', in Turkey. These handsome terraces are built up slowly as one thin film of calcite is deposited on another.

Sandstone

Sandstone is another popular stone used by architects and sculptors. In medieval times, it was used for all kinds of structures, ranging from Europe's famous cathedrals to small village churches, which are often outstandingly beautiful.

Sandstone is a sedimentary rock, composed largely of sand grains. The rock formed mostly in water and so marine fossils often occur in it. But some sandstones formed millions of years ago on land, from sand dunes in deserts. The sand grains are mostly quartz and so the surface of the stone resists weathering. The rock is also relatively soft and easy to carve.

In sandstone, the small, more or less rounded grains of quartz are cemented together by various binding materials. These natural cements are precipitated from water that seeps between the grains. The binding materials may be calcareous (calcite), clayey (clay minerals), ferruginous (haematite or limonite), or siliceous (silica). The colour of sandstone ranges from white, grey and yellow to brown or red. The colour depends on the nature of the binding material. The binding material also determines the hardness and resistance of the sandstone. In nature, sandstone is

Sandstone

a common rock and large outcrops of sandstone have a distinctive character. The rock is found in quantity in all continents. Sandstone is a porous rock and water can seep through the pores in it.

Underground sandstone layers are often saturated with water.

Labradorite

Labradorite is a mineral, which is one of the rock-forming feldspars. It was named after the Labrador peninsula, in Canada, where it was found about 200 years ago. This dark grey mineral attracted attention because, when it is turned around, it shows a lovely play of blues and greens on its cleavage surface. Labradorite can, therefore, brighten up the sometimes uniform look of buildings.

Labradorite is one of a series of minerals called plagioclase feldspars. They are aluminosilicates containing sodium and calcium, but the amounts of these two elements vary. This chemical composition distinguishes them from potassium feldspars, such as orthoclase, which has already been described. Labradorite is generally coarsely granular. The rare, tabular crystals are very small. It occurs in various igneous rocks, such as basalt and gabbro, and tiny crystals have been found in volcanic ash.

Labradorite (plagioclase feldspar, calcium and sodium aluminosilicates); Formula for the series: $CaAl_2Si_2O_8 \rightarrow NaAlSi_3O_8$; Crystal system: triclinic; Hardness: 6; Specific gravity: 2.6; Colour and transparency: greyish-blue, greenish-blue with play of colours, transparent to opaque; Streak: white; Lustre: vitreous, pearly; Cleavage: perfect.

Labradorite

Gems and Precious Stones

What gives gemstones their special magic? Why have people risked death in searching for them? What properties must a mineral have for it to be called a precious stone?

First, it must be beautiful. Geologists may hunt for a rare, grey mineral, which is lustreless, cracked and coarse, simply because it has useful properties. But such minerals are never described as precious stones. Beauty is determined by visual properties, particularly colour, lustre, translucence or transparency. A mineral's beauty is enhanced by grinding and polishing and also by being mounted in a suitable setting. Another feature of most precious stones is hardness, which makes their beauty long-lasting. Rarity is another factor which determines their value.

Diamond

Last of all, there is fashion, which affects the popularity of a given stone at a particular time.

Many highly-prized precious stones are also used in modern industry. In recent years, the demand has reached immense proportions and so many of them are now manufactured artificially. Such stones are called synthetic gems. Synthetic diamond, ruby, rutile, sapphire and spinel are among the stones that are now manufactured. But the cost of producing them is high, especially because synthetic gems are small and lack the quality of natural gems.

Diamond

The diamond is often called the 'king' of precious stones. Its name comes from the Greek word *adamas*, which means 'adamant' or 'unbreakable'. When cut and polished, it has a brilliant lustre and a magnificent play of colours. Its properties are important in industry. And yet diamond, chemically, is carbon, the same substance as the black lead in pencils or the purest forms of coal. The arrangement of the atoms in diamond accounts for its special properties.

Diamond, the birthstone for April, is the hardest mineral. It also possesses the greatest powers of refraction and the greatest brilliance. The best known specimens are clear and colourless. But diamonds may also be blue, red, green, yellow, brown, grey or even black. Diamonds were formed at great depths by tremendous heat and pressure in a rock called *kimberlite*. This rock fills volcanic pipes, called 'diamond pipes'.

The largest diamond ever found is called the Cullinan. It weighed 621.2 grams (more than 1 1/3 lb) and was found in 1905 in the Premier Mine, at Pretoria, in South Africa. It was later cut into 105 gemstones.

To bring out the lustre and brilliance of diamonds, they are cut in various ways. The most common cut is the 'brilliant', which is ground

Hexoctahedron crystal

on both the top and bottom into numerous small, inclined, flat surfaces, called *facets*. The facets reflect and refract light, increasing the sparkle and giving the diamond its unique play of colours.

Attempts at manufacturing diamonds were finally successful in 1955. However, synthetic crystals are extremely small. They cannot rival natural diamonds as gems, but they are used in industry. Synthetic and natural diamonds which lack gem quality are used to machine metals and cut glass. They are also made into abrasives to grind and polish hard materials and used for drilling.

Africa is the leading diamond-producing continent. The most valuable mines, for gem-quality diamonds, are at Kimberley, in South Africa. Other important mines are in Angola, Botswana, Ghana, Namibia, Sierra Leone, Tanzania and Zaire. The oldest known deposits are in Brazil and India. And, since 1954, diamonds have been obtained from rich deposits at Yakutsk, in the USSR.

Diamond (carbon), C; Crystal system: cubic, crystals are commonly octahedral, forming eight-faced double pyramids; Hardness: 10; Specific gravity: 3.5; Colour and transparency: colourless, red, blue, green, brown, grey, black, translucent to transparent; Lustre: adamantine, greasy; Cleavage: perfect.

Brilliant cut

Ruby, Sapphire (Corundum, aluminium oxide), Al_2O_3; Crystal system: trigonal; Hardness: 9; Specific gravity: 3.9-4.1; Colour and transparency: red (ruby), blue (sapphire), transparent to translucent; Streak: white; Lustre: adamantine to vitreous; Cleavage: none.

Ruby and synthetic boule

Ruby and Sapphire

The ruby and sapphire are also prized gems, although they are not as hard as diamond. Both are rare, coloured varieties of the common, dull-looking mineral corundum. Chemically, they are a crystalline form of pure aluminium oxide.

The ruby, the birthstone for July, has been used as a gem since ancient times and perfect specimens are now possibly the most expensive of all gemstones. Its rich red colour is due to the presence of a small amount of chromium. Its hardness and other properties have also proved useful in industry. It is used in some watches as one of the 'jewels' denoting the watch's quality, as an abrasive and in lasers. The sapphire, the September birthstone, is a beautiful relative of the ruby. It has a deep blue colour, possibly caused by iron or titanium.

The high price and great demand for gem varieties of corundum has-tened the manufacture of synthetic stones. Synthetic rubies, sapphires and other coloured varieties have

Sapphire

been produced by the Verneuil process since 1902. In this process, pure alumina powder is fused in an oxygen-hydrogen flame in a special furnace. The powder melts and fuses into single crystals, called *boules.*

The oldest known ruby deposits are found in Mogok, in northern Burma, where rubies are embedded in metamorphosed limestones. They are also obtained from alluvial deposits in Borneo, Cambodia, India, Sri Lanka and Thailand.

Emerald

The emerald, the birthstone for May, is a great rival of the diamond. It has a beautiful, vivid green colour.

The emerald is a precious variety of the mineral *beryl.* The vivid green colour is due to the presence of chromium in the internal structure. Other varieties of beryl which are used as gems include the green or bluish-green *aquamarine,* yellow *heliodor* and pink *morganite.* The mineral is found in such rocks as mica schists, limestones, coarse-grained granites and granite pegmatites.

The most famous deposits are found in Brazil, Colombia, India, Transvaal in South Africa and in Rhodesia (Zimbabwe).

Emerald (beryllium-aluminium silicate), $Be_3Al_2Si_6O_{18}$; Crystal system: hexagonal; Hardness: 7.5-8; Specific gravity: 2.6-2.8; Colour and transparency: green, transparent to translucent; Streak: white; Lustre: vitreous; Cleavage: poor.

Emerald

Topaz

Topaz

Topaz is prized chiefly for its attractive colours. Although many people think of topaz as a delicate, honey-yellow gem, it may also be golden-brown, pink, red, pale blue or clear and colourless. Another important property is its hardness. While it is softer than corundum, it is harder than quartz. Other valued properties are the mineral's transparency and clarity. The perfect cleavage of topaz in one direction makes it easy to divide.

Topaz is not as rare as many other precious stones and, hence, its price is now lower than it was in the past. It often occurs as large, perfect crystals. Sometimes, these column-like crystals may be enormous. Giant crystals weigh up to 300 kg (660 lb). Although synthetic topaz is not produced, amethyst (violet quartz) is sometimes treated to make imitation topaz. This may appear in shops as 'Spanish' or 'Madeira' topaz. Topaz and citrine (a yellow variety of quartz) are the birthstones for November.

The mineral occurs in cavities in granite, granite pegmatites, rhyolites and in quartz veins. It is also found in alluvial deposits. In the Middle Ages, Saxony was an important producer of topaz. Later, rich deposits were discovered in Brazil. Other valuable deposits of large crystals are found in the Ural region and Siberia, in the USSR.

Topaz (aluminium fluorsilicate), $Al_2SiO_4(F,OH)_2$; Crystal system: orthorhombic; Hardness: 8; Specific gravity: 3.4-3.6; Colour and transparency: colourless, yellow, brown, pink, red, blue, transparent to translucent; Streak: white; Lustre: vitreous; Cleavage: perfect.

Tourmaline

Tourmaline, one of the birthstones for October, is the name for a group of similar minerals with many interesting properties. Chemically, the minerals are highly complex and their composition varies greatly.

Varieties of tourmaline may be colourless or black, or they may occur in a wide range of colours. Common black tourmaline, called *schorl,* is a widely distributed mineral. It was sometimes used in jewellery which was worn during periods of mourning. Colourless tourmaline is called *achroite,* brown tourmaline is *dravite,* the blue variety is *indicolite* and the green is *verdelite.* Most highly prized, however, is the pink or red gemstone variety *rubellite.* A single crystal may have several colours.

Tourmaline and cross-section of a crystal

For instance, the base may be green or colourless, while the top is red and black. Such crystals are called 'Moors' heads'. And, in a thin slice of some tourmaline crystals, the core may have a different colour from the outside zones. But this is not all. For example, if we examine a green tourmaline against the light, it will display different colours when looked at from different directions. This property is called *pleochroism.*

When heated, tourmaline becomes positively charged at one end and negatively at the other. Because of this electrical charge, it attracts dust particles.

Tourmaline occurs in granites and granite pegmatites. The columnar crystals are often grooved parallel to their length. Superb, large crystals of the prized rubellite variety, which are the pride of many collections, come from California, in the USA.

Tourmaline (highly complex borosilicate of a number of elements), (Na,Ca)(Li,Mg,Al)(Al,Fe,Mn)$_6$(BO$_3$)$_3$(Si$_6$O$_{18}$)(OH)$_4$; Crystal system: trigonal; Hardness: 7-7.5; Specific gravity: 3-3.2; Colour and transparency: very varied colours, transparent to opaque; Streak: white; Lustre: vitreous; Cleavage: poor, conchoidal to uneven fracture.

Garnets

Garnets, the birthstones for January, are a group of closely re-

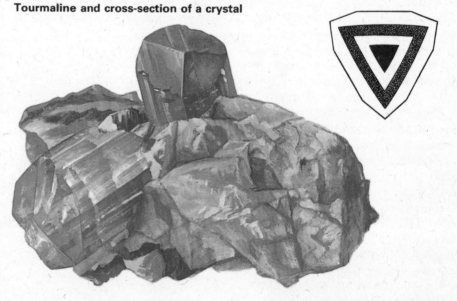

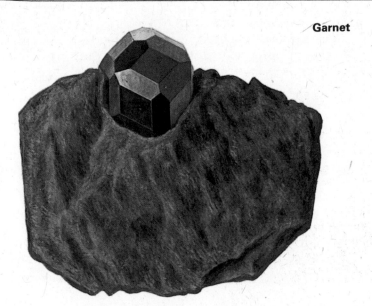

Garnet

lated minerals. They are all compounds of silica and two other elements, notably aluminium, calcium, chromium, iron, magnesium and manganese. They all have similar internal structures and similar properties. Their colour, however, differs according to their chemical composition and each variety has its own name.

Almandine is a garnet containing iron and aluminium. It has an attractive, dark red colour, with a purplish tinge. It has been used as a gem since ancient times. It occurs in metamorphic rocks and is found throughout the world. *Rhodolite* is a rose-red to violet variety, which is also often used as a gem. It is found in California and North Carolina in the USA.

Also popular as a gem is *pyrope*, a variety of garnet containing magnesium and aluminium. It is noted for its sparkling, ruby-red colour. It is associated with diamond and is found in the Kimberley region of South Africa and in the Yakutsk diamond deposits of the USSR.

Other garnets include *andradite*, emerald-green *demantoid*, black *melanite*, yellow and brown *hessonite*, *spessartine* and green *uvarovite*.

Garnets resist chemical weathering and eroded fragments are carried down into alluvial deposits, where they are mined. Small grains, which are hard and sharp-edged, also make excellent abrasives for grinding and polishing

other substances. Many numerous synthetic garnets are manufactured for use in jewellery and in industry. *Garnet group (compounds of silica and two other elements), Fe₃Al₂Si₃O₁₂ (almandine), Mg₃Al₂Si₃O₁₂ (pyrope); Crystal system: cubic; Hardness: 6.5-7.5; Specific gravity: 3.5-4.3; Colour and transparency: red, brown, yellow, orange, violet, green, black, transparent to translucent; Streak: white; Lustre: vitreous; Cleavage: none.*

Opal

Precious opal, a variety of the mineral opal, has been admired and prized for its bewitching beauty for 2,500 years. The ancient Romans ranked this mineral, which is noted for its iridescent play of colours, on a par with diamond. The electron microscopes show that the opal is made up of tiny, closely-packed spheres of silica, with air and water occupying the spaces between them. Viewed under an electron microscope, the spheres look like ping-pong balls piled on top of each other. Light strikes the planes of the gaps between the spheres and is bent and broken into the colours of the spectrum. Opal is cut in convex shapes, called *cabochons*.

Opal, a birthstone for October, is an amorphous mineral with water combined in its structure. It never forms crystals. It is a secondary mineral, formed by chemical weathering, and it forms masses in rock fissures or veinlets in igneous rocks. It also occurs in stalactite-like and rounded forms. Precious opal is white or yellowish. Other varieties are grey, milky-blue, red, brown or black. Black opal is also iridescent. It was once more costly than diamond. Precious opal was mined from the days of the Roman Empire in what is now Czechoslovakia, but this deposit has been exhausted. The richest deposits were found in the late 1 800s in New South Wales, Queensland and Victoria in Australia.

Opal (hydrous silicon dioxide), SiO₂.nH₂O; Crystal system: none, amorphous; Hardness: 5-6; Specific gravity: 1.9-2.2; Colour and transparency: varied colours, noted for its iridescence, transparent to subtranslucent; Streak: white; Lustre: vitreous, greasy; Cleavage: none, conchoidal fracture.

Opal

usually uneven. Stones of the deep rose pink variety are used in jewellery. The mineral is also carved into statuettes and other ornaments.

The mineral occurs in granite pegmatites in coarsely granular masses, which do not contain well-developed crystals. For a long time, geologists thought that it occurred only in a massive form. However, in 1960, crystals were found in Brazil. The chief producers of rose quartz are Brazil and Madagascar.

Rose quartz (silicon dioxide), SiO_2; Crystal system: trigonal; Hardness: 7; Specific gravity: 2.65; Colour and transparency; rose, translucent; Streak: white; Lustre: vitreous, greasy; Cleavage: none, conchoidal fracture.

Amethyst

The ancient Greeks believed that people who wore amethyst amulets would not become drunk. This explains how the mineral got its name, which comes from the Greek word *amethystos,* meaning 'not intoxicated'. Amethyst, the birthstone for February, is a popular variety of common quartz. It has a light to deep violet colour, which is caused by the presence of iron and by irradiation.

Amethyst forms from hot solutions in cavities and fissures in igneous rocks. As the solutions cool, crystals develop. Beautiful amethyst crystals are fairly common. The richest and best known deposits are in Brazil and Uruguay.

Amethyst (silicon dioxide), SiO_2; Crystal system: trigonal; Hardness: 7; Specific gravity: 2.65; Colour and transparency: violet, transparent to translucent; Streak: white; Lustre: vitreous; Cleavage: none, conchoidal fracture.

Rose quartz

Rose quartz

Rose quartz is a pale pink to deep rose pink variety of quartz. The colour is probably caused by the presence of a tiny quantity of manganese oxide. However, some specimens fade when exposed to light.

Rose quartz is translucent. It has a greasy lustre and the colour is

Tiger's eye

Quartz often contains inclusions of other minerals within it. One such quartz is called *tiger's eye.* It contains fine bands of a fibrous mineral called crocidolite or blue asbestos. Crocidolite is a blue to greyish-blue variety of riebeckite, a mineral which belongs to the common amphibole group which has already been described.

Tiger's eye has a banded appearance, caused by the parallel fibres, an unusual sparkle and a silky lustre. The stones are cut in a convex shape, called *cabochons.* When a cut stone is tilted, a light-coloured band skims over the surface. Stoneworkers called minerals which have this property, *chatoyancy,* 'eyes'. Some quartz contains bluish-grey inclusions of crocidolite. These are called fal-

Tiger's eye

con's eyes or pheasant's eyes. But the blue crocidolite, which contains iron, is often chemically weathered. As a result, it is partly or completely oxidized (combined with oxygen). The colour then changes to yellow or golden-brown. These yellow to brown stones are called tiger's eyes. Tiger's eye cannot be produced synthetically. At one time, tiger's eye was a highly prized precious stone. It was first brought to Europe from South Africa in the 1890s. It was rare, difficult to obtain and, therefore, it was used with the most precious gems and mounted in gold. However, interest in tiger's eye waned, both because of changing fashions and also because other sources were discovered and it became less rare. It was then used mainly for such things as paper weights, pen and ink sets and other ornamental objects. Today, however, interest in tiger's eye has revived. It is now used to make cameos and engravings.

The best known and longest worked deposits are in South Africa, where the mineral occurs as tabular fillings in rock fissures.

Chalcedony

Chalcedony is a form of silica, with the same chemical formula as quartz. At first glance, it appears to be a compact, shapeless mineral. But, viewed under a microscope, it is seen to be composed of tiny crystals of quartz. It is mostly translucent, with a greasy to vitreous lustre and it is slightly softer than quartz. Colours include grey, white, yellowish or bluish. *Carnelian* is a red to reddish-brown variety. *Jasper* is opaque and usually red. An apple-green variety is called *chrysoprase*. All these kinds of chalcedony are essentially evenly coloured.

However, the banded variety of chalcedony, called *agate,* has lost none of its appeal. It is a beautiful mineral, containing fine layers of differing colours, including shades of black, blue, brown, green, grey, red and white. The bands may be straight, circular or irregular in shape. Varieties of agate include the straight-banded *onyx* and *sardonyx, moss agate,* which contains delicate, moss-like, branching lines, and *mocha stone,* which contains

fern-like shapes. Today, the hardness and resistance of agate also make it valuable in industry. Agates are used to make bearings and pivots in precision instruments, tips for grinding and so on.

In nature, chalcedony fills cavities in certain rocks or it may form massive, rounded coatings on rocks. The mineral is precipitated from solutions that are rich in silica and it is quite common. It sometimes replaces the cells in buried logs, producing fossilized petrified wood.

Chalcedony and agate are found in many places throughout the world. The best known deposits in Europe are in Austria, Czechoslovakia, Germany, Iceland and the USSR. There are also large deposits in Brazil, China, India, Uruguay and in the USA.

Agate (a form of chalcedony, silicon dioxide), SiO_2; Crystal system: none; Hardness: about 6.5; Specific gravity: about 2.6; Colour and transparency: varied colours in bands, usually translucent; Streak: white; Lustre: vitreous to greasy; Cleavage: none, conchoidal fracture.

Agate

Chalcedony

Index

Page numbers in *italics* refer to illustrations

This book
belongs
TO
Colin Angel